Advances in Multiphase Flow and Related Problems

Proceedings of the Workshop on Cross Disciplinary Research in Multiphase Flow, Leesburg, Virginia, June 2–4, 1986.

This workshop was sponsored by SIAM and supported in part by the National Science Foundation Grant No. DMS8517142.

Advances in Multiphase Flow and Related Problems

EDITED BY
George Papanicolaou
Courant Institute of Mathematical Sciences
New York University

siam
Philadelphia 1986

Library of Congress Catalog Card Number 87-60050
ISBN 0-89871-212-2

Preface

A workshop on multiphase flow was held at the Xerox Training Center in Leesburg, Virginia on June 2-4, 1986. The purpose of the workshop was to bring together a cross disciplinary group of researchers from the universities, government and industry who are concerned with basic theoretical and computational problems in multiphase flow. The topic--already broad, since it includes much of fluid dynamics--was taken in an even broader sense at this workshop. There were lectures on composites, random media, continuum phase transitions, fluid-particle systems, bubbly liquids, computational methods for conservation laws, and many other topics.

This workshop was sponsored by the Applied Mathematics division of the National Science Foundation, directed at the time by M. Ciment. The divisions of Cross Disciplinary programs and Mechanical and Chemical Engineering also contributed to the workshop. A scientific organizing committee was put together in the Spring of 1985 under the chairmanship of George Papanicolaou of the Courant Institute, New York University. It consisted of the following, in addition to G. Papanicolaou: A. Acrivos (Chemical Engineering, Stanford University), D. Drew (Mathematics, Renssselaer Polytechnic Institute), J. Glimm (Courant Institute), G. Homsy (Chemical Engineering, Stanford University), D. Joseph (Aerospace and Mechanical Engineering, University of Minnesota), B. Levich (Chemical Engineering, City University of New York), L. E. Scriven (Chemical Engineering, University of Minnesota) and S. Soo (Mechanical Engineering, University of Illinois). This committee selected the topics that were to be covered in this broad area and invited the speakers. The workshop was advertised in SIAM publications and many more scientists attended as a result.

Twenty-three papers appear in this volume. The diversity of techniques and points of view that one sees in this collection is on doubt indicative of the state of exploratory research that basic questions in multiphase flow are in today.

George Papanicolaou
Courant Institute of Mathematical Sciences
New York University

List of Contributors

A. Acrivos, *Stanford University, Department of Chemical Engineering, Stanford, CA 94305*

H. Brenner, *Massachusetts Institute of Technology, Department of Chemical Engineering, Cambridge, MA 02139*

R. Burridge, *Schlumberger-Doll Research, Ridgefield, CT 06877*

Russel E. Caflisch, *Courant Institute of Mathematical Sciences, New York University, New York, NY 10012*

E. Chang, *Stanford University, Department of Chemical Engineering, Stanford, CA 94305*

Donald Drew, *Rensselaer Polytechnic Institute, Department of Mathematical Sciences, Troy, NY 12180*

Aaron L. Fogelson, *Courant Institute of Mathematical Sciences, New York University, New York, NY 10012*

A. F. Fortes, *University of Minnesota, Department of Aerospace Engineering and Mechanics, Minneapolis, MN 55455*

James Glimm, *Courant Institute of Mathematical Sciences, New York University, New York, NY 10012*

S. Haber, *Technion-Israel Institute of Technology, Faculty of Mechanical Engineering, Haifa, Israel*

R. Jackson, *Princeton University, Department of Chemical Engineering, Princeton, NJ 08544*

D. Joseph, *University of Minnesota, Department of Aerospace Engineering and Mechanics, Minneapolis, MN 55455*

Robert V. Kohn, *Courant Institute of Mathematical Sciences, New York University, New York, NY 10012*

Robert Lipton, *Cornell University, Mathematical Sciences Institute, Ithaca, NY 14853*

Jonathan H. C. Luke, *Courant Institute of Mathematical Sciences, New York University, New York, NY 10012*

T. S. Lundgren, *University of Minnesota, Department of Aerospace Engineering and Mechanics, Minneapolis, MN 55455*

Graeme W. Milton, *California Institute of Technology, Department of Physics, Pasadena, CA 91125*

A. Nadim, *Massachusetts Institute of Technology, Department of Mathematics, Cambridge, MA 02139*

Charles M. Newman, *University of Arizona, Department of Mathematics, Tucson, AZ 85721*

G. Papanicolaou, *Courant Institute of Mathematical Sciences, New York University, New York, NY 10012*

Charles S. Peskin, *Courant Institute of Mathematical Sciences, New York University, New York, NY 10012*

E. Bruce Pitman, *Courant Institute of Mathematical Sciences, New York University, New York, NY 10012*

Yuriko Renardy, *Virginia Polytechnic Institute and State University, Department of Mathematics, Blacksburg, VA 24061*

David G. Schaeffer, *Duke University, Department of Mathematics, Durham, NC 27706*

Leonard Schwartz, *Exxon Corporate Research, Annandale, NJ 08801*

William T. Sha, *Argonne National Laboratory, Argonne, IL 60439*

Michael Shearer, *North Carolina State University, Department of Mathematics, Raleigh, NC 27695*

P. Sheng, *Exxon Research Corporation, Annandale, NJ 08801*

P. Singh, *University of Minnesota, Department of Aerospace Engineering and Mechanics, Minneapolis, MN 55455*

M. Slemrod, *Rensselaer Polytechnic Institute, Department of Mathematical Sciences, Troy, NY 12180*

S. L. Soo, *University of Illinois, Department of Mechanical and Industrial Engineering, Urbana, IL 61801*

S. Torquato, *North Carolina State University, Department of Mechanical and Aerospace Engineering and of Chemical Engineering, Raleigh, NC 27695*

Grétar Tryggvason, *University of Michigan, Department of Mechanical Engineering, Ann Arbor, MI 48109*

Stephen Whitaker, *University of California, Department of Chemical Engineering, Davis, CA 95616*

B. White, *Exxon Research Corporation, Annandale, NJ 08801*

S. Yendler, *Stanford University, Department of Chemical Engineering, Stanford, CA 94305*

Contents

Sedimentation and Dispersion of Flexible Chains and Clusters of Hydrodynamically Interacting Brownian Particles

H. BRENNER*, A. NADIM** AND S. HABER†

Abstract. Generalized Taylor dispersion theory, incorporating so-called coupling effects, is used to calculate the transport properties of a single deformable 'chain' composed of hydrodynamically-interacting rigid Brownian particles bound together by internal potentials, and moving through an unbounded quiescent viscous fluid. The individual rigid particles comprising the flexible chain or cluster may each be of arbitrary shape, size and density, and are assumed to be 'joined' together to form the chain by a configuration-dependent internal potential V. Each particle separately undergoes translational and rotational Brownian motions; together, their relative motions give rise to a conformational or vibrational Brownian motion of the chain (in additional to a translational motion of the chain as a whole). Sufficient time is allowed for all accessible chain configurations to be sampled many times in consequence of this internal Brownian motion. In consequence, an internal Boltzmann probabilistic distribution of conformations derived from V effectively obtains.

In contrast with prior analyses of such chain transport phenomena, no ad hoc pre-averaging hypotheses are invoked to effect the averaging of the input conformation-specific hydrodynamic mobility data. Rather, the calculation is effected rigorously within the usual (quasistatic) context of configuration-specific Stokes-Einstein equations.

---

* Department of Chemical Engineering, Massachusetts Institute of Technology, Cambridge, MA 02139.

** Present Address: Department of Mathematics, Massachusetts Institute of Technology, Cambridge, MA 02139.

† Faculty of Mechanical Engineering, Technion-Israel Institute of Technology, Haifa 32000, Israel.

Explicit numerical calculations serving to illustrate the general scheme are performed for the simplest case of dumbbells composed of identically-sized spheres connected by a slack tether.

1. Introduction. Transport mechanics [10,11] in systems composed of isolated rigid particles moving through a continuous fluid phase has been the subject of extensive theoretical studies for well over a century. Building upon the pioneering hydrodynamic investigations of Stokes [27], who examined the 'slow' viscous translational motion of a spherical particle through a quiescent Newtonian fluid, a field has emerged which includes a strikingly rich variety of phenomena. Classified under the general title of "low-Reynolds-number hydrodynamics" [16], the latter field incorporates such diverse areas as suspension rheology, sedimentation processes, translational and rotational Brownian motions, and colloid science -- as well as a multitude of other nonequilibrium fluid-particle transport phenomena.

In circumstances where the suspended objects are flexible rather than rigid, progress has been limited, owing to the existence of several impediments. Not the least of these is the essentially pragmatic problem of dealing rigorously with the large numbers of degrees of freedom required to completely specify the instantaneous geometrical configuration of the flexible entity. A second related problem arises from the need to incorporate hydrodynamic interactions among the constituent rigid bodies making up the flexible body, and moving relative to one another. Usually, the first of these is dealt with within the more general framework of statistical mechanics [24] and kinetic theory [3], while the second is circumvented by either the complete neglect of hydrodynamic interactions, or by invoking lower-order approximations, such as the Burgers-Oseen interaction tensors with preaveraging [22]. This apparent necessity for introducing approximate hydrodynamic interaction calculations into the requisite analysis has not only hindered quantitative progress in calculations pertaining to specific models, but also the actual conceptual development of existing theories. Thus, a major aim of the present study is to provide a fresh impetus to the rigorous theoretical development of macromolecular (flexible body) transport mechanics, by utilizing the newly developed framework of generalized Taylor dispersion theory [6,8,9] to complement classical kinetic treatments [3] of macromolecular hydrodynamics. Our proposed framework allows both of the aforementioned difficulties to be surmounted (at least conceptually); specifically, all translational and orientational degrees of freedom of the individual constituent rigid particles comprising the cluster are retained, as too are all the requisite, many-body, configuration-specific, hydrodynamic phenomenological coefficients (grand resistance and mobility matrices).

The flexible-body model addressed herein is assumed to consist of a chain or cluster of rigid particles, not unlike the classical 'bead-spring' models of Rouse [26] and Zimm [31]. However, in our treatment the constituent rigid particles are taken to be of finite size and to be of arbitrary shape; moreover, the configuration-specific internal potential that serves to join them together -- thereby permitting collective identification of the cluster or chain as a _single_ entity moving through physical space -- is assumed arbitrary (rather than being limited, for example, by such restrictions as pairwise additivity) so long as the potential is sufficiently attractive at large particle separations to assure convergence of any subsequent integrals that arise in the theory. Other common models, such as the 'bead-rod' models of Kramers [23] and Hassager [18,19], or those of 'segmentally flexible macromolecules' [17,14,29], can presumably be treated with appropriate choices of the potential, although quantum mechanical effects may unexpectedly arise [25] in effecting the transition from flexible to rigid form for the constraining potential.

In the realm of kinematics, any arbitrary motion of a rigid particle can be decomposed into a translation (of a locator point affixed to the particle) and a rigid-body rotation (about an axis through that point); however, the same is not true of the arbitrary motion of a flexible cluster of rigid bodies. Indeed, it is not _a priori_ obvious which, if any, body-fixed geometrical point can best serve as a locator point for the chain 'position' in physical space. Points such as the centers of mass, volume, reaction [4], diffusion [30], or even an arbitrary point affixed to any one of the constituent rigid particles, all appear to constitute equally reasonable candidates, although the ultimate physical results characterizing the long-time transport properties [13] of the cluster as a whole must necessarily show themselves to be independent of the explicit choice made for the body-fixed chain locator point.

Another element of interest, particularly in problems pertaining to the sedimentation of flexible chains [32], is that although _on average_ such a chain may possess a definite 'mean configuration', the chain may instantaneously exist in any one of an infinite number of other accessible geometrical configurations (with the probability of a specific configuration governed by a Boltzmann distribution, entailing the configuration-specific internal potential). For example, although on time average the flexible body may possess some definite symmetric shape, it does not generally possess this symmetry at all times or indeed at any single instant of time. Since such deviations from the 'average' configuration normally create long-time secular or cumulative effects, the long-time physical properties of such a body can be expected in general to differ from those of its symmetric, pre-averaged _rigid_ counterpart. To rigorously analyze secular effects arising from instantaneous deviations from the average, generalized Taylor dispersion theory [8,13] can be employed. Indeed this paradigm has already been successfully used to investigate comparable sedimentation-dispersion phenomena in systems of _rigid_ nonspherical particles [5,7]. Upon incorporation of "coupling" effects [9], the generalized theory will be shown to be equally applicable to the macrotransport analysis of _flexible_ clusters too.

Prediction of the conventional molecular diffusivity of flexible macromolecules [30,15], free of any sedimentation effects, is itself a challenging goal. Recently, Wegener [30] and Haber and Brenner [15] have independently recognized the important role of coupling between the translational, rotational and internal motions of flexible macromolecules. The former has shown via a perturbation analysis that the long-time macroscopic translational dispersivity of a flexible body is equal to the mean diffusivity of its (unique) center of diffusion. Haber and Brenner [15] had independently examined the same general problem within the Taylor-Aris dispersion [28,1,20] framework, performing detailed calculations for the case of a flexible dumbbell.

In the next section, a summary of the pertinent results of generalized coupled Taylor dispersion theory are furnished. Then, in Section 3, transport mechanics of flexible chains and clusters -- composed of rigid Brownian particles and sedimenting through an otherwise quiescent viscous fluid -- are reviewed. The results obtained through application of generalized coupled Taylor dispersion theory are outlined, and specific numerical calculations pertaining to simple tethered dumbbells presented. Our general aim in the presentation of this work is to highlight and outline the principal methodology of our scheme. The details, which have deliberately been omitted, may be found in a concurrent contribution [12], of which the present paper constitutes a mere gloss.

2. Generalized coupled Taylor dispersion theory. A significant portion of the effort required to understand the non-equilibrium behavior of suspensions of flexible chains must necessarily be devoted to first describing the transport of a single one of the flexible bodies within the phase-space codifying its internal and external configurations. In addition to the conventional convective and (direct) diffusive contributions, the transport of isolated flexible objects also entails indirect or coupling contributions, whereby probability density gradients in the internal space give rise to fluxes in the external space and conversely. In order to treat the effects of such coupling upon the macrotransport properties of the isolated body (and ultimately upon the macroscale properties of the suspension as a whole), the rigorous results of generalized Taylor dispersion theory [6,8] -- including coupling effects [9] -- will be employed. Accordingly, this section is devoted to reviewing the relevant results of the latter theory. Subsequently, in the next section, the theory will be applied to the important problem of Brownian diffusion and sedimentation of a flexible chain or cluster, composed of hydrodynamically-interacting rigid Brownian particles, and moving through a quiescent viscous fluid.

The starting point of generalized Taylor dispersion theory is the canonical continuity equation [9]

$$\partial P/\partial t + \nabla_Q \cdot \mathbf{J} + \nabla_q \cdot \mathbf{j} = \delta(t)\, \delta(\mathbf{Q}-\mathbf{Q}')\delta(\mathbf{q}-\mathbf{q}') \tag{2.1}$$

governing conservation and transport of the conditional probability density $P(\mathbf{Q},\mathbf{q},t|\mathbf{Q}',\mathbf{q}')$ for a flexible 'tracer' to possess internal coordinates $\mathbf{q}$ and external (physical-space) coordinates $\mathbf{Q}$ at the instant t, given that at time $t = 0$ its respective $(\mathbf{Q},\mathbf{q})$ coordinates were $(\mathbf{Q}',\mathbf{q}')$. The local (internal) and global (external) fluxes $\mathbf{j}$ and $\mathbf{J}$ are respectively assumed to be given by the constitutive expressions

$$\mathbf{j} = \mathbf{u}(\mathbf{q})P - \mathbf{D}^{qQ}(\mathbf{q}) \cdot \nabla_Q P - e^{-E}\mathbf{D}^{qq}(\mathbf{q}) \cdot \nabla_q(e^{E}P), \tag{2.2}$$

$$\mathbf{J} = \mathbf{U}(\mathbf{q})P - \mathbf{D}^{QQ}(\mathbf{q}) \cdot \nabla_Q P - e^{-E}\mathbf{D}^{Qq}(\mathbf{q}) \cdot \nabla_q(e^E P). \tag{2.3}$$

In addition to the usual direct contributions arising from convection and diffusion in each subspace (i.e., local and global), coupling effects between these subspaces have been included, as directly embodied in the existence of the coupling diffusivity $\mathbf{D}^{qQ}$ (or its transpose $\mathbf{D}^{Qq}$). The Fickian diffusivities $\mathbf{D}^{QQ}$ and $\mathbf{D}^{qq}$ are each assumed to be both symmetric and positive definite, as too is also the combination $\mathbf{D}^{QQ} - \mathbf{D}^{Qq} \cdot (\mathbf{D}^{qq})^{-1} \cdot \mathbf{D}^{qQ}$. Inclusion of the dimensionless local-space potential $E(\mathbf{q})$ in the form given above summarizes the tranport effects due to conservative forces [6,8]; $\mathbf{U}$ and $\mathbf{u}$ are the respective global- and local-space velocity vectors of the tracer.

The conservation equation (2.1) is to be solved subject to appropriate attenuation-rate boundary conditions imposed as $|\mathbf{Q}| \to \infty$ to assure convergence, a no-flux condition $\mathbf{n}\cdot\mathbf{j} = 0$ at the boundaries $\partial\mathbf{q}_0$ of the local space, and the (pre-) initial condition $P = 0$ for $t < 0$. Application of a Lagrangian moment analysis to these equations reveals that the purely global transport of the tracer (i.e., from which transport process the internal degrees of freedom $\mathbf{q}$ have been eliminated via integration) is characterized by two macrotransport coefficients, namely the mean tracer velocity vector through physical space,

$$\bar{\mathbf{U}}^* = \int_{\mathbf{q}_0} d\mathbf{q}\ [P_0^\infty \mathbf{U} - e^{-E}\mathbf{D}^{Qq} \cdot \nabla_q(e^E P_0^\infty)\ ], \tag{2.4}$$

and its physical-space dispersivity dyadic,

$$\bar{\mathbf{D}}^* = \int_{\mathbf{q}_0} d\mathbf{q}\ \{P_0^\infty[\ \mathbf{D}^{QQ} - \mathbf{D}^{Qq} \cdot \nabla_q \mathbf{B}\ ] + [P_0^\infty(\mathbf{U}-\bar{\mathbf{U}}^*) - e^{-E}\mathbf{D}^{Qq} \cdot \nabla_q(e^E P_0^\infty)]\mathbf{B}\}. \tag{2.5}$$

The pair of coefficients $\bar{\mathbf{U}}^*$ and $\bar{\mathbf{D}}^*$ serves to describe the $\mathbf{Q}$-space kinematical properties of the flexible cluster for times sufficiently long to have established a terminal state (equilibrium or quasi-steady) within the internal space, a state that is independent of the initial internal configuration $\mathbf{q}'$.

Calculation of these long-time macrotransport coefficients requires knowledge of the internal equilibrium density P_0^∞, satisfying [9]

$$\nabla_q \cdot [\ \mathbf{u}P_0^\infty - e^{-E}\mathbf{D}^{qq} \cdot \nabla_q(e^E P_0^\infty)] = 0\ , \tag{2.6a}$$

$$\mathbf{n} \cdot [\ \mathbf{u}P_0^\infty - e^{-E}\mathbf{D}^{qq} \cdot \nabla_q(e^E P_0^\infty)] = 0 \text{ on } \partial\mathbf{q}_0, \tag{2.6b}$$

(2.6c) $$\int_{q_0} d\mathbf{q}\, P_0^{\infty} = 1 \quad .$$

Also required in (2.5) is the solution $\mathbf{B}(\mathbf{q})$ of the equations

(2.7a) $$\nabla_q \cdot \left[\, \mathbf{u}\, P_0^{\infty}\mathbf{B} - e^{-E}\mathbf{D}^{qq} \cdot \nabla_q(e^{E}P_0^{\infty}\mathbf{B}) + P_0^{\infty}\mathbf{D}^{qQ} \,\right] =$$

$$= P_0^{\infty}(\mathbf{U}-\bar{\mathbf{U}}^{*}) - e^{-E}\mathbf{D}^{Qq} \cdot \nabla_q(e^{E}P_0^{\infty}) \quad ,$$

(2.7b) $$P_0^{\infty}\, \mathbf{n} \cdot \left[\, \mathbf{D}^{qq} \cdot \nabla_q\mathbf{B} - \mathbf{D}^{qQ} \,\right] = \mathbf{0} \quad \text{on } \partial\mathbf{q}_0 \quad ,$$

uniquely defining $\mathbf{B}$ to within an irrelevant arbitrary additive constant vector [9].

Since only the long-time **Q**-space transport of the tracer is generally of interest in practice, the simplification afforded by being able to characterize the macrotransport process via only the two position-independent phenomenological coefficients (2.4) and (2.5) -- in place of the much larger set of **q**-dependent phenomenological coefficients appearing in (2.2)-(2.3) -- represents a major achievement of the general theory [albeit the not inconsiderable effort perhaps necessary to actually solve (2.6) and (2.7)].

3. Brownian diffusion and sedimentation of flexible chains and clusters. Consider a flexible cluster, synthesized by joining together n+1 rigid particles of arbitrary shapes via interparticle (internal) potentials. These potentials, which can be as elementary as simple tethers connecting pairs of particles, or as complex as one may wish to imagine, serve to permit collective identification of the n+1 rigid particles as a single identifiable entity -- namely, a "flexible chain". Its 'flexibility' arises from the fact that its constituent rigid particles are free to translate and rotate relative to one another, subject to any configurational constraints imposed by the internal potential; as such, the conformation of the chain can (and does) vary with time.

Label the individual rigid particles with the index A (A = 0, 1, 2,...,n, for a total of n+1 particles), and denote by O_A an arbitrarily-positioned particle 'locator point', rigidly affixed to particle A. At each instant the complete configuration of the cluster is entirely determined by specification of the n+1 position vectors $\mathbf{R}_A$ of the locator points O_A relative to an arbitrary space-fixed origin, and a comparable set of n+1 orientational triplets $\boldsymbol{\phi}_A$ of each of the constituent particles.

Denote by

(3.1) $$P(\mathbf{R}_0,\ldots,\mathbf{R}_n;\, \boldsymbol{\phi}_0,\ldots,\boldsymbol{\phi}_n,\, t\,|\,\mathbf{R}_0',\ldots,\mathbf{R}_n';\, \boldsymbol{\phi}_0',\ldots,\boldsymbol{\phi}_n')$$

the conditional probability density for finding the chain configuration in an infinitesimal neighborhood of $(\mathbf{R}_0,\ldots\mathbf{R}_n;\boldsymbol{\phi}_0,\ldots,\boldsymbol{\phi}_n)$ at time t, given that at time t=0 the chain possessed the corresponding primed configuration. The probability density P is chosen to satisfy the dual requirements

(3.2a) $$\int d\mathbf{R}_0\ldots d\mathbf{R}_n\, d\boldsymbol{\phi}_0\ldots d\boldsymbol{\phi}_n\, P = 1 \quad (t > 0) \quad ,$$

(3.2b) $$P = 0 \quad (t < 0) \quad ,$$

in which the limits of integration extend over the entire physical- and orientational-space domains available for the configurational transport.

The conservation equation governing the detailed configurational transport of the flexible cluster through a fluid continuum is of the general form

(3.3)
$$\frac{\partial P}{\partial t} + \sum_{A=0}^{n}\left(\frac{\partial}{\partial \mathbf{R}_A}\cdot \mathbf{J}[\mathbf{R}_A] + \frac{\partial}{\partial \boldsymbol{\phi}_A}\cdot \mathbf{j}[\boldsymbol{\phi}_A]\right) = \delta(t)\prod_{A=0}^{n}\delta(\mathbf{R}_A - \mathbf{R}'_A)\,\delta(\boldsymbol{\phi}_A - \boldsymbol{\phi}'_A) \ .$$

The configuration-specific physical- and orientational-space vector flux densities $\mathbf{J}[\mathbf{R}_A]$ and $\mathbf{j}[\boldsymbol{\phi}_A]$ of particle A will be assumed to possess the respective convective-diffusive constitutive forms

(3.4a) $$\mathbf{J}[\mathbf{R}_A] = \dot{\mathbf{R}}_A P - \sum_{B=0}^{n}\{\mathbf{D}[\mathbf{R}_A|\mathbf{R}_B]\cdot\frac{\partial P}{\partial \mathbf{R}_B} + \mathbf{D}[\mathbf{R}_A|\boldsymbol{\phi}_B]\cdot\frac{\partial P}{\partial \boldsymbol{\phi}_B}\} \quad ,$$

(3.4b) $$\mathbf{j}[\boldsymbol{\phi}_A] = \dot{\boldsymbol{\phi}}_A P - \sum_{B=0}^{n}\{\mathbf{D}[\boldsymbol{\phi}_A|\mathbf{R}_B]\cdot\frac{\partial P}{\partial \mathbf{R}_B} + \mathbf{D}[\boldsymbol{\phi}_A|\boldsymbol{\phi}_B]\cdot\frac{\partial P}{\partial \boldsymbol{\phi}_B}\} \quad ,$$

in which

(3.5a) $$\dot{\mathbf{R}}_A = -\sum_{B=0}^{n}\{\mathbf{M}[\mathbf{R}_A|\mathbf{R}_B]\cdot\frac{\partial V}{\partial \mathbf{R}_B} + \mathbf{M}[\mathbf{R}_A|\boldsymbol{\phi}_B]\cdot\frac{\partial V}{\partial \boldsymbol{\phi}_B}\} \quad ,$$

(3.5b) $$\dot{\boldsymbol{\phi}}_A = -\sum_{B=0}^{n}\{\mathbf{M}[\boldsymbol{\phi}_A|\mathbf{R}_B]\cdot\frac{\partial V}{\partial \mathbf{R}_B} + \mathbf{M}[\boldsymbol{\phi}_A|\boldsymbol{\phi}_B]\cdot\frac{\partial V}{\partial \boldsymbol{\phi}_B}\} \quad .$$

The configuration-specific mobility dyadics $\mathbf{M}[A|B]$ and diffusivity dyadics $\mathbf{D}[A|B]$ are functionally dependent only upon the internal configuration of the cluster; they are closely related to one another via the multibody configuration-specific Stokes-Einstein relations [4]

(3.6) $$\mathbf{D}[A|B] = kT\,\mathbf{M}[A|B] \quad .$$

The potential $V(\mathbf{R}_0,\ldots,\mathbf{R}_n;\ \boldsymbol{\phi}_0,\ldots,\boldsymbol{\phi}_n)$ appearing in (3.5) is also assumed to be a known function of the internal configuration, to which potential is added a further contribution arising from a constant external force $\mathbf{F}$ (e.g., gravity) causing sedimentation of the chain [cf. (3.9)] as a whole.

Subject to appropriate boundary and initial conditions, equation (3.3) for P may be solved to obtain an exact description of the configurational transport process; however, such a detailed resolution of the problem is not ordinarily the ultimate objective of interest. Indeed, for large n, such a description would be overwhelmingly detailed. Rather, if the flexible object is to be viewed as an entity unto itself, a much more physically useful and concise description is that of the transport through physical space of the flexible body as a whole, viewed as the sole object of interest. Attainment of this goal requires that we assign a single locator point to the flexible cluster as a definable entity, and focus exclusively on the stochastic trajectory of that point through ordinary three-dimensional (i.e., physical) space. All other coordinates, aside from the three required to specify the position of that locator point in physical space, are then assigned the role of specifying the internal conformation of the chain. Our eventual goal is thus to eliminate from the transport equation all the internal degrees of freedom, at least for sufficiently long times to assure that equilibrium with respect to the internal conformation has effectively been attained.

Equations (3.3) - (3.5), by themselves, effect no definite decomposition of the internal variables into respective local (internal) and global (external) sets. As this classification is prerequisite in applying to particular problems the results of coupled generalized Taylor dispersion theory given in Section 2, we will presently effect an appropriate transformation permitting this classification to be made.

Local/global form of the equations. Our ansatz consists of recognizing that for long times the choice of locator point for the flexible cluster is irrelevant; that is, any and all choices of chain locator point suffice equally well for identifying the location of the flexible body in physical space. In the long run, all such points behave alike as regards their net translational motions in physical space. As such, we are free to choose any point O_0 rigidly affixed to particle A=0 as the locator point of the flexible cluster. Denote by **Q** the position vector of this point, so that

$$\mathbf{Q} \equiv \mathbf{R}_0 \quad , \tag{3.7}$$

spanning the entire global (physical) space available for chain transport. Consider the local variables **q** to consist of the set

$$\mathbf{q} \equiv (\mathbf{r}_1,\mathbf{r}_2,\ldots,\mathbf{r}_n \ , \ \boldsymbol{\phi}_0,\boldsymbol{\phi}_1,\ldots,\boldsymbol{\phi}_n) \equiv (\mathbf{r}^n,\boldsymbol{\phi}^{n+1}) \quad , \tag{3.8a}$$

in which

$$\mathbf{r}_a \equiv \mathbf{R}_a - \mathbf{R}_0 \qquad (a = 1,2, \ldots, n) \tag{3.8b}$$

are the position vectors of the remaining points O_a relative to the designated point O_0.

Effecting the transformation of coordinates from $(\mathbf{R}_0,\ldots,\mathbf{R}_n;$ $\boldsymbol{\phi}_0,\ldots,\boldsymbol{\phi}_n)$ to $(\mathbf{Q},\mathbf{q})$ -- as represented by (3.7) and (3.8) -- into (3.3) - (3.5), in conjunction with the assumed decomposition of V into the global⊕local form

$$(3.9) \qquad V = -\mathbf{F}\cdot\mathbf{R}_0 + kT\, E(\mathbf{r}^n,\boldsymbol{\phi}^{n+1}) \quad ,$$

(with $\mathbf{F}$ a constant external force vector), ultimately yields the global/local form

$$(3.10) \qquad \partial P/\partial t + \nabla_Q\cdot\mathbf{J} + [\![\, \nabla_q \,]\!]^\dagger \cdot [\![\, \mathbf{j} \,]\!] = \delta(t)\delta(\mathbf{Q}-\mathbf{Q}')(\delta\mathbf{q}-\mathbf{q}') \ ,$$

of the conservation equation for P, in which

$$(3.11) \qquad [\![\, \nabla_q \,]\!]^\dagger \equiv [\![\, \partial/\partial\mathbf{r}_1^\dagger\ldots\partial/\partial\mathbf{r}_n^\dagger \quad \partial/\partial\boldsymbol{\phi}_0^\dagger\ldots\partial/\partial\boldsymbol{\phi}_n^\dagger \,]\!] \quad ,$$

$$(3.12) \qquad [\![\, \mathbf{j} \,]\!] = [\![\, \mathbf{u} \,]\!]P - [\![\, \mathbf{D}^{qQ} \,]\!]\cdot\nabla_Q P - e^{-E}[\![\, \mathbf{D}^{qq} \,]\!]\cdot[\![\, \nabla_q \,]\!](e^E P) \ ,$$

$$(3.13) \qquad \mathbf{J} = \mathbf{U}\,P - \mathbf{D}^{QQ}\cdot\nabla_Q P - e^{-E}[\![\, \mathbf{D}^{Qq} \,]\!]\cdot[\![\, \nabla_q \,]\!](e^E P) \quad .$$

In the above, the double-bracketed quantities are partitioned matrices, whose individual elements are themselves vectors or dyadics. The phenomenological coefficients appearing in (3.12) and (3.13) are defined as follows:

(i) Local velocity matrix $[\![\, \mathbf{u} \,]\!]$:

$$(3.14a) \qquad [\![\, \mathbf{u} \,]\!] = \left[\!\!\left[\begin{matrix} \mathbf{u}[\mathbf{r}_a] \\ \mathbf{u}[\boldsymbol{\phi}_A] \end{matrix} \right]\!\!\right] \quad ,$$

with

$$(3.14b) \qquad \mathbf{u}[\mathbf{r}_a] = \{\mathbf{M}[\mathbf{R}_a|\mathbf{R}_0] - \mathbf{M}[\mathbf{R}_0|\mathbf{R}_0]\}\cdot\mathbf{F} \quad ,$$

$$(3.14c) \qquad \mathbf{u}[\boldsymbol{\phi}_A] = \mathbf{M}[\boldsymbol{\phi}_A|\mathbf{R}_0]\cdot\mathbf{F} \quad ;$$

(ii) Coupling diffusivity matrices $[\![\, \mathbf{D}^{qQ} \,]\!]$ and $[\![\, \mathbf{D}^{Qq} \,]\!]$:

$$(3.15a) \qquad [\![\, \mathbf{D}^{qQ} \,]\!] = [\![\, \mathbf{D}^{Qq} \,]\!]^\dagger = \left[\!\!\left[\begin{matrix} \mathbf{D}^{qQ}[\mathbf{r}_a] \\ \mathbf{D}^{qQ}[\boldsymbol{\phi}_A] \end{matrix} \right]\!\!\right] \quad ,$$

with

(3.15b) $$\mathbf{D}^{qQ}[\mathbf{r}_a] = \mathbf{D}[\mathbf{R}_a|\mathbf{R}_0] - \mathbf{D}[\mathbf{R}_0|\mathbf{R}_0] \quad ,$$

(3.15c) $$\mathbf{D}^{qQ}[\phi_A] = \mathbf{D}[\phi_A|\mathbf{R}_0] \quad ;$$

(iii) Local diffusivity matrix $[\![\mathbf{D}^{qq}]\!]$:

(3.16a) $$[\![\mathbf{D}^{qq}]\!] = \left[\!\!\left[\begin{matrix} \mathbf{D}^{qq}[\mathbf{r}_a|\mathbf{r}_b] & \mathbf{D}^{qq}[\mathbf{r}_a|\phi_A] \\ \mathbf{D}^{qq}[\phi_A|\mathbf{r}_a] & \mathbf{D}^{qq}[\phi_A|\phi_B] \end{matrix} \right]\!\!\right] \quad ,$$

in which

(3.16b) $$\mathbf{D}^{qq}[\mathbf{r}_a|\mathbf{r}_b] = \mathbf{D}[\mathbf{R}_a|\mathbf{R}_b] - \mathbf{D}[\mathbf{R}_a|\mathbf{R}_0] - \mathbf{D}[\mathbf{R}_0|\mathbf{R}_b] + \mathbf{D}[\mathbf{R}_0|\mathbf{R}_0] \quad ,$$

(3.16c) $$\mathbf{D}^{qq}[\mathbf{r}_a|\phi_A] = \mathbf{D}[\mathbf{R}_a|\phi_A] - \mathbf{D}[\mathbf{R}_0|\phi_A] \quad ,$$

(3.16d) $$\mathbf{D}^{qq}[\phi_A|\mathbf{r}_a] = \mathbf{D}[\phi_A|\mathbf{R}_a] - \mathbf{D}[\phi_A|\mathbf{R}_0] = \mathbf{D}^{qq\dagger}[\mathbf{r}_a|\phi_A] \quad ,$$

(3.16e) $$\mathbf{D}^{qq}[\phi_A|\phi_B] = \mathbf{D}[\phi_A|\phi_B] \quad ;$$

(iv) Global velocity vector $\mathbf{U}$:

(3.17) $$\mathbf{U} \equiv \mathbf{M}[\mathbf{R}_0|\mathbf{R}_0] \cdot \mathbf{F} \quad ;$$

(v) Global diffusivity dyadic $\mathbf{D}^{QQ}$:

(3.18) $$\mathbf{D}^{QQ} \equiv \mathbf{D}[\mathbf{R}_0|\mathbf{R}_0] \quad .$$

Equations (3.10), (3.12) and (3.13) are identical to the canonical forms of the equations of generalized coupled Taylor dispersion theory (2.1) - (2.3), albeit the more complex appearance of their partitioned-matrix phenomenological coefficients. As such, the results (2.4) and (2.5) are directly applicable in this circumstance.

Results. For completeness, explicit expressions for the mean velocity vector $\overline{\mathbf{U}}^*$ and dispersivity dyadic $\overline{\mathbf{D}}^*$ of the flexible chain sedimenting through an otherwise quiescent viscous fluid are now presented. These correspond identically to equations (2.4) and (2.5) of Section 2, namely

(3.19) $$\overline{\mathbf{U}}^* = \int_{\mathbf{q}_0} d\mathbf{q}\{P_0^\infty \mathbf{U} - e^{-E} [\![\mathbf{D}^{Qq}]\!] \cdot [\![\nabla_q]\!] (e^E P_0^\infty)\}$$

and

$$\overline{\mathbf{D}}^{*} = \int_{q_0} d\mathbf{q}\{P_0^{\infty} [\ \mathbf{D}^{QQ} - [\![\mathbf{D}^{Qq}]\!] \cdot [\![\nabla_q]\!] \mathbf{B}\ ] + [P_0^{\infty}(\mathbf{U} - \overline{\mathbf{U}}^{*}) - e^{-E} [\![\mathbf{D}^{Qq}]\!] \cdot [\![\nabla_q]\!] (e^{E} P_0^{\infty})]\ \mathbf{B}\} \quad . \tag{3.20}$$

The requisite local fields P_0^{∞} and **B** appearing above are to be obtained by solving equations equivalent to (2.6) and (2.7).

Expressions (3.19) and (3.20) represent the major results of our analysis. Together with identification from existing data of the phenomenological coefficients required therein, as given by (3.14) - (3.18), they permit rigorous calculation of the long-time mean sedimentation velocity vector $\overline{\mathbf{U}}^{*}$ and dispersivity dyadic $\overline{\mathbf{D}}^{*}$ for any flexible chain or cluster composed of rigid Brownian constituent particles bound together via internal potentials.

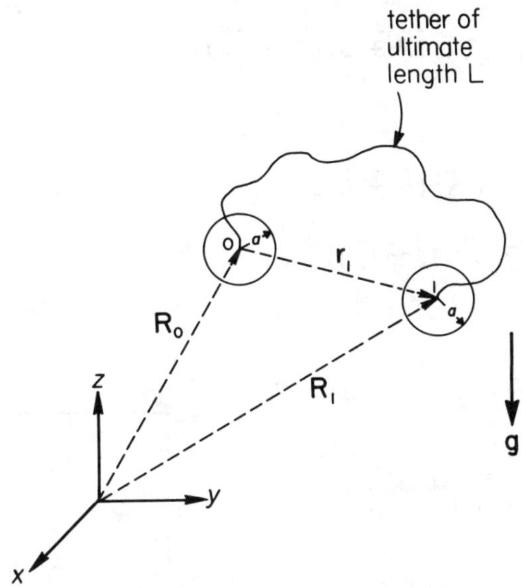

FIG. 1. Tethered dumbbell sedimenting under the influence of gravity (**g**) in an otherwise quiescent viscous fluid

Example. As an example consider the sedimentation-dispersion problem for the tethered dumbbell depicted in Fig 1. The latter consists of two identical rigid spherical particles of radii $\underline{a}$, each denser than its surrounding fluid by an amount corresponding to a mass difference $|\Delta m|$. A tether of ultimate extensibility L is assumed to connect their geometric centers. The total sedimentation force acting on the dumbbell is here

$$\mathbf{F} = 2|\Delta m|\mathbf{g} = F\hat{\mathbf{F}} \quad , \tag{3.21}$$

with $\hat{\mathbf{F}}$ a unit vector along gravity field vector $\mathbf{g}$. If the dimensionless tether-length/sphere-radius ratio is denoted by

$$\chi = L/a \quad , \tag{3.22}$$

and the Langevin parameter (embodying the ratio of gravitational to thermal energies of the dumbbell) designated as

$$\gamma = Fa/kT \quad , \tag{3.23}$$

the final results of the foregoing analysis may be summarized by the explicit relations

$$\overline{\mathbf{U}}^*/(M_\infty F) = \hat{\overline{M}}\,\hat{\mathbf{F}} \tag{3.24}$$

and

$$\overline{\mathbf{D}}^*/D_\infty = \hat{D}_\parallel\,\hat{\mathbf{F}}\hat{\mathbf{F}} + \hat{D}_\perp(\mathbf{I} - \hat{\mathbf{F}}\hat{\mathbf{F}}) \quad , \tag{3.25a}$$

in which

$$\hat{D}_\parallel = \hat{\overline{M}} + \gamma^2(\hat{\alpha} + \hat{\beta}/5) \quad , \tag{3.25b}$$

$$\hat{D}_\perp = \hat{\overline{M}} + \gamma^2(3\hat{\beta}/20) \quad . \tag{3.25c}$$

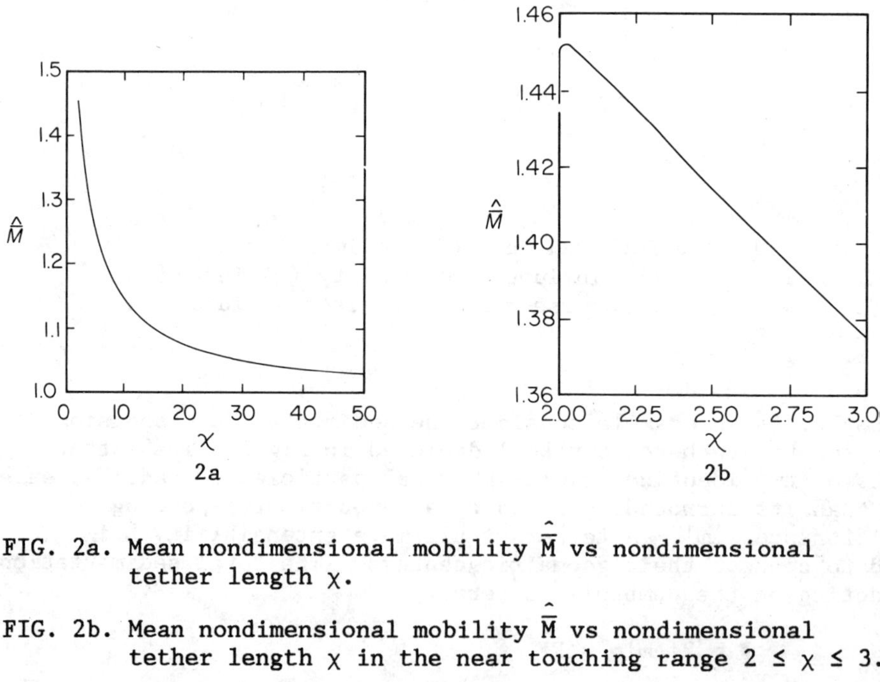

FIG. 2a. Mean nondimensional mobility $\hat{\overline{M}}$ vs nondimensional tether length χ.

FIG. 2b. Mean nondimensional mobility $\hat{\overline{M}}$ vs nondimensional tether length χ in the near touching range $2 \le \chi \le 3$.

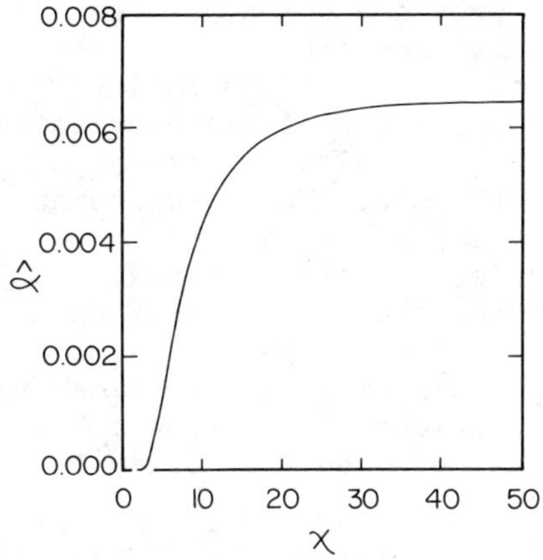

3

FIG. 3. Taylor dispersivity coefficient $\hat{\alpha}$ vs nondimensional tether length χ.

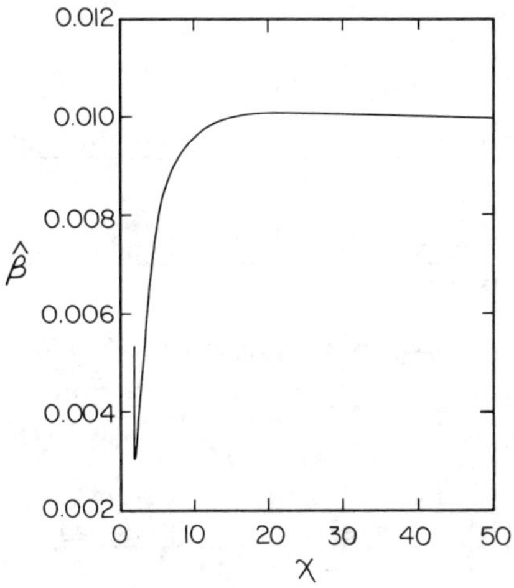

4a

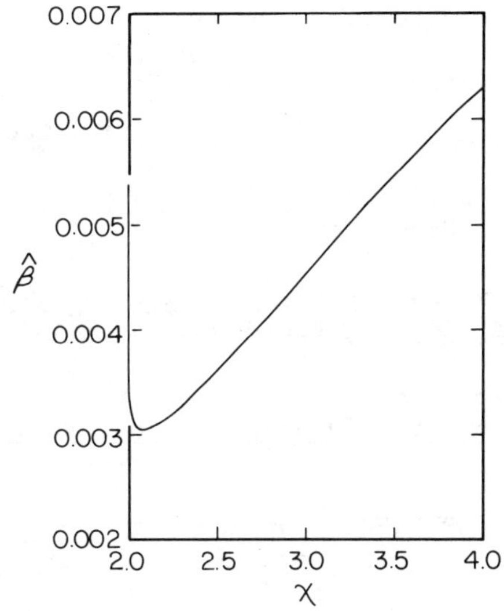

4b

FIG. 4a. Taylor dispersivity coefficient $\hat{\beta}$ vs nondimensional tether length χ.

FIG. 4b. Taylor dispersivity coefficient $\hat{\beta}$ vs nondimensional tether length χ in the near touching range $2 \le \chi \le 4$.

Here,

(3.26a,b) $$M_\infty = (12\pi\mu a)^{-1} \quad , \quad D_\infty = kTM_\infty$$

with μ the fluid viscosity.

The functions $\hat{\overline{M}}$, $\hat{\alpha}$ and $\hat{\beta}$ appearing in (3.24) - (3.25) are dependen only upon χ. These have been calculated numerically, using as input data the hydrodynamic calculations of Jeffrey and Onishi [21] and Batchelor [2]. Calculated results for these coefficients are embodied in Figs. 2-4.

4. Summary. A general scheme has been provided which permits rigorous calculations of the long-time macrotransport coefficients characterizing the mean translational velocity vector and Taylor dispersivity dyadic of flexible chains and clusters of rigid Brownian particles, sedimenting under the influence of a uniform external force within an otherwise quiescent viscous fluid. The constituent rigid particles may be of arbitrary shapes and sizes, and may mutually interact via arbitrary configuration-specific internal potentials. Full account is taken of the nonzero sizes and of the hydrodynamic interactions among the individual rigid particles comprising the flexible chain. No *ad hoc* pre-averaging approximations have been invoked to bring the calculations to fruition. Explicit numerical results are given for tethered dumbbells.

Acknowledgment. Howard Brenner is grateful to the Microgravity Sciences and Applications Program of NASA for support of this research.

REFERENCES

[1] R. ARIS, *On the dispersion of a solute in a fluid flowing through a tube*, Proc. Roy. Soc., A235 (1956), pp. 67-77.

[2] G.K. BATCHELOR, *Brownian diffusion of particles with hydrodynamic interaction*, J. Fluid Mech., 74 (1976), pp.1-29.

[3] R.B. BIRD, O. HASSAGER, R.C. ARMSTRONG, and C.F. CURTISS, *Dynamics of Polymeric Liquids, Vol. 2, Kinetic Theory*, Wiley & Sons, 1977.

[4] H. BRENNER, *Coupling between the translational and rotational Brownian motions of rigid particles of arbitrary shape II. General theory*, J. Colloid Interface Sci., 23 (1967), pp. 407-436.

[5] H. BRENNER, *Taylor dispersion in systems of sedimenting nonspherical Brownian particles I. Homogeneous, centrosymmetric, axisymmetric particles*, J. Colloid Interface Sci., 71 (1979), pp. 189-208.

[6] H. BRENNER, *A general theory of Taylor dispersion phenomena*, PhysicoChem. Hydrodyn., 1 (1980), pp. 91-23.

[7] H. BRENNER, *Taylor dispersion in systems of sedimenting nonspherical Brownian particles II. Homogeneous ellipsoidal particles*, J. Colloid Interface Sci., 80 (1981), pp. 548-588.

[8] H. BRENNER, *A general theory of Taylor dispersion phenomena II. An extension*, PhysicoChem. Hydrodyn., 3 (1982), pp. 139-157.

[9] H. BRENNER, *A general theory of Taylor dispersion phenomena IV. Direct coupling effects*, Chem. Eng. Comm., 18 (1982), pp. 355-379.

[10] H. BRENNER, and D.W. CONDIFF, *Transport mechanics in systems of orientable particles III. Arbitrary particles*, J. Colloid Interface Sci., 41 (1972), pp.228-274.

[11] H. BRENNER, and D.W. CONDIFF, Transport mechanics in systems of orientable particles IV. Convective transport, J. Colloid Interface Sci., 47 (1974), pp.199-264.

[12] H. BRENNER, A. NADIM and S. HABER, Long-time molecular diffusion, sedimentation and Taylor dispersion of a fluctuating cluster of interacting particles, J. Fluid Mech., submitted (1986).

[13] H. BRENNER, and M. PAGITSAS, Macrotransport Processes, to appear, 1987.

[14] J. GARCIA DE LA TORRE, P. MELLADO, and V. RODES, Diffusion coefficients of segmentally flexible macromolecules with two spherical subunits, Biopolymers, 24 (1985), pp. 2145-2164.

[15] S. HABER, and H. BRENNER, Taylor-Aris dispersion of N particles with internal constraints I. Molecular dispersion, in "Recent Developments in Structured Continua", (P.N. Kaloni and D. DeKee, eds.), Longmans, 1986.

[16] J. HAPPEL, and H. BRENNER, Low Reynolds Number Hydrodynamics, Nijhoff, 1983.

[17] S.C. HARVEY, P. MELLADO, and J. GARCIA DE LA TORRE, Hydrodynamic resistance and diffusion coefficients of segmentally flexible macromolecules with two subunits, J. Chem. Phys., 78 (1983), pp. 2081-2090.

[18] O. HASSAGER, Kinetic theory and rheology of bead-rod models for macromolecular solutions I. Equilibria and steady flow properties, J. Chem. Phys., 60 (1974), pp. 2111-2124.

[19] O. HASSAGER, Kinetic theory and rheology of bead-rod models for macromolecular solutions II. Linear unsteady flow properties, J. Chem. Phys., 60 (1974), pp. 4001-4008.

[20] F.J.M. HORN, Calculation of dispersion coefficients by means of moments, AIChE J., 17 (1971), pp. 613-620.

[21] D.J. JEFFREY and Y. ONISHI, Calculation of the resistance and mobility functions for two unequal spheres in low-Reynolds-number flow, J. Fluid Mech., 139 (1984), pp. 261-290.

[22] J.G. KIRKWOOD and J. RISEMAN, The intrinsic viscosities and diffusion constants of flexible macromolecules in solution, J. Chem. Phys., 16 (1948), pp. 565-573.

[23] H.A. KRAMERS, The behavior of macromolecules in inhomogeneous flow, J. Chem. Phys., 14 (1946), pp. 415-424.

[24] L.D. LANDAU and E.M. LIFSHITZ, Course of Theoretical Physics Vol. 5, Statistical Mechanics, Pergamon, 1980.

[25] J.M. RALLISON, The role of rigidity constraints in the rheology of dilute polymer solutions, J. Fluid Mech., 93 (1979), pp. 251-279.

[26] P.E. ROUSE, A theory of linear viscoelastic properties of coiling polymers, J. Chem. Phys., 21 (1953), pp. 1272-1280.

[27] G.G. STOKES, On the effect of the internal friction of fluids on the motion of pendulums, Trans. Cambr. Phil. Soc., 9 (1851), pp. 8-106.

[28] G.I. TAYLOR, Dispersion of soluble matter in solvent flowing slowly through a tube, Proc. Roy. Soc., A219 (1953), pp. 186-203.

[29] W.A. WEGENER, Bead models of segmentally flexible macromolecules, J. Chem. Phys., 76 (1982), pp. 6425-6430.

[30] W.A. WEGENER, Center of diffusion of flexible macromolecules, Macromolecules, 18 (1985), pp. 2522-2530.

[31] B.H. ZIMM, Dynamics of polymer molecules in dilute solution: visco-elasticity, flow birefringence and dielectric loss, J. Chem. Phys., 24 (1956), pp. 269-278.

[32] B.H. ZIMM, Sedimentation of asymmetric elastic dumbbells and the rigid-body approximation in the hydrodynamics of chains, Macromolecules, 15 (1982), pp. 520-525.

Direct and Inverse Problems for Pulse Reflection from a Stratified Random Halfspace

R. BURRIDGE*, G. PAPANICOLAOU**, P. SHENG† AND B. WHITE†

When the intensity of disturbances is small, acoustic wave propagation in inhomogeneous media can be analyzed using the wave equation as for example in bubbly liquids [1,2]. In a one-dimensional medium the pressure $p(x,t)$ and velocity $u(x,t)$ satisfy the equations

$$\rho(x)u_t(x,t)+p_x(x,t)=0 \tag{1}$$

$$\frac{1}{K(x)}p_t(x,t)+u_x(x,t)=0$$

where $\rho(x)$ and $K(x)$ are the density and bulk modulus. We shall present some recent work [3,4] on the propagation of pulses in one-dimensional media described by equations (1) when ρ and K are random functions. The randomness of the density and bulk modulus can model inhomogeneities at a microscopic level or at a more macroscopic one. In the latter case inhomogeneities reflect the complexity of the medium arising from its composite or multiphase character, for example.The methods of analysis that we use rely on the fact that the medium has plane parallel structure so that (1) is one-dimensional.

There are many interesting initial-boundary value problems that can be posed and analyzed for (1). To fix ideas we assume that the positive x axis is occupied by a homogeneous medium with density ρ_0 and bulk modulus K_0 and that a pulse disturbance is incident from the right on the inhomogeneous halfspace $x<0$. Thus we shall consider (1) with p and u satisfying

$$u(x,t)=\frac{1}{2}f(t+\frac{x}{c_0}) \tag{2}$$

$$p(x,t)=\frac{-1}{2}\rho_0 c_0 f(t+\frac{x}{c_0})$$

for $t<0$ and large and all x. Here $f(t)$ is a smooth function that vanishes for $t<0$ and t large and it represents the shape of the incident pulse and $c_0=\sqrt{K_0/\rho_0}$ is the sound speed. The objec-

* Schlumberger-Doll Research, Ridgefield, CT 06877

** Courant Institute, New York University, 251 Mercer Street, New York, NY 10012

† Exxon Research & Engineering Company, Route 22 East, Clinton Township, Annandale, NJ 08801

tive of our analysis is to calculate the velocity u and the pressure p for $t>0$ and all x ; in particular at $x=0$ which is the reflected signal.

The main assumption that defines the scope of our results is that the width of the incident pulse is long compared to the time it takes to travel one correlation length of the inhomogeneities but short compared to the scale of mean variation in ρ and K. Thus, the pulse width can resolve the large scale features of the medium but feels only an avereged effect from the random inhomogeneities.

To make this precise we shall rewrite (1-2) in dimensionless form as follows. Let x and t be space and time measured in macroscopic units and let ε^2 be a small parameter characterizing the correlation length of ρ and K. The scaled form of (1-2) is then

$$\rho_0(x)(1+\gamma(x)\eta(\frac{x}{\varepsilon^2}))u_t+p_x=0 \tag{3}$$

$$\frac{1}{K_0(x)}(1+\zeta(x)\nu(\frac{x}{\varepsilon^2}))p_t+u_x=0$$

for $t>0$ and $x<0$ and

$$u(x,t)=\frac{1}{2\sqrt{\varepsilon}}f(\frac{t+x}{\varepsilon}) \tag{4}$$

$$p(x,t)=\frac{-1}{2\sqrt{\varepsilon}}f(\frac{t+x}{\varepsilon})$$

for $t<0$. Here $\rho_0(x)$ and $K_0(x)$ are deterministic functions that define the mean density and bulk modulus of the inhomogeneous half space and $\eta(\xi)$ and $\nu(\xi)$ are zero mean, stationary, bounded random functions characterizing the fluctuations. The slow variation of the fluctuations is given by $\gamma(x)$ and $\zeta(x)$. Note that in (3-4) the pulse width is ε which is small compared to the scale of variation of the mean characteristics but large compared to ε^2 which is the correlation length. Thus there are three scales in this problem: one, ε and ε^2. We have also assumed that the density and bulk modulus are equal to one in the right half space and we have divided by $\sqrt{\varepsilon}$ on the right in (4) to make the energy of the incident pulse independent of ε.

The problem now is to study the solution $u(x,t)$, $p(x,t)$ of (3-4) for $t>0$ as a stochastic process when ε is small. A rather complete analysis is given in [3] for the case of a half space whose density and bulk modulus are random but have no macroscopic variations. This means that

$$\rho_0(x)=K_0(x)=1 \tag{5}$$

$$\gamma(x)=\zeta(x)=1$$

for $x<0$. The aim of the analysis in [3] is to characterize the reflected signal under condition (5). To describe the results we let $A(x,t)$ denote the right travelling wave disturbance

$$A(x,t)=u(x,t)+p(x,t) \tag{6}$$

We also let

$$n(x)=\frac{(\eta(x)-\nu(x))}{2} \tag{7}$$

and define the reflected process by

$$A_{t,f}^{\varepsilon}(\sigma)=A(0,t+\varepsilon\sigma) \tag{8}$$

The notation indicates dependence on the pulse form f, the time $t+\varepsilon\sigma$ at which the reflected signal is observed at x=0 and the parameter ε. We call this parametrization of the time a windowing. We anticipate that the reflected process has this windowed structure with t being the center of the window and σ the time variable within this window. We have now the following Theorem.

Assume that the processes $\eta(x)$ and $\nu(x)$ have rapidly decaying correlations and let t and the pulse form f be fixed in (8). Then the process $A_{t\,f}^{\varepsilon}(\sigma)$, $-\infty<\sigma<+\infty$, goes to a zero mean, stationary Gaussian process $A_{t\,f}(\sigma)$ in the limit $\varepsilon\to 0$. The power specral density $S_{t\,f}(\omega)$ of this process is given by

$$S_{t\,f}(\omega)=|\hat{f}(\omega)|^2\frac{1}{t}\mu(\sqrt{\alpha t}\,\omega) \tag{9}$$

where

$$\alpha=\int_0^{\infty}E\{n(x)n(0)\}dx \tag{10}$$

$\hat{f}(\omega)$ is the Fourier transform of $f(t)$ and μ is a universal function that is given by

$$\mu(\omega)=\frac{\omega^2}{(1+\omega^2)^2} \tag{11}$$

The material properties affect the statistics of the reflected pulse only through the parameter α defined by (10).

To understand the content and possible applications of this result we should make some remarks in addition to the ones above regarding the scaling. From (9) we see that the reflected signal decays in amplitude roughly like one over the square root of time. Roughly, because time enters also through the unversal function μ. On the time scale of the pulse width the reflected signal has a canonical structure as given by its power spectrum. We see that the canonical reflected signal has the structure of a time-windowed process since it depends not only on t, the center of the time window, but also on σ, the time scale within the window that picks up the fluctuations.

How can we apply this result? Before any other considerations we must understand how it fares when tested with direct numerical simulations. This has been done [5] and the conclusion is that our theory is not only accurate but also robust. This means that even when ε, the small parameter, is not very small the structure of the reflected signal is still as we described it. Regarding applications, we need the solution of the more general problem (3-4) without the homogeneity hypothesis (5).This is described in the following paragraphs and in [7]. With it we have a structure theory for reflected pulses for sufficiently general random media to attempt the statistical inverse problem: to construct likelihood functionals and tests that are able to determine the medium from the way it reflects pulses. The Gelfand-Levitan theory in the time domain [6] can be used for the inverse problem directly. The present theory however is able to use the power spectral density of the reflected pulse to extract information about the macroscopic features of the scattering medium after first eliminating the effects of the fluctuations in a canonical way.

We now describe briefly the analysis of the inhomogeneous problem (3),(4) i.e. when condition (5) does not hold. We need to assume however that (i) the functions ρ_0 , K_0 , γ and ζ describing the mean properties of the halfspace are smooth and (ii) that for all $-x$ large, ρ_0 , K_0 , γ and ζ are constant. We comment on the case when the differentiability condition (i) does not hold after stating the results.

We again define A by (6) but at $x=0$ only. We also define $A_{t,f}^{\varepsilon}(\sigma)$ by (8) as before. Our Theorem is now that again for each t,f fixed the process $A_{t,f}^{\varepsilon}(\sigma)$ converges weakly as ε tends to zero to a mean zero, stationary, Gaussian process with power spectral density

$$S_{t,f}(\omega)=|\hat{f}(\omega)|^2\mu(t,\omega) \tag{12}$$

Now however, the function $\mu(t,\omega)$ does not scale as in (9) and depends on the medium. It is obtained as follows.

Let

$$\alpha(x)=\frac{1}{4}\gamma^2(x)\int_0^\infty E\{\eta(s)\eta(0)\}ds-\frac{1}{4}\zeta(x)\gamma(x)\int_0^\infty E\{\eta(s)\nu(0)\}ds \tag{13}$$

$$-\frac{1}{4}\zeta(x)\gamma(x)\int_0^\infty E\{\nu(s)\eta(0)\}ds+\frac{1}{4}\zeta^2(x)\int_0^\infty E\{\nu(s)\nu(0)\}ds$$

and let

$$c_0(x)=\sqrt{\frac{K_0(x)}{\rho_0(x)}}\ ,\ x<0 \tag{14}$$

be the mean speed of propagation in the inhomogeneous half space. Let $u(x,\phi;h,\omega)$ be the solution of

$$\frac{\partial u}{\partial x}+\frac{h}{c_0(x)}\frac{\partial u}{\partial \phi}+\frac{4\omega^2\alpha(x)}{c_0^2(x)}(1-\cos\phi)\frac{\partial^2 u}{\partial \phi^2}=0\ ,\ \ x<0 \tag{15}$$

$$u(0,\phi;h,\omega)=e^{-i\phi}$$

Since c_0 and α tend to finite, nonzero values as $x\to-\infty$, the limit

$$U(h,\omega)=\lim_{x\to-\infty}u(x,\phi;h,\omega) \tag{16}$$

exists and is independent of ϕ. Then

$$\mu(t,\omega)=\int_{-\infty}^{\infty}e^{iht}U(h,\omega)dh \tag{17}$$

The canonical spectral density $\mu(t,\omega)$ can also be obtained by carrying out the h integration in (15) and by using a Fourier series to represent u. Let V be defined by

$$V=\int_{-\infty}^{\infty}e^{iht}u \tag{18}$$

and

$$V^N(x,t,\omega)=\frac{1}{2\pi}\int_0^{2\pi}e^{iN\phi}Vd\phi \tag{19}$$

From (15) we obtain the following equation for V^N

$$\frac{\partial V^N}{\partial x}-\frac{N}{c_0(x)}\frac{\partial V^N}{\partial t}+\frac{4\omega^2\alpha(x)}{c_0^2(x)}\left[\frac{(N+1)^2}{2}V^{N+1}-N^2V^N+\frac{(N-1)^2}{2}V^{N-1}\right]=0 \tag{20}$$

This holds for $x<0$ and $N=1,2,3\cdots$ with

$$V^N(0,t,\omega)=2\pi\delta(t)\delta_{N1} \tag{21}$$

Then we have that

$$\mu(t,\omega)=\int_{\tau^{-1}(t)}^{0}\frac{4\omega^2\alpha(s)}{c_0^2(s)}V^1(s,t,\omega)ds \tag{22}$$

where

$$\tau(x)=\int_x^0\frac{1}{c_0(s)}ds\ ,\ x<0 \tag{23}$$

is the travel time into the left half space and τ^{-1} is the inverse function of τ. Note that the V^N are nonnegative, that τ^{-1} is negative and hence that $\mu(t,\omega)$ given by (22) is nonnegative (actually positive). Moreover, the hyperbolic nature of system (20) and formula (22) conform to the expected domain of dependence conditions that $\mu(t,\omega)$ must satisfy.

When the mean functions ρ_0, K_0, γ and ζ are piecewise differentiable functions the above theory is extended easily. In between discontinuities one solves the equation (15) or (20) and across the points of discontinuity the solutions satisfy interface conditions in accordance with the nature of the discontinuity.

In the special case (5) of a homogeneous random half space we recover the results (9)-(11)

$$\mu(t,\omega)=\frac{1}{t}\mu(\sqrt{\alpha t}\ \omega) \tag{24}$$

with $\mu(\omega)$ given by (11).

Let us now consider inverse problems associated with the pulse reflection problem. Inverse problems associated with (3)-(4) are of little interest before the limit $\varepsilon\to 0$ because the usual procedures (c.f. [6]) are overwhelmed by the fluctuations. So we want to pose inverse problems after the limit, i.e. for the reflected process $A_{t,f}(\sigma)$. Perhaps the simplest such problem is this: what can we say about the slowly varying properties of the medium if we know the power spectral density $\mu(t,\omega)$ of the windows given by (22)? We have the following Theorem.

The limit

$$\lim_{\omega\to 0}\frac{1}{\omega^2}\mu(t,\omega)=\frac{8\pi\alpha(\tau^{-1}(t))}{c_0(\tau^{-1}(t))} \tag{25}$$

exists and is given by the right side of (25).

This Theorem is an immediate and elementary consequence of (20)-(23). Let us consider a simple application of it. Suppose that $\gamma(x)=\zeta(x)=1$ so that $\alpha(x)$ in (13) equals the constant α in (10). Let us assume that α is known and let

$$\Theta(t)=\frac{1}{8\pi\alpha}\lim_{\omega\to 0}\frac{1}{\omega^2}\mu(t,\omega)=\frac{1}{c_0(\tau^{-1}(t))} \tag{26}$$

be the quantity that is measured. From (23) we see that

$$\frac{d\tau^{-1}(t)}{dt}=-c_0(\tau^{-1}(t))=-\frac{1}{\Theta(t)}$$

Thus, the distance travelled up to time t, $\tau^{-1}(t)$, is obtained from the reflection statistics by

$$\tau^{-1}(t)=-\int_0^t\frac{1}{\Theta(s)}ds \tag{27}$$

Since $\tau(x)$ and $\tau^{-1}(t)$ are monotone inverse functions, (27) also determines the travel time $\tau(x)$ and hence the mean propagation speed $c_0(x)$ as a function of distance.

This result is in striking contrast to what the usual inverse methods give when applied to (3)-(4). It is known for example that quantities such as impedance can only be computed as functions of the (in general unknown from normal incidence data) travel time. In the stochastic case, at least when $\alpha(x)=\alpha$, we are able to determine the mean speed c_0 as a function of distance into the medium from the power spectra via (25)-(27). This appears paradoxical at first because we get more out of (the limit of) the noisy problem. We get more because we have access to more, namely power specra which are ensemble averages and contain therefore information from more than one realization. In many applied contexts one has only a single realization of the reflected signal available. Then $\mu(t,\omega)$ must be estimated statistically from this sample and so $\Theta(t)$ in (26) is known only approximately and depends on the realization i.e., it is random. But we know in addition that the reflected signal is a Gaussian process (in the limit $\varepsilon \to 0$). This information is now used to get sharp estimates for the travel time statistics.

It is clear that there are many interesting problems, direct and inverse, that can be posed for the pulse reflection problem. Our framework and results provide a theoretical basis for solving some of them.

The proofs of the above Theorems follow the setup of [3] which relies heavily on the asymptotic analysis of stochastic equations. The Theorem characterizing the reflection statistics from the inhomogeneous-in-the-mean medium is given in [7] along with applications to direct and inverse problems.

Acknowledgment

The work of G. Papanicolaou was supported by the National Science Foundation and by the Office of Naval Research.

References.

[1] L. van Wijngaarden, One-dimensional flow of liquid containing small gas bubbles, Ann. Rev. Fluid Mech. 4, 1972, 369-394

[2] R. Caflisch, M. Miksis, G. Papanicolaou and Lu Ting, Effective equations for wave propagation in bubbly liquids, J. Fluid Mech. 153, 1985, 259-273 and Wave propagation in bubbly liquids at finite volume fraction, 160, 1985, 1-14.

[3] R. Burridge, G. Papanicolaou and B. White, Statistics for pulse reflection from a randomly layered medium, Siam J. Appl. Math. 1987, to appear.

[4] R. Burridge, G. Papanicolaou and B. White, One dimensional wave propagation in a highly discontinuous medium, to appear.

[5] P. Sheng, Z.-Q. Zhang, B. White and G. Papanicolaou, Multiple scattering noise in one dimension: universality through localization lenght scales, Phys. Rev. Letters 57, number 8, 1986, 1000-1003.

[6] R. Burridge, The Gelfand-Levitan, the Marchenko and the Gopinath-Shondi integral equations of the inverse scattering theory regarded in the context of inverse impulse-response problems, Wave Motion, 2, 1980, 305-323.

[7] R. Burridge, G. Papanicolaou, P. Sheng and B. White, Direct and inverse problems for pulse reflection from inhomogeneously random halfspaces, to appear.

Statistical Properties of a Sedimenting Suspension and the Point Particle Approximation

RUSSEL E. CAFLISH*,† AND JONATHAN H. C. LUKE**,†

Abstract

For a sedimenting suspension of hard sphere in a viscous liquid, bulk properties (such as average sedimentation speed), are sensitive to the statistics of the particle positions. Since hydrodynamic interaction of two spheres does not change their relative positions, determination of particle statistics must involve interactions of at least three particles. We use the results of interactions of three point particles to approximately determine the two particle distribution function. The results are valid only for interparticle distances much larger than a particle radius and much smaller than the average interparticle distance.

I. Introduction

Consider a monodisperse suspension of solid spheres in a viscous liquid. The spheres fall through the fluid due to gravity and interact with the fluid and with each other through hydrodynamic forces; no other interparticle forces are included. The Reynolds number is assumed to be small enough that Stokes equations are valid, i.e. inertial forces are neglected both for particles and fluid.

The most important properties of such a sedimenting suspension are bulk properties, such as the average sedimentation speed and the coefficient of hydrodynamic dispersion (particle diffusion driven by

---

*Research partially supported by the National Science Foundation under grant number NSF-MCS-83-01260

**Research supported by a graduate fellowship from the Fannie and John Hertz Foundation

†Courant Institute of Mathematical Sciences, New York University

paricle interactions). If Brownian motion were included in the particle dynamics, the effective diffusion coefficient would be another important bulk paramter.

The objective of this paper is partial determination of the statistics of particle positions, namely the two-particle distribution function. The mechanism for this determination (analogous to a virial expansion) is calculation of three particle interactions using a point particle approximation. As described in section 2, two-particle interactions by themselves do not determine the particle statistics. The resulting two particle distribution function is only valid for interparticle distances much larger than a particle radius (because of the point particle approximation) and much smaller than the average interparticle distance (because of only three particle interaction). Interactions with the many distant particles are neglected, but are not expected to cause qualitative changes in the results.

Motivation for a careful analysis of particle statistics comes from the apparent sensitivity of a suspension's bulk properties to the particle distribution. The average sedimentation speed v_{sed} has been calculated for a periodic array of spheres by Hasimoto [7] and for a random, well-stirred (independent except for volume exclusion) array by Batchelor [1]. For the periodic array,

$$\underset{\sim}{v}_{sed} = (1 - c\beta^{1/3})\underset{\sim}{v}_{st} \tag{1.1}$$

while for the well-stirred array,

$$\underset{\sim}{v}_{sed} = (1 - 6.55\beta)\underset{\sim}{v}_{st} \tag{1.2}$$

in which $\beta = (4\pi/3)a^3n$ is the volume fraction of spheres and $\underset{\sim}{v}_{st} = mg6\pi\mu a$ is Stokes velocity for a single sphere in an unbounded fluid. The constant c in (1.1) depends on the type of periodic lattice. The particle radius is a; another important length scale is the average interparticle spacing $\ell = n^{-1/3}$, in which n is particle number density. The major difference between these formulae, the exponent 1 or 1/3, shows the sensitivity of v_{sed} to the particle distribution.

2. Point Particle Model

If the particles are very small compared to their inter-particle spacing, i.e. $a << \ell$, they may be treated as points. The total force exerted by a particle on the fluid is the same as the total external force on the particle, which is $F = mg = 6\pi\mu a v_{st}$. We scale the problem so that as $a \to 0$, F remains constant. Thus each particle acts as a point force in the fluid. Stokes equations for a suspension of point particles are then

$$\mu\nabla^2\underset{\sim}{u} - \underset{\sim}{\nabla}p = \underset{\sim}{F}\sum_{i=1}^{N}\delta(\underset{\sim}{x} - \underset{\sim}{x}_i) \tag{2.1}$$
$$\underset{\sim}{\nabla}\cdot\underset{\sim}{u} = 0$$

which has the solution

$$\text{(2.2)} \qquad \underset{\sim}{u}(\underset{\sim}{x}) = |F| \sum_{i=1}^{N} \underset{\sim}{U}(\underset{\sim}{x} - \underset{\sim}{x}_i)$$

in which $\underset{\sim}{U}$ is the Stokeslet given by

$$\text{(2.3)} \qquad \underset{\sim}{U}(\underset{\sim}{x}) = \frac{\underset{\sim}{e}}{|\underset{\sim}{x}|} + \frac{\underset{\sim}{x}(\underset{\sim}{e}\cdot\underset{\sim}{x})}{|\underset{\sim}{x}|^3}$$

and $\underset{\sim}{e} = \underset{\sim}{F}/|\underset{\sim}{F}|$.

According to Faxen's laws [3,6] the velocity of the ith particle in such a flow is (for a = 0)

$$\text{(2.4)} \qquad \dot{\underset{\sim}{x}}_i = \underset{\sim}{v}_i = \underset{\sim}{v}_{st} + |\underset{\sim}{F}| \sum_{j\neq i} \underset{\sim}{U}(\underset{\sim}{x}_i - \underset{\sim}{x}_j) \ .$$

The system of equations (3.4) describe the motion of a suspension of point particles. Simplify equation (3.4) by a galilean transformation to remove the term $\underset{\sim}{v}_{st}$ and by setting $|\underset{\sim}{F}| = 1$, to get

$$\text{(2.4')} \qquad \dot{\underset{\sim}{x}}_i = \sum_{j\neq i} \underset{\sim}{U}(\underset{\sim}{x}_i - \underset{\sim}{x}_j) \ .$$

If there are only two particles, N = 2, then

$$\text{(2.5)} \qquad \begin{aligned} \underset{\sim}{v}_1 &= \dot{\underset{\sim}{x}}_1 = \underset{\sim}{U}(\underset{\sim}{x}_1 - \underset{\sim}{x}_2) \\ \underset{\sim}{v}_2 &= \dot{\underset{\sim}{x}}_2 = \underset{\sim}{U}(\underset{\sim}{x}_2 - \underset{\sim}{x}_1) \ . \end{aligned}$$

Since $\underset{\sim}{U}$ is even, then $\underset{\sim}{v}_1 = \underset{\sim}{v}_2$, that is two isolated particles fall togther without changing their separation. Thus if only two particle interactions are included, any two-particle distribution function is an equilibrium.

3. Variance in the Sedimentation Speed

The result (1.2) for the average sedimentation speed was calculated under the assumption of a well-stirred distribution of particles, i.e. the two particle distribution function is

$$\text{(3.1)} \qquad P(\underset{\sim}{x}_1,\underset{\sim}{x}_2) = \begin{cases} P & \text{if } |\underset{\sim}{x}_1 - \underset{\sim}{x}_2| > 2a \\ 0 & \text{if } |\underset{\sim}{x}_1 - \underset{\sim}{x}_2| < 2a \ . \end{cases}$$

This distribution function is not exactly realizable, but replacement of (3.1) by the actual hard-spheres distribution function would not significantly change (1.2) or the results of this section. In this section it is shown that for a sedimenting suspension of point particles in a fixed container with a distribution function (3.1), the sedimentation speed has an infinite variance in the continuum limit.

Let the container be Ω_R in which $(4\pi/3)R^3 = \mathrm{vol}(\Omega_R)$, and for any R let Ω_R be a scaled version of Ω_1. Let N + 1 be the total number of particles in Ω_R. The continuum limit is $N \to \infty$, $a/R \to 0$ with $\beta = N(a/R)^3$ constant.

For a point particle at x_0, the velocity is

$$(3.2) \qquad \underset{\sim}{v} = \frac{|F|}{\mu} \sum_{i=1}^{N} \underset{\sim}{U}(\underset{\sim}{x}_0, \underset{\sim}{x}_i, R)$$

in which $|F| = 6\pi\mu a|\underset{\sim}{v}_{st}|$ and $\underset{\sim}{U}(\underset{\sim}{x}_0, \underset{\sim}{x}_i, R)$ is the velocity at x_0 due to a vertical unit point force at $\underset{\sim}{x}_i$ in the container Ω_R. Thus $\underset{\sim}{U}(\underset{\sim}{x}_1, \underset{\sim}{y}_1, \underset{\sim}{R})$ satisfies

$$(3.3) \qquad \nabla_x^2 \underset{\sim}{U} - \underset{\sim}{\nabla}_x \Pi = \hat{\underset{\sim}{e}}\,\delta(\underset{\sim}{x} - \underset{\sim}{y})$$

$$(3.4) \qquad \underset{\sim}{\nabla}_x \cdot \underset{\sim}{U} = 0 \quad \text{for } \underset{\sim}{x} \in \Omega_R$$

$$(3.5) \qquad \underset{\sim}{U} = 0 \quad \text{for } \underset{\sim}{x} \in \partial\Omega_R$$

with $\hat{\underset{\sim}{e}}$ the unit vertical vector.

It is easy to see that $\underset{\sim}{U}$ satisfies

$$(3.6) \qquad \underset{\sim}{U}(\underset{\sim}{x}, \underset{\sim}{y}, R) = R^{-1}\underset{\sim}{U}(R^{-1}\underset{\sim}{x}, R^{-1}\underset{\sim}{y}, 1) .$$

Moreover averaging over $\underset{\sim}{y}$, with $\langle\cdot\rangle = |\Omega_R|^{-1} \int_{\Omega_R} \cdot\, d\underset{\sim}{y}$ results in

$$(3.7) \qquad \nabla_x^2 \langle \underset{\sim}{U} \rangle - \underset{\sim}{\nabla}_x \langle \Pi \rangle = \hat{\underset{\sim}{e}}$$

$$(3.8) \qquad \underset{\sim}{\nabla}_x \cdot \langle \underset{\sim}{U} \rangle = 0 \quad \text{for } \underset{\sim}{x} \in \Omega_R$$

$$(3.9) \qquad \langle \underset{\sim}{U} \rangle = 0 \quad \text{for } \underset{\sim}{x} \in \partial\Omega_R .$$

The only solution of (3.7) - (3.9) is

$$(3.10) \qquad \langle \underset{\sim}{U} \rangle = 0 .$$

Next define for $\underset{\sim}{x} \in \Omega_1$

$$\sigma(\underset{\sim}{x}) = (6\pi)^{-2} \langle |\underset{\sim}{U}(\underset{\sim}{x}, \cdot, 1)|^2 \rangle$$

$$(3.11) \qquad = (6\pi)^{-2} |\Omega_1|^{-1} \int |\underset{\sim}{U}(\underset{\sim}{x}, \underset{\sim}{y}, 1)|^2 dy .$$

With these preliminaries finished it is easy to calculate $\langle|\underset{\sim}{v}|^2\rangle$. Since the $\underset{\sim}{x}_i$'s are independent and identically distributed, $\langle\underset{\sim}{U}(\underset{\sim}{x}_0,\underset{\sim}{x}_i)\cdot\underset{\sim}{U}(\underset{\sim}{x}_0,\underset{\sim}{x}_j)\rangle = \delta_{ij}\langle|\underset{\sim}{U}(\underset{\sim}{x}_0,\underset{\sim}{x}_1)|^2\rangle$ so that

$$\begin{aligned}\langle|\underset{\sim}{v}|^2\rangle &= (6\pi a|\underset{\sim}{v}_{st}|)^2 N\langle|\underset{\sim}{U}(\underset{\sim}{x}_0,\underset{\sim}{x}_1,R)|^2\rangle \\ &= (6\pi a|\underset{\sim}{v}_{st}|)^2\, NR^{-2}\langle|\underset{\sim}{U}(\underset{\sim}{x}_0/R,\underset{\sim}{x}_1/R,1)|^2\rangle \\ &= |\underset{\sim}{v}_{st}|^2 N(a/R)^2\ \sigma(x_0/R)\ . \end{aligned} \tag{3.12}$$

The result (3.12) can also be written as

$$\langle|\underset{\sim}{v}|^2\rangle = |\underset{\sim}{V}_{st}|^2 N^{1/3}\beta^{2/3}\sigma(x_0/R)\ . \tag{3.13}$$

Since σ is independent of N, a and R and $|\underset{\sim}{v}_{st}|$ sets the scale for $\underset{\sim}{v}$, the variance $\langle|\underset{\sim}{v}|^2\rangle \to \infty$ in the continuum limit. This result was first presented in [2].

4. Three Particle Problem

The equations for three particles are

$$\begin{aligned}\dot{\underset{\sim}{x}}_1 &= \underset{\sim}{U}(\underset{\sim}{x}_1 - \underset{\sim}{x}_2) + \underset{\sim}{U}(\underset{\sim}{x}_1 - \underset{\sim}{x}_3) \\ \dot{\underset{\sim}{x}}_2 &= \underset{\sim}{U}(\underset{\sim}{x}_2 - \underset{\sim}{x}_1) + \underset{\sim}{U}(\underset{\sim}{x}_2 - \underset{\sim}{x}_3) \\ \dot{\underset{\sim}{x}}_3 &= \underset{\sim}{U}(\underset{\sim}{x}_3 - \underset{\sim}{x}_1) + \underset{\sim}{U}(\underset{\sim}{x}_3 - \underset{\sim}{x}_2)\ . \end{aligned} \tag{4.1}$$

If the three particles are located on a horizontal equilateral triangle, then they fall together without changing their relative positions. However, this equilibrium configuration can be shown to be unstable, although only algebraically so.

For any other 3-particle configuration, we believe that the particle trajectories can be characterized as follows: At t near $-\infty$ there is a pair of particles very far above the third single particle. As time progresses the pair falls toward the single particle (since two particles fall faster than one). As they get closer the three particles interact, but eventually a pair of them (not necessarily the original pair) will pull away from the third. At t near $+\infty$, the pair will be very far below the single particle. A particle pair may be described by the distance r between the pair, and the cosine of the angle that the pair makes with the vertical. Thus the three particle interaction may be thought of as a scattering event, by which an initial pair described by (r_i,α_i) (near $t=-\infty$) goes to a final pair described by (r_f,α_f) (near $t=+\infty$). This description is consistent with previous studies of three particle interactions [5,8].

In this collision between a pair and a single particle, the relative position of the single particle can be described by impact parameters, just as in the kinetic theory of gases. The impact parameters describe the limiting (as $t \to -\infty$) position of the third particle, in a plane perpendicular to the relative velocity between the pair the the third particle. Because of the long range of the hydrodynamic interaction, there is also a logarithmic correction to the impact parameters. However no logarithmic correction is needed in the pair-separation vector. A graphical description of the impact parameter, with or without logarithmic correction, is depicted in figure 1.

Precise definitions of the pair separation radius vector $\mathbf{z}_0$ before the collision and $\mathbf{z}_1$ after the collision are

$$\text{(4.2)} \qquad \mathbf{z}_0 = \lim_{t\to-\infty} \frac{1}{2}(\mathbf{x}_1(t) - \mathbf{x}_2(t))$$

$$\text{(4.3)} \qquad \mathbf{z}_1 = \lim_{t\to\infty} \frac{1}{2}(\mathbf{x}_1(t) - \mathbf{x}_2(t)) \ .$$

These limits are approached at the rate $|t|^{-1}$. Note that the labels 1,2,3 may need to be permuted in (4.2) and (4.3), since the particles in the pair after the collision may be different from those in the pair before the collision. A precise definition of the impact vector $\mathbf{y}_\perp$, involving a logarithmic correction, is calculated in [4] to be

$$\text{(4.4)} \qquad \mathbf{y}_\perp = \lim_{t\to-\infty} \mathbf{y}(t) - (t - t_0)\mathbf{v}_0 - \operatorname{sgn}(t - t_0)\log|1 - t/t_0|\mathbf{v}_1$$

in which the pair-single particle separation y(t), the pair-single particle relative velocity $\mathbf{v}_0$ and the logarithmic correction velocity $\mathbf{v}_1$ are defined by

$$\text{(4.5)} \qquad \mathbf{y}(t) = \frac{1}{2}(\mathbf{x}_1 + \mathbf{x}_2) - \mathbf{x}_3$$

$$\text{(4.6)} \qquad \mathbf{v}_0 = \mathbf{U}(2\mathbf{z}_0)$$

$$\mathbf{v}_1 = -(\mathbf{u}_1 \cdot \nabla)\mathbf{U}(\mathbf{x}_0) - \mathbf{U}(\mathbf{v}_0) \text{(4.7)}$$

in which $\mathbf{u}_1 = 4(\mathbf{z}_0 \cdot \nabla)\mathbf{U}(\mathbf{v}_0)$. The constant t_0 in (4.4) is chosen so that $\mathbf{y}_\perp \cdot \mathbf{v}_0 = 0$. The limit in (4.4) is approached at the rate $|t|^{-1}\log|t|$. It is important to note that $\mathbf{v}_0, \mathbf{v}_1$ and t_0 depend only on the pair separation and not on the location of the third particle; otherwise it might not have been possible to define an impact vector.

It is convenient to use spherical coordinates for the pair separations $\mathbf{z}_0$ and $\mathbf{z}_1$ and polar coordinates for the impact vector $\mathbf{y}_\perp$. Define

$$r_0 = 2|\mathbf{z}_0| \qquad\qquad r_1 = 2|\mathbf{z}_1|$$

$$\alpha_0 = (\mathbf{e} \cdot \mathbf{z}_0)/|\mathbf{z}_0| \qquad\qquad \alpha_1 = (\mathbf{e} \cdot \mathbf{z}_1)/|\mathbf{z}_1|$$

$$\delta = |\mathbf{y}_\perp| \qquad\qquad \psi = \cos^{-1}(\mathbf{y}_\perp \cdot \kappa/\delta)$$

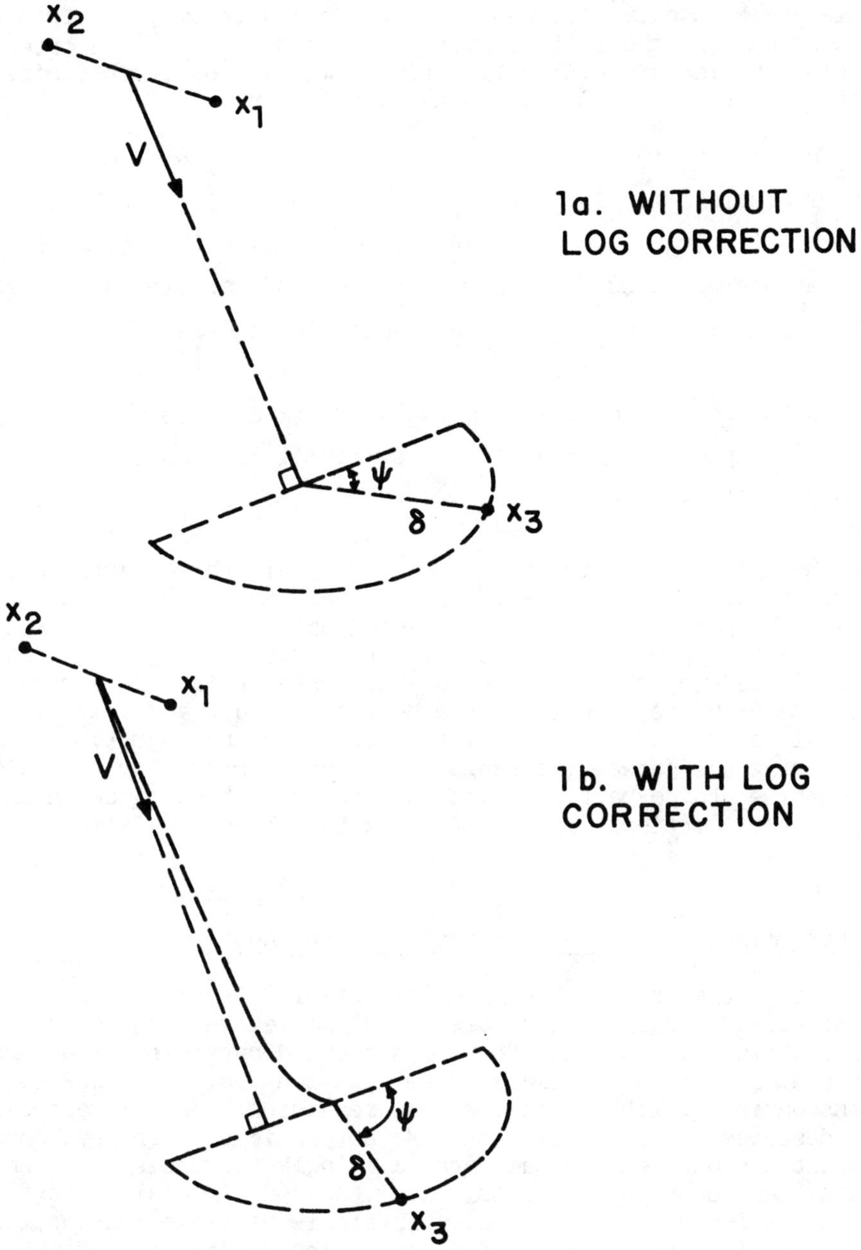

FIGURE 1 IMPACT PARAMETERS (ψ, δ)

in which $\underset{\sim}{\kappa}$ is a fixed unit vector in the plane perpendicular to v_0. Since the suspension is assumed to be homogeneous and the pair and third particle are infinitely far apart at $t = -\infty$, the impact parameters δ and ψ are assumed to be uniformly distributed and independent of the initial pair parameters (r_0,α_0). In order to make the distribution of (δ,ψ) well-defined, we put on (artificial) upper limit D on δ, so that $0 < \delta < D$ and $0 < \psi < 2\pi$.

The scattering from (r_0,α_0) to (r_1,α_1) can be written as a (deterministic) mapping $(r_1,\alpha_1) = (R,A)(r_0,\alpha_0,\delta,\psi)$. This mapping scales with r_0, i.e. $(r^{-1}R,A)(r,\alpha_0,r\delta,\psi) = (R,A)(1,\alpha_0,\delta,\psi)$. The computational results of a large number of single scatterings for various values of α_0 $(r_0=1)$ are graphed in figures 2-5. In any one of the graphs the points given the outcomes (r_1,α_1) for 10,000 scatterings each with the same value of α_0 and with different, randomly chosen values of impact paramters δ,ψ. The initial separation $(r_0=1,\alpha_0)$ is indicated with a cross.

One conlcusion from this study of single three-particle collsions is that a vertical pair $(\alpha_0=1)$ is the most stable configuration and a horizontal pair $(\alpha_0=0)$ is the least stable. This is evidenced by the small changes in pair parameter r, α for a vertical pair and the much larger possible changes for a horizontal pair. This result, which contrasts with the stability of a horizontal pair and instability of a vertical pair in a convective flow, can be partly understood by considering the parametric range of the scattering: for a vertical pair the entire three-particle interaction will take place in the (time-independent) vertical plane containing the three particles.

5. Determination of the Pair-Distribution Function

Using the results of three-particle interactions, which are numerically calculated as described above, we wish to determine the pair distribution function for the suspended particles as accurately as possible. This approximate determination is analogous to a virial expansion in equilibrium statistical mechanics. In this approximation, we describe the evolution of a particle pair by considering interactions between the pair and a single particle for which the impact parameter δ is less than the (artificial cutoff) D. Two physical effects are neglected: pair-single particle interactions with $\delta > D$ (collisions with a distant third particle) and the possible shearing motion in the net fluid flow caused by the distribution of distant particles. Our neglect of these two effects is unjustified but we believe that their inclusion would not qualitatively change the results.

For a very dilute suspension the time between collisions for a given pair will be long compared to the duration of the interaction during any collision. So the collisions may be considered as instantaneous, as in the kinetic theory of gases. Thus the pair

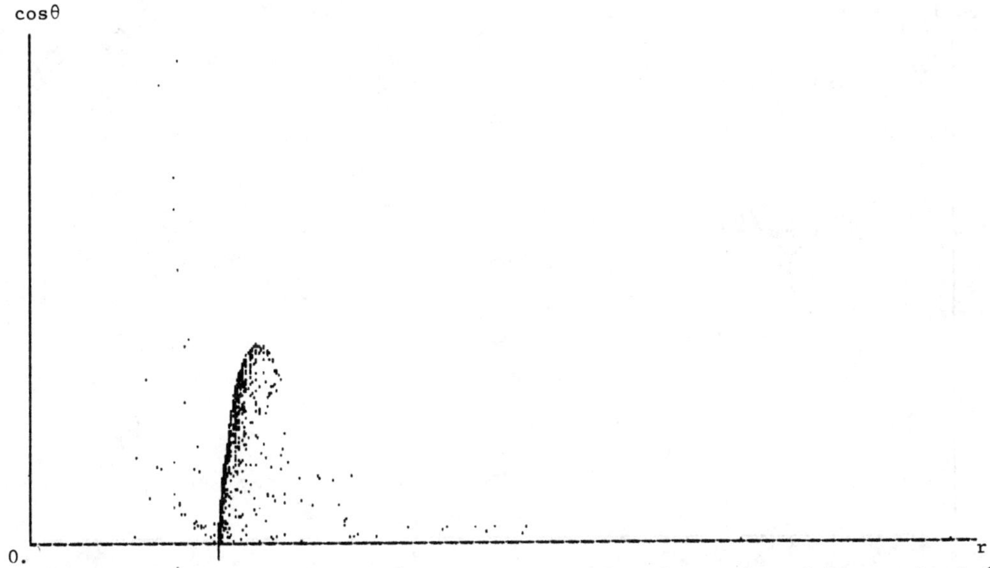

Figure 2. Values of pair separation r and pair angle with vertical θ before and after interaction with third particle. + gives (r,θ) before collision; . gives (r,θ) after collision. For all interactions r=1 and cosθ=0 before collision. Different dots correspond to 1000 random choices of different positions for third particle.

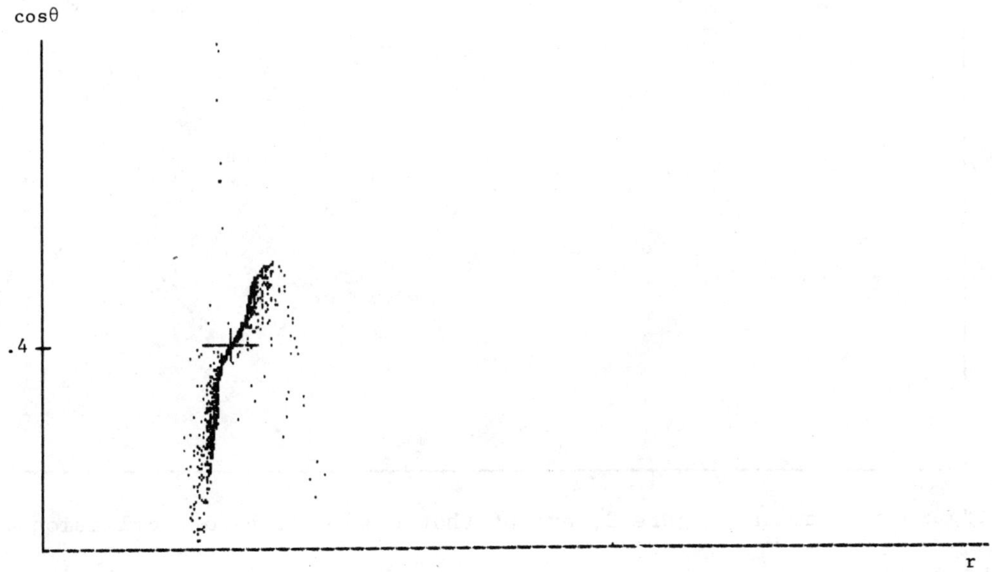

Figure 3. Same as figure 2, except that cosθ = .4 before collision.

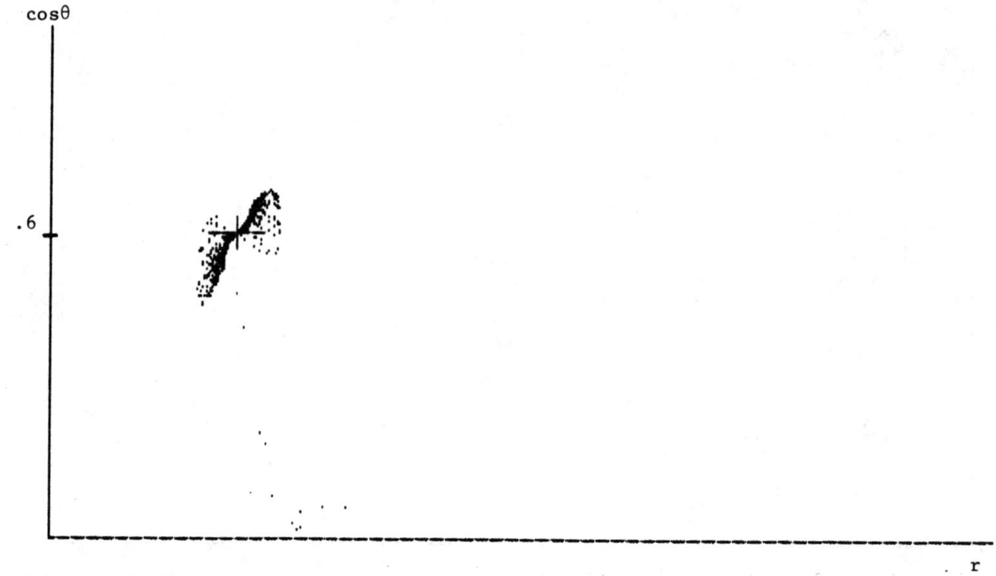

Figure 4. Same as figure 2, except that cosθ = .6 before collisions.

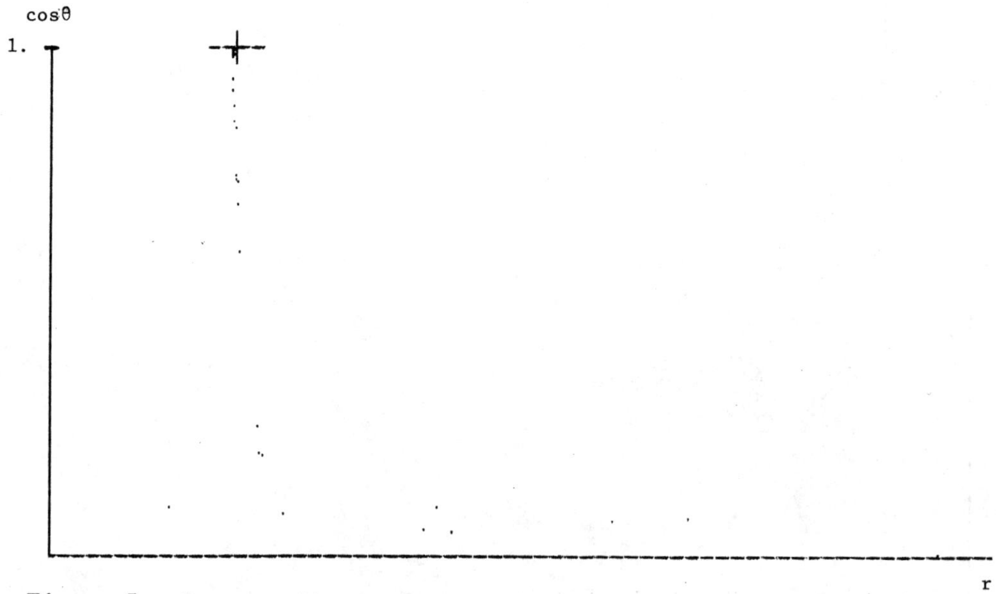

Figure 5. Same as figure 2, except that cosθ = 1. before collisions.

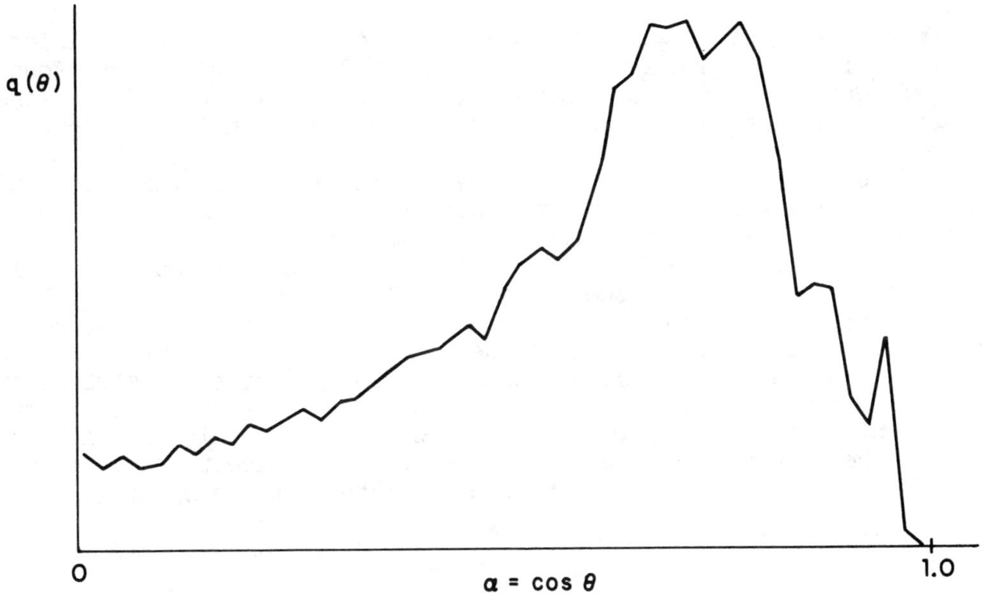

FIGURE 6 ANGULAR DISTRIBUTION
30,000 TRIALS, D = 160r

separation vector (r,α) may be thought of as a stochastic process that undergoes discrete instantaneous jumps at random times. The rate of jumps is proportional to the velocity $U((r,\alpha,0)) = r^{-1}v(\alpha)$ of the pair relative to a distant single particle. The outcome of a jump is $(r',\alpha') = (R,A)(r,\alpha,\delta,\psi)$ in which the impact parameters δ,ψ are chosen randomly. This stochastic process is described in details in [4].

Here we are interested in the equilibrium distribution $p(r,\alpha)$ for this stochastic process; $p(r,\alpha)$ is the approximate pair distribution function. Because the jump rate and the jump outcome both scale with r, the equilibrium distribution may be sought in separated form $p(r,\alpha) = r^m q(\alpha)$ [4]. Simulation of this stochastic process with $D = 160r$ and with 30,000 trials has yielded the angular distribution plotted in figure 6. This result should be regarded as tentative. The exponent m is more difficult to accurately determine, but preliminary results indicate at least that $m > 0$. It would be most interesting to determine whether $m < 2$, $m = 2$ or $m > 2$ corresponding to an overpopulation, even population or underpopulation of nearby pairs.

6. Conclusion

The infinite variance result of section 3 suggests that the well-stirred distribution for a fluid-particle suspension may be incorrect. In order to determine the correct dynamically determined equlibrium distribution we have considered interactions of 3 point particles, which can be described as a collision between a pair of particles and a single particle.

Our first result was calculation of the range of such collisions between a given pair and a randomly chosn third particle. We found that a vertical pair is fairly stable to collision, while a more horizontal pair may be more greatly changed as shown in figures 2-5.

Our second result is a preliminary approximate determination of the equilibrium pair-distribution function $p(r,\alpha)$, taking account only of three particle interactions. We found that p factors as $p(r,\alpha) = r^m q(\alpha)$. Determination of m and q is only tentative. If $m > 0$, as we believe, then the point particle approximation is valid. By contrast, if m were negatve a pair separations would become continually closer due to collisions, so that the point particle approximation would be invalid eventually. Note that the distribution $p(r,\alpha)$ is not normalizable. This is because it is not valid for large r, at which multiple particle interactions must dominate. At such large r, the distribution p will be cutoff. We expect that length scale for this cutoff to be ℓ, the average interparticle spacing.

7. References

[1] G.K. Batchelor, "Sedimentation in a dilute suspension of spheres." JFM 52 (1972) 245-268.

[2] R.E. Caflisch and J.H.C. Luke, "Variance in the sedimentation speed of a suspension." Physics of Fluids 28 (1985) 759-760.

[3] R.E. Caflisch and J. Rubinstein, Lectures on the Mathematical Theory of Multi-Phase Flow (1986) CIMS.

[4] R.E. Caflisch and J.H.C. Luke, "Stochastic modelling of a dilute fluid-particle suspension." Proc. Workshop Stochastic Processes and Hydrodynamics, Inst. Math. and Its Appl., Minneapolis (1985), to appear.

[5] P. Ganatos, R. Pfeffer and S. Weinbaum, "A numerical solution technique for three-dimensional Stokes flows, with applications to the motion of strongly interacting spheres in a plane." JFM 84 (1978) 79-111.

[6] J. Happel and H. Brenner, Low Reynolds Numbers Hydrodynamics (1983) Nÿhoff.

[7] H. Hasimoto, "On the periodic fundamental solutions of the Stokes equations and their application to viscous flow past a cubic array of spheres." JFM 5 (1959) 317-328.

[8] L.M. Hocking, "The behavior of clusters of spheres falling in a viscous fluid. Part 2 Slow motion theory." JFM 20 (1964) 129-139.

A Model for Estimating the Effective Thermal Conductivity of a Random Suspension of Spheres

E. CHANG*, B. S. YENDLER* AND A. ACRIVOS*

Abstract. We present two different effective medium approximations, referred to as methods A and B, and apply them to the problem of finding the effective thermal conductivity of a composite consisting of a matrix containing a random dispersion of spheres whose thermal conductivity is different from that of the matrix. This problem is equivalent to that of finding the average dipole strength of any one of the spheres, the so-called test sphere, which, according to the aforementioned effective medium approximations, is then computed by replacing the composite surrounding this sphere with a continuum having a position dependent effective thermal conductivity. The ensuing calculations require only knowledge of a pair distribution function in order to characterize the statistics of the composite. Although the two methods are similar in the way they replace the composite surrounding the test sphere with an effective continuum, they differ as to how they treat the bulk and local fields.

We then compare the values of the effective thermal conductivity thus obtained with other theoretical and experimental results and find that method B gives the closest agreement with the experimental data when the volume fraction is large and the inclusion conductivity is high.

Finally, via the method of averaged equations, we present a theoretical justification of method A to second order in the inclusion volume fraction. Unfortunately, we were unable to-date to obtain a comparable justification for method B, even though it is similar to method A and appears to be superior.

* Department of Chemical Engineering, Stanford University
Stanford, California 94305

Introduction. The problem of determining the effective thermal conductivity of a two-phase material, in which one of the phases is finely dispersed within the other, has attracted the attention of numerous investigators for over a century. These, primarily theoretical, studies can be conveniently grouped into three categories: a) exact calculations which are generally restricted to systems containing either i) small amounts of the dispersed phase, or, ii) identically shaped inclusions, e.g. spherical particles, arranged in spatially periodic configurations, e.g. cubic lattices; b) variational methods leading to exact bounds which, however, diverge whenever the thermal conductivities of the two phases are widely dissimilar; and c) approximate, and generally **ad hoc**, techniques whose reliability is often impossible to ascertain **a priori**.

The purpose of this paper is to describe yet another approximate method which, although also **ad hoc** and therefore of **a priori** unknown reliability and accuracy, has been found to predict values for the effective thermal conductivity, as well as for other effective parameters, in suprisingly close agreement with those obtained experimentally or from exact calculations.

To this end, consider a macroscopically homogeneous composite which consists of a continuous matrix and a random dispersion of identical spherical inclusions of unit radius and volume concentration c. The thermal conductivity of the matrix is set equal to unity and that of the inclusions is denoted by α. Also suppose that a bulk temperature gradient of unit strength is applied across the material in the +x direction. Under these conditions, the effective thermal conductivity k^* is given exactly by [1]

$$k^* = 1 + (\alpha - 1)\, c \left\langle \left(\frac{\partial T}{\partial x} \right)_0 \right\rangle \tag{1}$$

where $(\partial T/\partial x)_0$ is the x-component of the temperature gradient evaluated at the origin which is fixed at the center of a given inclusion henceforth referred to as the **test particle.** In addition, the brackets $\langle\ \rangle$ denote the conditional average of the indicated quantity over all possible locations of the other spheres within the composite given that the test particle is at the origin.

In the limit $c \to 0$ the temperature profile within the test particle is of course uninfluenced by all the other spheres, and hence, by solving a conduction problem for a single sphere, one obtains readily that

$$\left\langle \frac{\partial T}{\partial x} \right\rangle = \frac{3}{\alpha + 2} \qquad \text{for} \quad 0 \le r \le 1$$

which, in conjunction with (1), leads directly to Maxwell's well-known expression

$$k^* = 1 + 3\beta c + O(c^2) \;, \qquad \beta \equiv \frac{\alpha - 1}{\alpha + 2} \tag{2}$$

The extension of eq. (2) to $O(c^2)$ requires both the solution of a conduction problem involving two spheres, for all values of their center-to-center distance R, plus knowledge of the pair correlation function g(R) which, for a well-mixed suspension of hard spheres, is, as $c \to 0$,

$$g(R) = H(R - 2) \tag{3}$$

with H denoting the Heaviside function. The resulting exact expression for k* is then [1]

$$k^* = 1 + 3\beta c + 3\beta^2 c^2 [\, 1 + 2\delta(\alpha)\beta \,] + O(c^3) \tag{4}$$

where $\delta(\alpha)$ was found to depend weakly on α and to equal 0.22, 0.21, and 0.25 for, respectively, $\alpha = 0$, 1, and ∞. Although, in principle, the expansion in c could be continued to $O(c^3)$ and beyond, this would require the solution of a three-sphere conduction problem for an arbitrary arrangement of the three particles, followed by the determination of the corresponding ensemble average using the appropriate 3-particle statistics, clearly an almost unsurmountable task. Hence the need to construct tight bounds and to develop approximate techniques.

An Effective Medium Approximation. In what follows, we shall present two approximate methods, to be referred to henceforth as methods A and B, both of which involve the solution of a heat conduction problem in which a single sphere is viewed as being embedded in an effective medium with a position-dependent effective thermal conductivity.

a) Method A We begin by considering the extension of eq. (2), Maxwell's expression for the effective thermal conductivity, to higher order in c. According to method A, a preliminary version of which has already been presented [2], we replace the composite surrounding the test sphere with an effective medium having a position-dependent effective thermal conductivity κ of the form

$$\kappa(r,\alpha,c) = 1 + (k^* - 1)\, \rho(r,c)/c \quad , \qquad r \geq 1 \tag{5}$$

where $k^*(\alpha,c)$ is the unknown effective thermal conductivity of the mixture, and $\rho(r,c)$ is the probability that any given point, which is measured with respect to the center of the test sphere, lies within any of the other spheres comprising the composite. Moreover, since for impenetrable spheres it can be shown that

$$\rho(r,c) = \frac{3c}{4r} \int_{r-1}^{r+1} g(R,c)\ (-R^2 - r^2 + 1 + 2Rr)\ R\ dR \qquad (6)$$

it is easy to see that the probability density function ρ is given uniquely in terms of the statistics of the particle configuration. In particular for a well-mixed suspension of hard spheres, we have in view of eq.(3) that, in the dilute limit,

$$\rho = \frac{c}{16r} (-r^4 + 30r^2 - 56r + 27) + O(c^2) \qquad (7)$$

Thus, the boundary-value problem to be solved is:

$$\nabla^2\langle T\rangle = 0 \text{ for } 0 \leqq r \leqq 1 \quad ; \quad \nabla\cdot\kappa\nabla\langle T\rangle = 0 \text{ for } r \geqq 1, \qquad (8)$$

together with the boundary conditions:

$$\langle T\rangle \to x \text{ as } r \to \infty, \qquad (9a)$$

$$\langle T_-\rangle = \langle T_+\rangle \text{ and } \alpha \left(\frac{\partial\langle T\rangle}{\partial r} \right)_- = \left(\frac{\partial\langle T\rangle}{\partial r} \right)_+ \text{ at } r = 1 \qquad (9b)$$

where the subscripts - and + refer, respectively, to the interior and exterior faces of the surface $r = 1$. Finally, eq. (1) relates the unknown effective thermal conductivity k^*, which appears in the expression for κ given by eq. (5), to the value of $\partial\langle T\rangle/\partial x$ at $r = 0$ as computed from the solution of eqs. (8) or (9). It is easily seen moreover that, since κ is assumed to be radially symmetric, the system given above reduces to an ordinary differential equation via separation of variables.

It is clear from eq.(1) that the expansion for k^* in terms of c can be carried out by successive approximations. Thus, to obtain k^* to $O(c^2)$ it suffices to know $(\partial\langle T\rangle/\partial x)_0$ correct to $O(c)$. In turn, the latter can be computed from the solution of eqs. (8) and (9) with κ set equal to $1 + 3\beta\rho$ and $\rho(r)$ given by eq. (7). It can be shown readily that the expression for k^* thus obtained is

$$k^* = 1 + 3\beta c + 3\beta^2 c^2 (1 + 0.42\beta) + O(c^3) \qquad (10)$$

which is in excellent agreement with the exact result given by eq. (4). Higher order terms can then be derived by substituting the above in eqs. (5) and (8) and then determining $(\partial\langle T\rangle/\partial x)_0$ correct to $O(c^2)$. Before this can be done, however, it is necessary that the expression for ρ given by (7) also be extended to $O(c^2)$ which in turn requires that the pair-correlation function $g(r,c)$ be known to $O(c)$.

Although the exact expression for this function will of course depend on the actual mode of preparation of the composite material, especially at higher values of c, we shall adopt an assumption commonly made and suppose throughout this work that, for a well-mixed suspension, g(r,c) can be set equal to the pair-correlation function of a hard sphere fluid in equilibrium which, as $c \to 0$, takes on the form [3]

$$g(r,c) = H(r-2)\left[1 + c\left(8 - 3r + \frac{r^3}{16} \right) H(4-r) + O(c^2) \right] \tag{11}$$

Inserting the above into eq. (6) and evaluating eqs. (8) and (9) to $O(c^2)$, we obtain that

$$k^*=1+3\beta c+3\beta^2c^2(1+0.42\beta)+3\beta^3c^3(0.12+0.76\beta+0.36\beta^2)+O(c^4) \tag{12}$$

Although, in principle, this expansion could have been continued further, this proved to be impractical and, instead, the effective thermal conductivity was evaluated for selected values of c by first calculating the pair-probability function -- in a manner to be described in the next section -- and then solving numerically, by iteration, the mathematical system consisting of eqs. (1), (5), (8) and (9), as simplified using separation of variables.

b) Method B This technique was developed in the U.S.S.R., primarily by Buyevich and his collaborators, and has already been described in a number of publications [4],[5],[6]. The essence of the method lies in the assumption that, in the region external to the test sphere, the heat flux **q** is given by

$$\mathbf{q} = -k^* \nabla\bar{T} - \kappa \nabla t \tag{13}$$

where $\bar{T}$ is the unconditional bulk temperature, which varies over a length scale comparable to the macroscale, and $t \equiv \langle T\rangle - \bar{T}$ is the disturbance temperature due the the presence of the test sphere which of course varies over distances comparable to the microscale, i.e. the sphere's radius. Also, k* is the unknown constant effective conductivity, corresponding to the volume fraction c, and κ is its position dependent counterpart as given by eq. (5). Consequently, since for the problem at hand, $\bar{T} = x$, the boundary value problem to be solved is, in lieu of eqs. (8) and (9),

$$\nabla^2\langle T\rangle = 0 \text{ for } 0 \leq r \leq 1 \;;\quad \nabla\cdot\kappa\nabla t = 0 \text{ for } r \geq 1 \tag{14}$$

$$t \to 0 \text{ as } r \to \infty , \tag{15a}$$

$$\langle T_-\rangle = x + t_+ \quad\text{and}\quad \alpha\left(\frac{\partial\langle T\rangle}{\partial r} \right)_- = k^*x + \left(\frac{\partial t}{\partial r} \right)_+ \quad \text{at } r = 1 \tag{15b}$$

plus eq.(1). Once again this system reduces to an ordinary differential equation by separation of variables and can be solved in a manner identical to that described earlier in our discussion of method A.

Suprisingly, both methods yield the same expression for k* to $O(c^2)$, viz. eq. (10), but thereafter they diverge as seen from the formula for k* to $O(c^3)$ given by method B,

$$k^*=1+3\beta c+3\beta^2c^2(1+0.42\beta)+3\beta^3c^3(1.09+.66\beta+.57\beta^2)+O(c^4) \qquad (16)$$

This divergence of the two methods is to be expected since, if expressed in terms of $\bar{T}$ and t, eqs. (8) and (9) of method A reduce to

$$\nabla^2\langle T\rangle = 0 \text{ for } 0 \leqq r \leqq 1 \;; \quad \nabla\cdot\kappa\nabla t = -\frac{x}{r}\frac{d\kappa}{dr} \text{ for } r \geqq 1$$

$$t \to 0 \text{ as } r \to \infty$$

$$\langle T_-\rangle = x + t_+ \text{ and } \alpha\left(\frac{\partial\langle T\rangle}{\partial r}\right)_- = x + \left(\frac{\partial t}{\partial r}\right)_+ \text{ at } r = 1$$

which of course differ from (14) and (15). Nevertheless, although differing in detail, both methods are similar in spirit in that they combine an effective medium approximation with an exact self-consistent scheme, eq. (1), to yield estimates for the effective transport parameters, in this case the effective thermal conductivity, of two-phase materials. We also note that although numerous studies have already appeared in the literature on this subject in which effective medium theories, self-consistent schemes, as well as position dependent effective properties have all been employed in one form or another, none of them seem to have combined these three in quite the same manner as was done above. Hence, the approach adopted here appears to contain a certain degree of novelty.

c) Description of Numerical Method and Results We now return to describe, in more detail, the numerical technique that was used to obtain estimates for k* over the whole range of c under the assumptions of methods A and B. First of all, as shown in the Appendix, a closed form analytic expression for the pair-distribution function was derived via a solution of the Percus-Yevick integral equation [7], and the corresponding expression for the probability density function $\rho(r,c)$ was then obtained from eq. (6) using MACSYMA. In addition, the simplification $g(r,c) \equiv 1$ for $r \geqq 4$ was introduced which was found to affect $\rho(r,c)$ by no more than 4%, even for c as large as 0.6. Given, then, $\rho(r,c)$ for a chosen value of c, the ordinary differential equation, which resulted from eqs. (8) and (9) or (14) and (15) together with the consistency condition (1) through separation of variables, was solved numerically for an assumed value of the parameter k*, and the iteration was repeated until the remaining boundary condition at infinity was satisfied. The values for k* thereby determined for $\alpha = 0$ and $\alpha = \infty$ are listed in Tables 1 and 2, and are seen plotted in Figures 1 and 2. Also shown in Table 2 are the values of k*, as computed via method B but using, for the pair distribution function $g(r,c)$, eq. (3) for the entry denoted B_1, and eq. (11) for B_2. The numbers in the 3 columns, B, B_1, and B_2,

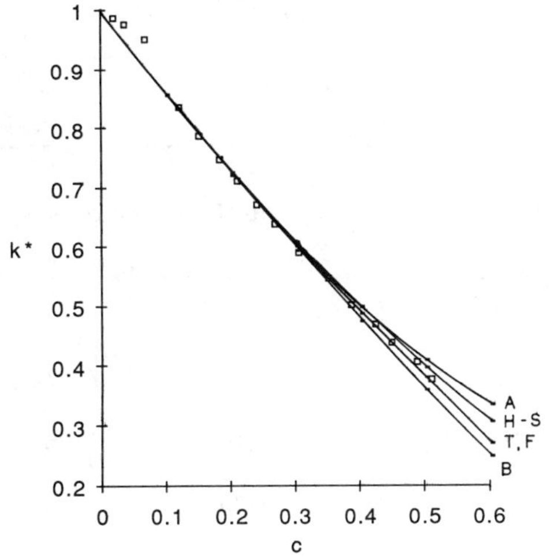

FIG. 1. k^* for $\alpha = 0$. The squares are the measurements of Meredith and Tobias [22]. The curves are labelled according to the text. The curve labelled H-S is the second order upper bound for k^* derived by Hashin and Shtrikman [15]. Note that curves T and F are indistinguishable.

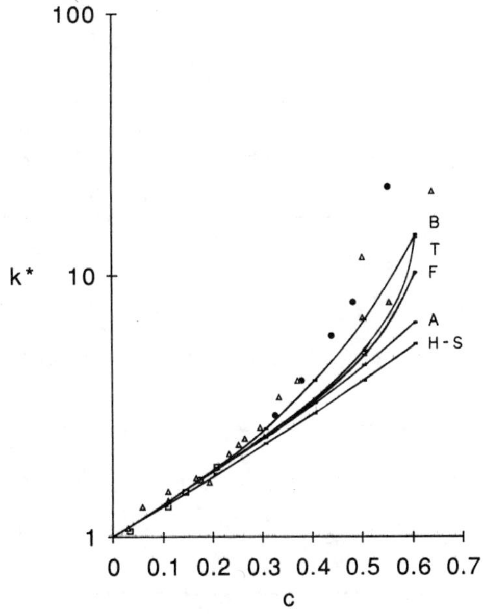

FIG. 2. k^* for $\alpha = \infty$. The squares are the measurements of Meridith and Tobias [22]; the circles are the measurements of Turner [16]; and the triangles are the measurements of Pierce [23]. The curve labelled H-S is the second order lower bound for k^* derived by Hashin and Shtrikman [15].

TABLE 1

$\alpha = 0$

c	A	B	B_1	B_2	Torquato	Felderhof	Hashin-Shtrikman upper bound
0.1	.856	.856	.856	.856	.856	.856	.857
0.2	.724	.721	.723	.721	.723	.723	.727
0.3	.606	.595	.596	.596	.602	.602	.609
0.4	.502	.476	.488	.478	.489	.490	.500
0.5	.412	.361	.385	.362	.383	.383	.400
0.6	.336	.250	.292	.279	.271	.273	.308

TABLE 2

$\alpha = \infty$

c	A	B	B_1	B_2	Torquato	Felderhof	Hashin-Shtrikman upper bound
0.1	1.35	1.35	1.35	1.35	1.348	1.347	1.33
0.2	1.81	1.85	1.82	1.85	1.815	1.811	1.75
0.3	2.42	2.63	2.49	2.59	2.457	2.442	2.29
0.4	3.28	3.99	3.48	3.79	3.402	3.358	3.00
0.5	4.55	6.82	5.03	5.88	5.174	4.982	4.00
0.6	6.64	14.5	7.70	9.92	14.06	10.36	5.50

give some indication regarding the degree to which k^* depends on the statistical arrangement of the particles comprising the composite.

A few additional calculations were also performed using the pair distribution function $g(r,c)$ given by Verlet and Weis [8] which is based on the results of molecular dynamics simulations under equilibrium conditions and is therefore considered to be superior at high concentrations to that derived from the Percus-Yevick equation. Although the new values for k^* thus obtained were found to be slightly larger (for $\alpha = \infty$) than those listed in Table 1, the maximum difference between the two sets was only 4% at $c = 0.6$. Hence, no attempt was made to recalculate the entries in Tables 1 and 2 using this presumably more accurate pair distribution function.

In addition, Tables 1 and 2 contain numerical values for k^* as computed from the following two expressions due, respectively, to Felderhof [9],[10] (F) and to Torquato [13],[14] (T)

$$k^* = \frac{1 + 2\beta c + 4\beta^3 c(1-c)\zeta_2}{1 - \beta c - 2\beta^3 c(1-c)\zeta_2} \quad \text{(Felderhof)} \qquad (17)$$

$$k^* = \frac{1 + 2\beta c - 2\beta^2 (1-c)\zeta_2}{1 - \beta c - 2\beta^2 (1-c)\zeta_2} \quad \text{(Torquato)} \qquad (18)$$

where ζ_2 is a microstrustural parameter which is uniquely related to the two-point and three-point correlation functions and has been tabulated as a function of c [13]. Its appearance in eq. (17) follows from Lado and Torquato [14]. Lastly, the last column refers to the well-known Hashin-Shtrikman bounds [15].

It is easy to show that, to $O(c^2)$, eqs. (17) and (18) reduce to eqs. (12) and (16), but thereafter all four diverge. In addition, as seen in Figure 2, all of these four expressions for k^* and $\alpha = \infty$ fall below Turner's data [16] which pertain to the experimentally determined effective electrical conductivity in a fluidized bed. It is worth noting, however, that, as pointed out by Turner [16], there is a tendency for the particles in a fluidized bed to line up in each others wakes, thereby leading to higher values of k^* than would have been the case if the spheres were randomly dispersed and therefore not in actual contact.

Finally, it is quite clear from Tables 1 and 2, as well as from Figures 1 and 2, that Method A, although quite accurate at the lower concentrations, underestimates the influence of the disperse phase in determining the effective conductivity at the higher concentrations and, in fact, the values of k^* thereby computed fall slightly outside the Hashin-Shtrikman bounds when $c \geq 0.4$ and $\alpha = 0$. In contrast, method B appears to retain its accuracy even up to $c = 0.6$ and, for the case $\alpha = \infty$, gives predictions in closer agreement with the experimental data than all the other expressions for k^* presented above. This would indicate, then, that it should be possible to apply method B with a certain degree of confidence and thereby determine a variety of other effective transport coefficients in macroscopically homogeneous two-phase materials having random structure.

Justification of Method A. Since methods A and B lead to accurate expressions for the effective thermal conductivity of a dispersion of spheres in a matrix, at least to $O(c^2)$, it appears reasonable to anticipate that the assumptions inherent in their derivation should have some physical basis. We begin by noting that, although these assumptions cannot be justified **a priori**, they contain the appealing feature of introducing an "excluded volume", of depleted inclusion density, whose thickness is set by the physical nature of the model and is not chosen in some **ad hoc** or empirical fashion as is the case in some cell models.

Chang and Acrivos [17] have already presented a partial justification of Model A under certain restricted conditions. The particular problem considered by these authors concerned the rate of heat conduction from a heated sphere with constant surface temperature which was surrounded by a matrix having unit thermal conductivity and containing equisized passive spheres of conductivity α. It was shown that the $O(c)$ correction to this rate, as calculated using Method A, was to $O(\alpha-1)$ equal to its exact numerical value, but, of course, this result does not justify the method since it involves only the limiting behavior in one particular system.

In what follows, we shall derive a justification of method A to $O(c^2)$ but over the whole range of α, by adapting to the problem under discussion a technique due originally to Hinch [18]. Recall that the first step of Hinch's method is to unconditionally average the conservation and constitutive equations over all possible locations of the spheres thereby obtaining that

$$\nabla \cdot \langle \mathbf{q} \rangle = 0 \tag{19}$$

$$\langle \mathbf{q} \rangle = -\langle \nabla T \rangle + (1 - \alpha)\langle I \nabla T \rangle \tag{20}$$

where $\mathbf{q}$ and T are the local heat flux and temperature, and I is a generalized function which is unity within a particle and zero elsewhere. Here, the brackets denote the unconditional average over all possible locations of the sphere centers in the ensemble, and the argument of each averaged quantity is $\mathbf{r}$, the position vector relative to a fixed origin. Suppressing the unnecessary averaging brackets and simplifying (20) leads to

$$\mathbf{q} = -\nabla T + (1-\alpha)\int_{|\mathbf{r}-\mathbf{r}'|\leq 1} \nabla T(\mathbf{r}|\mathbf{r}')P(\mathbf{r}')dV' \tag{21}$$

since all contributions to the last term only arise from those configurations in which a sphere overlaps $\mathbf{r}$. Here, $T(\mathbf{r}|\mathbf{r}')$ is the conditionally averaged temperature field with a single sphere having its center at $\mathbf{r}'$ held fixed, and $P(\mathbf{r}')$ is the unconditional probability of finding a sphere with its center at $\mathbf{r}'$. Eqs. (19) and (21) are solved by imposing the boundary conditions of continuity of temperature and normal heat flux at $r = 1$, together with the requirement that $T \to x$ as $r \to \infty$. We note, however, that to compute the integral in (21), it is necessary to obtain $\nabla T(\mathbf{r}|\mathbf{r}')$ which requires the solution of the conditionally averaged equations with one particle held fixed. These, in turn, involve the conditionally averaged temperature with two particles held fixed, and therefore an infinite hierarchy of equations emerges which needs to be terminated somehow. In the present case, we only require the solution for $\mathbf{T(r|r')}$, from which we are able to obtain the dipole strength of a single sphere and thereby the effective thermal conductivity of the composite via (1). In this vein, we seek the solution to

$$\nabla\cdot\mathbf{q}(\mathbf{r}|\mathbf{r}') = 0 \tag{22a}$$

$$\mathbf{q}(\mathbf{r}|\mathbf{r}') = -\nabla T(\mathbf{r}|\mathbf{r}') + (1-\alpha)\int_{|\mathbf{r}-\mathbf{r}''|\leq 1} \nabla T(\mathbf{r}|\mathbf{r}',\mathbf{r}'')P(\mathbf{r}''|\mathbf{r}')dV'' \tag{22b}$$

for $|\mathbf{r}-\mathbf{r}'| > 1$ and

$$\mathbf{q}(\mathbf{r}|\mathbf{r}') = -\alpha\nabla T(\mathbf{r}|\mathbf{r}') \quad \text{for} \quad |\mathbf{r}-\mathbf{r}'| < 1 \tag{22c}$$

with boundary conditions given by (9a) and (9b). Next, instead of continuing to the following hierarchy of equations in order to calculate $T(\mathbf{r}|\mathbf{r}',\mathbf{r}'')$, we apply the formula due to Brown [19] for the temperature gradient, $\nabla T(\mathbf{r}|\mathbf{r}',\mathbf{r}'')$ induced within the sphere by the external field, $T(\mathbf{r}|\mathbf{r}')$,

$$[(\alpha-1)\nabla T(\mathbf{r}|\mathbf{r}',\mathbf{r}'')]_{|\mathbf{r}-\mathbf{r}''|\leq 1} = 3\beta[\nabla T(\mathbf{r}|\mathbf{r}')]_{|\mathbf{r}-\mathbf{r}''|\leq 1} + O(\beta^3) \tag{23}$$

which may easily be derived from Maxwell's equations for the analogous electrostatic problem [20]. We note that in this case, eq. (23) is exact to $O(\beta^2)$ since it is exact for the $O(1)$ bulk temperature, $\bar{T}$, by Faxen's law [12], and furthermore since the disturbance temperature, t, is itself $O(\beta)$, only an $O(\beta^3)$ error is made in estimating its influence. Therefore, (22b) may be replaced by

$$\mathbf{q}(\mathbf{r}|\mathbf{r}') = -\left(1 + 3\beta c\rho(r,c)\right)\nabla T(\mathbf{r}|\mathbf{r}') \qquad |\mathbf{r}-\mathbf{r}'| \geq 1 \tag{24}$$

where we have also used eq. (6) and the relation $(3c/4\pi)g(|\mathbf{r}''-\mathbf{r}'|) = P(\mathbf{r}''|\mathbf{r}')$.

It is clear now that (22a) and (24) are precisely the same as (8), the conservation and constitutive equations predicted by Method A. Hence, the step of replacing the composite surrounding the test sphere with an effective continuum whose volume fraction varies with position according to $\rho(r,c)$ is equivalent to closing the hierarchy of averaged equations by estimating the thermal gradient within the second sphere (at $\mathbf{r}''$) to $O(\beta^2)$.

Since Method A leads to an $O(c^2)$ contribution to the effective thermal conductivity which has been shown to be correct to $O(\beta^3)$, it is not suprising that it is related to two other methods that have been proposed in the literature which yield similar results also correct to this order. The first of these is due to Felderhof [9],[10],[11],[12],[13] who begins by expressing the formal solution for the local electric field, in the presence of N polarizable spheres, as an operator, characterizing the microstructure and relative polarizability of the inclusions, acting on the bulk field imposed at infinity. The effective dielectric polarizability may then be derived by obtaining a relationship between the bulk average polarization and the electric field in terms of this operator averaged over all configurations in the ensemble. We pause to note that since this problem is analogous to that of heat conduction, this is equivalent to obtaining the effective thermal conductivity. The original operator involves an N-body problem and is therefore too complicated to evaluate. Instead, this operator is expanded in a series consisting of various simpler operators each involving n ($n = 1,2,\ldots,N$) spheres. Each of these n-body operators is expanded, in turn, into a series of yet simpler operators corresponding to reflective interactions between the spheres. This process leads to a so-called "weak-coupled" expansion in powers of $\alpha - 1$ for each n-body operator, yielding an $O(c^n)$ correction to the effective polarizability of the medium. The weak-coupled series is then rearranged into a "strong-coupled" one in powers of β from which, as expected, we recover (10), the result of method A.

The coefficient of the $O(c^3)$ correction obtained by Felderhof [9] for $\alpha = \infty$ is, however, 3.98. The difference between that and eq. (12) is partially due to the fact that Method A does not yield the exact coefficient of the $O(\beta^3c^3)$ term, which is given by Felderhof as equal to 1.45, since it incorrectly estimates the effect of the other spheres on the one located at $\mathbf{r}''$.

The procedure derived by Torquato [13],[14] differs in that, following Brown [19], it develops a formal relationship between the two fields

$$\mathbf{E}(\mathbf{r}) \equiv \frac{(\alpha - 1)I(\mathbf{r})}{4\pi}\nabla T \qquad \mathbf{G}(\mathbf{r}) \equiv \left(1 + \frac{(\alpha - 1)I(\mathbf{r})}{3}\right)\nabla T \qquad (25)$$

rather that between the polarization and the electric field.

Moreover, in view of the fact that

$$\bar{\mathbf{E}} = \frac{3}{4\pi} \frac{k^* - 1}{k^* + 2} \bar{\mathbf{G}} \tag{26}$$

we see that the proportionality constant between $\bar{\mathbf{E}}$ and $\bar{\mathbf{G}}$ is bounded for all α and therefore this procedure bypasses the need of rearranging a series from a weak-coupled to a strong-coupled form.

The analysis then proceeds in the following stages. First, the equation relating $\mathbf{E}$ to the imposed bulk temperature gradient $\nabla\bar{T}$ s solved iteratively to yield an explicit relation for $\mathbf{E}$ in terms of $\nabla\bar{T}$. The latter is next eliminated by combining the above with the equation relating $\mathbf{G}$ to $\nabla\bar{T}$. Finally, the resulting relation between $\mathbf{E}$ and $\mathbf{G}$ is averaged and leads, in conjunction with (26), to the desired expression for k^*. On this occasion, however, the microstructural parameters characterizing the statistics of the media, which appear in this series, are expressed in terms of point rather than body correlation functions. Thus, the series representation for the effective conductivity is already an expansion in powers of β, but because of the complexity of evaluating microstructural parameters of higher order, the series is truncated at the $O(\beta^3)$ term. Again, the results of Model A are recovered to $O(c^2)$, but the $O(c^3)$ coefficient of k^* for $\alpha = \infty$ is given by this analysis to be 4.51. The difference between this and the previous result is actually $O(\beta^4)$ and arises from the fact that extra higher order terms in β may be included in Torquato's method depending on whatever self-consistent quotient form is employed for expressing the answer. Evidence of this may be found by comparing the various self-consistent forms in which Chiew and Glandt [3] present their results for the effective thermal conductivity.

We can then see from these comments and comparisons that Method A rests on a sound fundamental basis for dilute systems, although it not clear at this stage why such a first order method should remain so accurate over the whole range of α. Unfortunately, we are unable at present to present a similarly comprehensive justification for Method B. In fact, we have found that, although the two methods predict the same $O(c^2)$ correction to the effective thermal conductivity, the application of Method B leads to a different, and therefore incorrect, $O(c)$ perturbation, at lowest order in β, to the dipole field induced by the test sphere. This indicates that methods A and B are fundamentally different, and that any attempts to link the latter with rigorously derived strong or weak-coupled expansions will probably prove to be unsuccessful.

Acknowledgment. This work was supported in part by the National Science Foundation under Grant No. 8117299 and by the Department of Energy under Grant No. DE13328.

Appendix. Analytic Expression for the Percus-Yevick Pair-Correlation function for a Hard-Sphere Potential. The pair-correlation function contains a considerable amount of information the internal structure of a dispersed medium and is therefore widely used in calculating the effective properties of dispersed media. One of the more accurate pair-correlation functions that describes the equilibrium structure arises from the Percus-Yevick integral equation, whose solution was given in the form [7]:

$$g(r) = \frac{1}{12\pi cri} \int_{\delta-i\infty}^{\delta+i\infty} \frac{tL(t)\exp(rt/2)dt}{[L(t)+ S(t)\exp(t)]} \qquad \delta > 0 \qquad (27)$$

where $S(t)=(1-c)^2t^3 +6c(1-c)t^2+18c^2t-12c(1+2c)$, $L(t)=12c\,[\,(1+0.5c)t + 1 + 2c\,]$, c is the volume concentration, and r is the radial distance scaled by the radius of the sphere.

The above integral was evaluated [7] by means of the residue theorem and the result was given as:

for $r < 2$, $g(r) = 0$;

for $2 < r < 4$ $g(r) = g_1(r)$,

$$g_1(r) = \frac{1}{6cr} \sum_1^3 \lim_{t\to t_i} \left\{ (t-t_i) \frac{tL(t)\exp[t(r/2-1)]}{S(t)} \right\} \qquad (28)$$

where t_i is a root of $S(t)$;

for $4 < r < 6$, $g(r) = g_1(r) + g_2(r)$,

$$g_2(r) = \frac{1}{6cr} \sum_1^3 \lim_{t\to t_i} \frac{d}{dt} \left\{ (t-t_i)^2 t \left| \frac{L(t)}{S(t)} \right|^2 \exp[t(r/2-2)] \right\}; \qquad (29)$$

for $6 < r < 8$, $g(r) = g_1(r) + g_2(r) + g_3(r)$,

$$g_3(r) = \frac{1}{6cr} \sum_1^3 \lim_{t\to t_i} \frac{d^2}{dt^2} \left\{ (t-t_i)^3 t \left| \frac{L(t)}{S(t)} \right|^3 \exp[t(r/2-3)] \right\} \qquad (30)$$

and so on.

Let us consider (28) and rewrite it in the form:

$$g_1(r) = \frac{2}{(1-c)^2 r} \sum_1^3 \lim_{t\to t_i} \left\{ (t-t_i) \frac{t[(1+0.5c)t+1+2c]\exp[t(r/2-1)]}{(t+2c/(1-c))^3 + p(t+2c/(1-c))+q} \right\}$$

$$p = \frac{6\,c^2}{(1-c)^2} \qquad q = \frac{4c(c^2-3c-3)}{(1-c)^3} \qquad (31)$$

In order to calculate this limit we must factor the denominator. It is a cubic equation with roots:

$$t_1 = A+B-K \qquad t_{2,3} = -\frac{1}{2}(A+B)-K \pm \frac{i\sqrt{3}}{2}(A-B) \tag{32}$$

where

$$A = \sqrt[3]{\left(-\frac{q}{2} + \sqrt{Q}\right)} \qquad B = \sqrt[3]{\left(-\frac{q}{2} - \sqrt{Q}\right)}$$

$$Q = 8\left(\frac{c}{1-c}\right)^6 + \frac{4c}{(1-c)^6}\left(c^2 - 3c - 3\right)^2$$

$$K = \frac{2c}{1-c}$$

Now we can transform (31) into:

$$\frac{(1-c)^2 r g_1(r)}{2} = \left\{ \frac{t_1[(1+0.5c)t_1+1+2c]\exp[t_1(r/2-1)]}{(t_1-t_2)(t_1-t_3)} + \frac{t_2[(1+0.5c)t_2+1+2c]\exp[t_2(r/2-1)]}{(t_2-t_1)(t_2-t_3)} + \frac{t_3[(1+0.5c)t_3+1+2c]\exp[t_3(r/2-1)]}{(t_3-t_1)(t_3-t_2)} \right\} \tag{33}$$

Following arduous manipulations, we obtain the final expression in the form:

$$g_1(r) = \frac{2}{3r(1-c)^2(A^2+AB+B^2)} * \{P_1\exp[h_1(r/2-1)] + [P_2\cos(\sqrt{3}f(r/2-1)) + P_3\sin(\sqrt{3}f(r/2-1))]\exp[-h_2(r/2-1)]\}$$

where the variables P_1, P_2, P_3, f, h_1 and h_2 depend only on the concentration c and are expressed in the following forms:

$$P_1 = \left(A+B-\frac{2c}{1-c}\right)\left[(1+0.5c)\left(A+B-\frac{2c}{1-c}\right)+1+2c\right]$$

$$P_2 = (1+0.5c)\left[2A^2+AB+2B^2+\frac{4c}{1-c}\left(A+B-\frac{c}{1-c}\right)\right]-(1+2c)\left(A+B-\frac{2c}{1-c}\right)$$

$$P_3 = 2\sqrt{3}\,(A-B)^{-1} * \left[(1+2c)\left(\frac{A^2+B^2}{2}+\frac{c}{1-c}(A+B)\right)-(1+0.5c)\left(\left(AB+\frac{4c^2}{(1-c)^2}\right)\frac{A+B}{2}+\frac{2c}{1-c}(A^2+B^2)\right)\right].$$

$$h_1 = A + B - \frac{2c}{1-c} \qquad h_2 = \frac{A+B}{2} + \frac{2c}{1-c} \qquad f = \frac{A-B}{2} \tag{34}$$

By inserting the pair-distribution function given by (33) and (34) we obtain for $\rho(r,c)$:

$$\rho = \begin{cases} \rho_1 & 1 < r < 3 \\ \rho_2 & 3 < r < 5 \\ 1 & r > 5 \end{cases}$$

where

$$\rho_1 = \frac{P_1 I_1 + P_2 II_1 + P_3 III_1}{3(1-c)^2(A^2 + AB + B^2)}$$

$$\rho_2 = \frac{P_1 I_2 + P_2 II_2 + P_3 III_2}{3(1-c)^2(A^2 + AB + B^2)}$$

$$I_1 = \frac{3}{rh_1^3}\left[4(h_1-2)\exp\left(\frac{h_1(r-1)}{2}\right) + h_1(3h_1-8)+8+rh_1(rh_1+4-4h_1)\right]$$

$$II_1 = \frac{1}{r(h_2^2+3f^2)^3} *$$

$$\Big\{ \Big[12(h_2^4-9f^4-2h_2(h_2^4-9f^2))\cos(a_1)+$$

$$+24\sqrt{3}f(h_2(h_2^2+3f^2)-3(h_2^2-f^2))\sin(a_1) \Big]\exp\left(\frac{h_2(r-1)}{2}\right)$$

$$+ 9h_2(h_2^2+3f^2)^2-24(h_2^4-9f^4)+24h_2(h_2^2-9f^2)+$$

$$+r[3rh_2(h_2^2+3f^2)^2 -12h_2(h_2^2+3f^2)^2 +12(h_2^4-9f^4)] \Big\}$$

$$III_1 = \frac{1}{r(h_2^2+3f^2)^3} *$$

$$\Big\{ \Big[12(h_2^4-9f^4-2h_2(h_2^2-9f^2))\sin(a_1)-$$

$$-24\sqrt{3}f(h_2(h_2^2+3f^2)-3(h_2^2-f^2))\cos(a_1) \Big]\exp\left(\frac{h_2(r-1)}{2}\right)$$

$$- \sqrt{3}f[9(h_2^2+3f^2)^3+72(h_2^2-f^2)-48h_2(h_2^2+3f^2)]-$$

$$-r^2 3\sqrt{3}f(h_2^2+3f^2)^2 +r12\sqrt{3}f(2h_2(h_2^2+3f^2)-(h_2^2+3f^2)^2) \Big\}$$

$$a_1 = \frac{\sqrt{3}}{2}f(r-1) \qquad a_2 = \frac{\sqrt{3}}{2}f(r-3)$$

$$I_2 = \frac{3\exp(h_1)}{rh_1^3}\left[-4(h_1+2)\exp\left(\frac{h_1(r-5)}{2}\right)+h_1(15h_1-16)+8+rh_1(rh_1+4-8h_1)\right]$$

$$
\begin{aligned}
II_2 = {} & \frac{-\exp(h_2)}{r(h^2+3f^2)^3} * \\
& \Big\{\ 12\big[\ (-h_2^4+9f^4-2h_2(h_2^2-9f^2))\cos(a_2)- \\
& \qquad -2\sqrt{3}f(h_2(h_2^2+3f^2)+3(h_2^2-f^2))\sin(a_2)\ \big]\exp\left(\frac{h_2(r-5)}{2}\right) + \\
& + \cos(\sqrt{3}f)[45h_2(h_2^2+3f^2)^2-48(h_2^4-9f^4)+24h_2(h_2^2-9f^2)]+ \\
& + \sqrt{3}f\sin(\sqrt{3}f)[45(h_2^2+3f^2)^2-96h_2(h_2^2+3f^2)+72(h_2^2-f^2)]+ \\
& +3r^2(h_2^2+3f^2)^2\ (h_2\cos(\sqrt{3}f)+\sqrt{3}f\sin(\sqrt{3}f))+ \\
& +12r[(-2h_2(h_2^2+3f^2)^2+h_2^4-9f^4)\cos(\sqrt{3}f)+ \\
& \qquad + 2\sqrt{3}f(h_2^2+3f^2)(h_2-h_2^2-3f^2)\sin(\sqrt{3}f)]\ \Big\}
\end{aligned}
$$

$$
\begin{aligned}
III_2 = {} & \frac{-\exp(h_2)}{r(h_2^2+3f^2)^3} * \\
& \Big\{\ 12\big[\ (-h_2^4+9f^4-2h_2(h_2^2-9f^2))\sin(a_2)+ \\
& \qquad +2\sqrt{3}f(h_2(h_2^2+3f^2)+3(h_2^2-f^2))\cos(a_2)\ \big]\exp\left(\frac{h_2(r-5)}{2}\right) \\
& + \sin(\sqrt{3}f)[45h_2(h_2^2+3f^2)^2-48(h_2^4-9f^4)+24h_2(h_2^2-9f^2)]- \\
& - \sqrt{3}f\cos(\sqrt{3}f)[45(h_2^2+3f^2)^2-96h_2(h_2^2+3f^2)+72(h_2^2-f^2)]+ \\
& +3r^2(h_2^2+3f^2)^2\ (h_2\sin(\sqrt{3}f)-\sqrt{3}f\cos(\sqrt{3}f))+ \\
& +12r[(-2h_2(h_2^2+3f^2)^2+h_2^4-9f^4)\sin(\sqrt{3}f)- \\
& \qquad - 2\sqrt{3}f(h_2^2+3f^2)(h_2-h_2^2-3f^2)\cos(\sqrt{3}f)]\ \Big\}
\end{aligned}
$$

We also list below the expression for the first derivative of ρ. It has the form:

$$\frac{d\rho}{dr} = \begin{cases} \rho_1' & 1 < r < 3 \\ \rho_2' & 3 < r < 5 \\ 0 & r > 5 \end{cases}$$

where

$$\rho_1' = \frac{P_1DI_1 + P_2DII_1 + P_3DIII_1}{3(1-c)^2(A^2 + AB + B^2)}$$

$$\rho_2' = \frac{P_1DI_2 + P_2DII_2 + P_3DIII_2}{3(1-c)^2(A^2 + AB + B^2)}$$

$$DI_1 = \frac{3}{r^2\, h_1^3}\{\ [2rh_1(h_1-2)+8-4h_1]\exp\left(\frac{h_1(r-1)}{2}\right)-h_1(3h_1-8)+8+r^2h_1^2\ \}$$

$$DII_1 = \frac{1}{r^2(h_2^2+3f^2)^3} *$$

$$\{\ [\ -12(h_2^4-9f^4+2h_2(-h_2^2+9f^2))\cos(a_1)-$$
$$-24\sqrt{3}f(-3(h_2^2-f^2)+h_2(h_2^2+3f^2))\sin(a_1)$$
$$+6r(\cos(a_1)(\ h_2(h_2^2+3f^2)^2-2(h_2^4-9f^4))$$
$$+\sqrt{3}f\sin(a_1)((h_2^2+3f^2)^2-4h_2(h_2^2+3f^2)))\ ]\exp\left(\frac{h_2(r-1)}{2}\right)$$
$$+r^2 3h_2(h_2^2+3f^2)^2+24(h_2^4-9f^4)-24h_2(h_2^2-9f^2)-9h_2(h_2^2+3f^2)^2\ \}$$

$$DIII_1 = \frac{1}{r^2(h_2^2+3f^2)^3}$$

$$\{\ [\ 12(h_2^4-9f^4\ -2h_2(h_2^2-9f^2))\sin(a_1)-$$
$$-24\sqrt{3}f(h_2(h_2^2+3f^2)-3(h_2^2-f^2))\cos(a_1)c\ +$$
$$+6r(\sin(a_1)(-h_2(h_2^2+3f^2)^2+2(h_2^4-9f^4))$$
$$+\sqrt{3}f\cos(a_1)((h_2^2+3f^2)^2-4h_2(h_2^2+3f^2)))\ ]\exp\left(\frac{h_2(r-1)}{2}\right)$$
$$+3\sqrt{3}fr^2(h_2^2+3f^2)^2+48\sqrt{3}f(h_2^2+3f^2)-24\sqrt{3}(h_2^2-f^2)-9\sqrt{3}f(h_2^2+3f^2)^2\ \}$$

$$DI_2 = \frac{-3\exp(h_1)}{3r^2h_1^3}[2(2-rh_1)(h_1+2)\exp\left(\frac{h_1(r-5)}{2}\right)+h_1(-15h_1+16)-8+r^2h_1^2]$$

$$DII_2 = \frac{\exp(h_2)}{r^2(h_2^2+3f^2)^3} *$$

$$\{\ [\ 12(-h_2^4+9f^4-2h_2(-h_2^2+9f^2))\cos(a_1)+$$
$$+24\sqrt{3}f(-3(h_2^2-f^2)-h_2(h_2^2+3f^2))\sin(a_1)+$$

$$+6r(\cos(a_1)(\ h_2(h_2^2+3f^2)^2+2(h_2^4-9f^4))$$

$$+\sqrt{3}f\sin(a_1)((h_2^2+3f^2)^2+4h_2(h_2^2+3f^2)))\ ]\exp\left(\frac{h_2(r-5)}{2}\right)$$

$$+\ \cos(\sqrt{3}f)[45h_2(h_2^2+3f^2)^2-48(h_2^4-9f^4)+24h_2(h_2^2-9f^2)]+$$

$$+\ \sqrt{3}f\sin(\sqrt{3}f)[45(h_2^2+3f^2)^2-96h_2(h_2^2+3f^2)+72(h_2^2-f^2)]+$$

$$-3r^2(h_2^2+3f^2)^2\ (h_2\cos(\sqrt{3}f)+\sqrt{3}f\sin(\sqrt{3}f))\ \}$$

$$DIII_2 = \frac{\exp(h_2)}{r^2(h_2^2+3f^2)^3} *$$

$$\{\ [\ 12(-h_2^4+9f^4-2h_2(h_2^2-9f^2))\sin(a_2)-$$

$$-24\sqrt{3}f(h_2(h_2^2+3f^2)+3(h_2^2-f^2))\cos(a_2)$$

$$-6r(\sin(a_1)(h_2(h_2^2+3f^2)^2+2(h_2^4-9f^4))$$

$$+\sqrt{3}f\cos(a_1)((h_2^2+3f^2)^2+4h_2(h_2^2+3f^2)))\ ]\exp\left(\frac{h_2(r-5)}{2}\right)$$

$$+\ \sin(\sqrt{3}f)[45h_2(h_2^2+3f^2)^2-48(h_2^4-9f^4)+24h_2(h_2^2-9f^2)]-$$

$$-\ \sqrt{3}f\cos(\sqrt{3}f)[45(h_2^2+3f^2)^2-96h_2(h_2^2+3f^2)+72(h_2^2-f^2)]-$$

$$-3r^2(h_2^2+3f^2)^2\ (h_2\sin(\sqrt{3}f)-\sqrt{3}f\cos(\sqrt{3}f))\ \}$$

REFERENCES

[1] D. J. Jeffrey, Conduction through a random suspension of spheres, Proc. R. Soc. Lond. A, 335 (1973), pp. 355-367.

[2] A. Acrivos and E. Chang, A model for estimating transport quantities in two-phase materials, Phys. Fluids, 29 (1986), pp. 3-4.

[3] Y. C. Chiew and E. D Glandt, The effect of structure on the conductivity of a dispersion, J. Colloid and Interface Sci., 94 (1983), pp. 90-104.

[4] Y. A. Buyevich and I. N Shchelchkova, Flow of dense suspensions, Prog. Aerospace Sci., 18 (1978), pp. 121-150.

[5] Y. A. Buyevich and Y.A. Korneev, Effective thermal conductivity of disperse media at small Peclet number, J. Eng. Phys., 31 (1976), p. 607-612.

[6] B. S. Endler, Effective viscosity and heat conductivity of a disperse media, J. Eng. Phys., 37 (1979), pp. 110-117.

[7] G. J. Throop and R. J. Bearman, Numerical solution of the Percus-Yevick equation, J. Chem. Phys., 42 (1965), pp. 2408-2411.

[8] L. Verlet and J-J. Weis, Equilibrium theory of simple liquids, Phys. Rev. A, 5 (1972), pp. 939-952.

[9] B. U. Felderhof, Bounds for the effective dielectric constant of a suspension of spherically symmetric particles, J. Phys. C, 15 (1982), pp. 3943-3951.

[10] B. U. Felderhof, Bounds for the effective dielectric constant of a suspension of uniform spheres, J. Phys. C, 15 (1982), pp. 3953-3966.

[11] B. U. Felderhof, G. W. Ford, and E. G. D. Cohen, Cluster expansion for the dielectric constant of a polarizable suspension, J. Stat. Phys., 28 (1982), pp. 135-164.

[12] B. U. Felderhof, Two particle cluster integral in the expansion of the dielectric constant of a polarizable suspension, J. Stat. Phys., 28 (1982), pp. 649-672.

[13] S. Torquato, Effective electrical conductivity of two phase disordered composite media, J. Appl. Phys., 58 (1985), pp. 3790-3797.

[14] F. Lado and S. Torquato, Effective properties of disordered composite media. I. Simplification of bounds on the conductivity and bulk modulus of dispersions of impenetrable spheres, Phys. Rev. B, 33 (1986), pp. 3370-3378.

[15] Z. Hashin and S. Shtrikman, A variational approach to the theory of the effective magnetic permeability of multiphase materials, J. Appl. Phys., 58 (1962), pp. 3125-.

[16] J. C. R. Turner, Two-phase conductivity. The electrical conductivity of liquid-fluidized beds of spheres, Chem. Eng. Sci., 31 (1976), pp. 487-492.

[17] E. Chang and A. Acrivos, Rate of heat conduction from a heated sphere to a matrix containing passive spheres of a different conductivity, J. Appl. Phys., 59 (1986), pp. 375-3382.

[18] E. J. Hinch, An averaged equation approach to particle interactions in a fluid suspension, J. Fluid Mech., 83 (1977), pp. 695-720.

[19] W. F. Brown, Jr., Dielectric constants, permeabilities, and conductivities of random media, Trans. Soc. Rheol. 9 (1965), pp. 357-380.

[20] J. D. Jackson, Classical Electrodynamics, vol. 2, Wiley, New York, 1975, p. 145.

[21] E. Thiele, Equation of state for hard spheres, J. Chem. Phys., 39 (1963), pp. 474-479.

[22] R. E. Meredith and C. W. Tobias, Conductivities in Emulsions, J. Electrochemical Soc., 108 (1961), pp. 286-290.

[23] C. A. R. Pierce, The electrical conductivity and permeability of mixtures, with special reference to emulsions of water in fuel oil, Brit. J. of Appl. Phys., 6 (1955), pp. 113-120.

A Flow Structure in the Poiseuille Flow of a Particle-Fluid Mixture

DONALD A. DREW*

Abstract. The shear flow of a particle-fluid mixture is studied using the two-fluid model, which is based on equations of conservation of mass and momentum for each material. The equations are applied to plane parallel flow, where the equations or mixture axial momentum and an equation for the concentration across the channel are derived. An asymptotic solution to the equations is derived which shows that the particles move to the center of the channel, forming a core which does not shear. The core is surrounded by clear fluid layers where all the shear occurs.

Introduction. The equations of motion for the flow of a particle-laden fluid mixture with both phasic densities constant are

$$(1) \qquad \frac{\partial \alpha}{\partial t} + \nabla \cdot \alpha \, \mathbf{v}_p = 0$$

$$(2) \qquad \frac{\partial (1 - \alpha)}{\partial t} + \nabla \cdot (1 - \alpha) \, \mathbf{v}_f = 0$$

$$(3) \qquad \alpha \, \rho_p \left[\frac{\partial \mathbf{v}_p}{\partial t} + \mathbf{v}_p \cdot \nabla \, \mathbf{v}_p \right] = - \alpha \nabla p_p + \left[p_{pi} - p_p \right] \nabla \alpha + \alpha \, S \left[\mathbf{v}_f - \mathbf{v}_p \right]$$

* Department of Mathematical Sciences
Rensselaer Polytechnic Institute
Troy, New York 12180-3590

$$+ \alpha \rho_f C_{vm} \left[\left[\frac{\partial \mathbf{v}_f}{\partial t} + \mathbf{v}_f \cdot \nabla \mathbf{v}_f \right] - \left[\frac{\partial \mathbf{v}_p}{\partial t} + \mathbf{v}_p \cdot \nabla \mathbf{v}_p \right] \right]$$

$$+ \alpha \rho_f L \left[\mathbf{v}_p - \mathbf{v}_f \right] \times \left[\nabla \times \mathbf{v}_f \right] + \mathbf{F}_F + \nabla \cdot \alpha \underline{\tau}_p$$

$$- \nabla \alpha \cdot \underline{\tau}_{pi} + \nabla \cdot \alpha \underline{\tau}_p^T - \alpha \rho_p \mathbf{g}$$

$$(4) \quad \left[1 - \alpha\right] \rho_f \left[\frac{\partial \mathbf{v}_f}{\partial t} + \mathbf{v}_f \cdot \nabla \mathbf{v}_f \right] = - \left[1 - \alpha\right] \nabla p_f$$

$$- \left[p_{fi} - p_f\right] \nabla \alpha + \alpha S \left[\mathbf{v}_p - \mathbf{v}_f\right]$$

$$+ \alpha \rho_f C_{vm} \left[\left[\frac{\partial \mathbf{v}_p}{\partial t} + \mathbf{v}_p \cdot \nabla \mathbf{v}_p \right] - \left[\frac{\partial \mathbf{v}_f}{\partial t} + \mathbf{v}_f \cdot \nabla \mathbf{v}_f \right] \right]$$

$$+ \alpha \rho_f L \left[\mathbf{v}_f - \mathbf{v}_p \right] \times \left[\nabla \times \mathbf{v}_f \right] - \mathbf{F}_F$$

$$+ \nabla \cdot \left[1 - \alpha\right] \underline{\tau}_f + \nabla \alpha \cdot \underline{\tau}_{fi}$$

$$+ \nabla \cdot \left[1 - \alpha\right] \underline{\tau}_f^T - \left[1 - \alpha\right] \rho_f \mathbf{g}$$

where α is the volume fraction of particles, ρ_p is the particle density, ρ_f is the fluid density, $\mathbf{v}_p$ is the particle velocity, $\mathbf{v}_f$ is the fluid velocity, p_p is the particle pressure, p_{pi} is the pressure at the particle interface, p_f is the fluid pressure, p_{fi} is the fluid pressure at the interface, S is the drag coefficient, C_{vm} is the virtual mass coefficient, L is the lift coefficient, $\mathbf{F}_F$ is the Faxen force, $\mathbf{g}$ is the acceleration due to gravity, $\underline{\tau}_p$ is the particle shear stress, $\underline{\tau}_{pi}$ is the particle shear stress at the interface, $\underline{\tau}_f$ is the fluid shear stress, $\underline{\tau}_{fi}$ is the fluid shear stress at the interface, $\underline{\tau}_p^T$ is particle turbulent shear stress, and $\underline{\tau}_f^T$ is the fluid turbulent shear stess.

The combination of virtual mass and lift is that derived by Drew and Lahey [1]. This force is calculated by considering the force on a single sphere accelerating relative to an inviscid fluid which is undergoing a pure shear plus a rotation. This results in $C_{vm} = 1/2$ and $L = 1/2$. This force is objective only if $L = C_{vm}$. For our present purposes, we shall assume that the virtual mass-lift combination is objective, hence $L = C_{vm}$, but we shall not assign a specific value for L at this point.

The Faxen force is usually neglected. Since we wish to treat viscous effects and the Faxen force is comparable to the viscous forces in the fluid phase, we retain this term here. We assume

(5) $\mathbf{F_F} = k\,\alpha\,\mu_f\,\nabla^2\,\mathbf{v}_f$.

The assumptions made about the pressures are known to have profound consequences for wave propagation in bubble flows. We wish to do something here which is consistent with the other assumptions made about other terms such as virtual mass and lift. Therefore, we assume

(6) $p_p = p_{pi} = p_{fi} = p_f - \xi\,\rho_f\,|\mathbf{v}_f - \mathbf{v}_p|^2.$

Stuhmiller [2] gives $\xi = 1/4$ for inviscid flow.

We next discuss models for the stress terms. If the particles are small and rigid, they are essentially stress transmitters on the microscale. Thus, we assume that the particle stress and the interfacial stresses are the same. That is,

(7) $\underline{\tau}_p = \underline{\tau}_{pi} = \underline{\tau}_{fi}$.

The stress which the particles are transmitting corresponds to the extra needed to make the mixture more viscous. To account for this effect, we take

(8) $\underline{\tau}_{pi} = \beta\,(\alpha)\,\underline{\tau}_f$.

For the viscous stress, we take Ishii's [3] form, which is derived from averaging the microscopic viscous stress tensor. We have

(9) $$\underline{\tau}_f = \mu\left\{\left[\nabla\,\mathbf{v}_f + \left(\nabla\,\mathbf{v}_f\right)^T\right] + \nabla\alpha\,\left(\mathbf{v}_f - \mathbf{v}_p\right) + \left(\mathbf{v}_f - \mathbf{v}_p\right)\,\nabla\,\alpha\right\}.$$

Forms for the turbulent, or Reynolds stresses are not considered standard in the two-phase flow literature, even though they are responsible for diffusive effects in the momentum balance. We take a simple form of a model proposed by Drew and Lahey [4] which has coefficients which can be calculated from inviscid flow around a single sphere [5]. This gives

(10a) $$\underline{\tau}_f^T = \alpha\,\rho_f\,a\,|\mathbf{v}_f - \mathbf{v}_p|^2\,\underline{I} + \alpha\,\rho_f\,b\,\left(\mathbf{v}_f - \mathbf{v}_p\right)\left(\mathbf{v}_f - \mathbf{v}_p\right)$$

and

(10b) $$\underline{\tau}_p^T = \rho_p\,c\,|\mathbf{v}_f - \mathbf{v}_p|^2\,\underline{I} + \rho_p\,d\,\left(\mathbf{v}_f - \mathbf{v}_p\right)\left(\mathbf{v}_f - \mathbf{v}_p\right)$$

This last assumption results from assuming that the particles follow closely the motions of the fluid.

We assume a symmetric flow in a channel of width 2h. We shall neglect gravitational forces. Then the no slip condition on the fluid gives $\mathbf{v}_f = 0$ and the condition of impenetrability of the wall to particles gives $\mathbf{n} \cdot \mathbf{v}_p = 0$ at $y = \pm h$. We also impose conditions that the total flow is given, and the concentration of the incoming fluid is known.

$$\int_{-h}^{h} \mathbf{i} \cdot \mathbf{v}_f \, dy = 2 h V_o \tag{11a}$$

$$\int_{-h}^{h} \mathbf{i} \cdot \mathbf{v}_p \, dy = 2 h U_o \tag{11b}$$

$$\int_{-h}^{h} \alpha(y) \, dy = 2 h \langle\alpha\rangle. \tag{11c}$$

We wish to study the plane Poiseuille flow of a particle-fluid mixture. For plane parallel flow, let

$$\mathbf{v}_p = U(y) \, \mathbf{i} \tag{12a}$$

$$\mathbf{v}_f = V(y) \, \mathbf{i} \tag{12b}$$

$$\alpha = \alpha(y). \tag{12c}$$

Then the continuity equations are satisfied automatically, and the momentum equations yield

$$0 = - \alpha \frac{\partial p_f}{\partial x} + \alpha S \left[V - U\right] + \alpha \mu \frac{d}{dy} \beta \frac{dV}{dy} + \alpha k \mu \frac{d^2 V}{dy^2} \tag{13}$$

$$0 = - \left[1 - \alpha\right] \frac{\partial p_f}{\partial x} + \alpha S \left[U - V\right] + \mu \frac{d}{dy} \left[\left[1 - \alpha\right] \frac{dV}{dy}\right] - \alpha k \mu \frac{d^2 V}{dy^2} + \frac{d\alpha}{dy} \beta \mu \frac{dV}{dy} \tag{14}$$

$$0 = - \alpha \frac{\partial p_f}{\partial y} + 2 \alpha \xi \rho_f \left[V - U\right] \frac{d}{dy} \left[U - V\right] + \alpha \rho_f L \left[U - V\right] \frac{dV}{dy} + \frac{d}{dy} \alpha \rho_p c \left[U - V\right]^2 \tag{15}$$

closely the motions of the fluid.

We assume a symmetric flow in a channel of width 2h. We shall neglect gravitational forces. Then the no slip condition on the fluid gives $\mathbf{v}_f = 0$ and the condition of impenetrability of the wall to particles gives $\mathbf{n} \cdot \mathbf{v}_p = 0$ at $y = \pm h$. We also impose conditions that the total flow is given, and the concentration of the incoming fluid is known.

(11a) $$\int_{-h}^{h} \mathbf{i} \cdot \mathbf{v}_f \, dy = 2 h V_o$$

(11b) $$\int_{-h}^{h} \mathbf{i} \cdot \mathbf{v}_p \, dy = 2 h U_o$$

(11c) $$\int_{-h}^{h} \alpha(y) \, dy = 2 h \langle\alpha\rangle .$$

We wish to study the plane Poiseuille flow of a particle-fluid mixture. For plane parallel flow, let

(12a) $$\mathbf{v}_p = U(y) \, \mathbf{i}$$

(12b) $$\mathbf{v}_f = V(y) \, \mathbf{i}$$

(12c) $$\alpha = \alpha(y) .$$

Then the continuity equations are satisfied automatically, and the momentum equations yield

(13) $$0 = - \alpha \frac{\partial p_f}{\partial x} + \alpha S \left[V - U\right] + \alpha \mu \frac{d}{dy} \beta \frac{dV}{dy} + \alpha k \mu \frac{d^2 V}{dy^2}$$

(14) $$0 = - \left[1 - \alpha\right] \frac{\partial p_f}{\partial x} + \alpha S \left[U - V\right] + \mu \frac{d}{dy} \left[\left[1 - \alpha\right] \frac{dV}{dy}\right] - \alpha k \mu \frac{d^2 V}{dy^2} + \frac{d\alpha}{dy} \beta \mu \frac{dV}{dy}$$

(15) $$0 = - \alpha \frac{\partial p_f}{\partial y} + 2 \alpha \xi \rho_f \left[V - U\right] \frac{d}{dy} \left[U - V\right] + \alpha \rho_f L \left[U - V\right] \frac{dV}{dy} + \frac{d}{dy} \alpha \rho_p c \left[U - V\right]^2$$

$$(16)\quad 0 = -\left[1-\alpha\right]\frac{\partial p_f}{\partial y} + \xi\,\rho_f\left[V-U\right]^2\frac{d\alpha}{dy} + \alpha\,\rho_f\,L\left[V-U\right]\frac{dV}{dy} + \frac{d}{dy}\,\alpha\,\rho_f\,a\left[U-V\right]^2$$

Differentiating (9) or (10) with respect to x, and using (6) gives

$$\frac{\partial p_f}{\partial x \partial y} = 0 ,$$

so

$$(17)\qquad p_f = f(x) + P(y).$$

Differentiating (7) or (8) with respect to x gives

$$f'(x) = 0.$$

Therefore

$$(18)\qquad p_f = -\frac{\Delta p}{\Delta x}\,x + P(y) ,$$

where Δp is the imposed pressure drop on the channel and Δx is its length. If we add (13) and (14), we have

$$(19)\qquad -\frac{\Delta p}{\Delta x} = \frac{d}{dy}\left[\left[1-\alpha\right]\mu\,\frac{dV}{dy} + \alpha\,\beta\,\mu\,\frac{dV}{dy}\right] .$$

Note that $\beta = 7/2$ gives the Einstein [6] formula for effective viscosity. A form which agrees with Einstein's for small α and with that of Frankel and Acrivos [7] for large concentrations is given by Graham [8] as

$$(20)\qquad \mu_{eff}/\mu = \left[1+\frac{5}{2}\,\alpha\right] + \frac{9}{4}\left[1+\left[h/2a\right]\right]^{-1}\left[\frac{1}{\left[h/a\right]} - \frac{1}{\left[1+\left[h/a\right]\right]} - \frac{1}{\left[1+\left[h/a\right]\right]^2}\right]$$

where, for a simple cubic packing,

$$h/a = 2\left[\left[1-\left[\alpha/\alpha_m\right]^{1/3}\right]/\left[\alpha/\alpha_m\right]^{1/3}\right] ,$$

where α_m is the experimentally determined maximum packing of spheres.

The relative motion between the particles and the fluid can be obtained from either of the remaining momentum equations in the x-direction. If we divide equation (13) by α S, we have

$$U - V = \frac{1}{S}\frac{\Delta p}{\Delta x} + \frac{\mu}{S}\frac{d}{dy}\left[\beta + k\right]\frac{dV}{dy}\ . \tag{21}$$

The momentum equations in the y-direction are instrumental in determining the distribution of particles across the channel. These equations involve the transverse pressure gradient dP/dy. The transverse pressure gradient can be eliminated from (15) and (16) by subtracting $\alpha/(1 - \alpha)$ times eq (16) from eq (15). This gives

$$0 = \left[\xi + c - a\frac{\alpha}{1-\alpha}\right] 2\,\alpha\,\rho_f\left[U - V\right]\frac{d\left[U - V\right]}{dy} \tag{22}$$

$$+ \frac{\alpha}{1-\alpha}\rho_f\,L\left[U - V\right]\frac{dV}{dy} +$$

$$+ \left[-\left[\xi + a\right]\frac{\alpha}{1-\alpha} + c\right]\rho_f\left[U - V\right]^2$$

We can further eliminate U - V by using equation (21). This gives

$$0 = \alpha\left[\xi + c - a\frac{\alpha}{1-\alpha}\right]\frac{2\,\mu}{S}\frac{d^2}{dy^2}\left[\beta + k\right]\frac{dV}{dy} - \frac{\alpha\,L}{1-\alpha}\frac{dV}{dy} \tag{23}$$

$$+ \left[-\left[\xi + a\right]\frac{\alpha}{1-\alpha} + c\right]\left[\frac{1}{S}\frac{\Delta p}{\Delta x} + \frac{\mu}{S}\frac{d}{dy}\left[\beta + k\right]\frac{dV}{dy}\right]\frac{d\alpha}{dy}$$

Equations (19) and (23) govern the fluid velocity profile and the concentration of particles in the channel. For these equations, the appropriate boundary conditions are

$$V(h) = 0 \tag{24a}$$

$$\int_{-h}^{h} V(y)\,dy = 2\,h\,V_o \tag{24b}$$

$$\int_{-h}^{h} \alpha(y)\,dy = 2\,h\,\langle\alpha\rangle. \tag{24c}$$

Let us nondimensionalize the problem by

$$w = V(y)/V(0) \tag{25a}$$

(25b)v $\quad \zeta = y/h\ .$

The equations become

$$\left[1 + \left[\beta - 1\right]\alpha\right]\frac{dw}{d\zeta} = -\,R\,\zeta \tag{26}$$

$$(27) \qquad 2\,\alpha \left[\xi + c - a\,\frac{\alpha}{1-\alpha}\right] \frac{d^2}{d\zeta^2}\,[\beta + k]\,\frac{dw}{d\zeta} + \alpha\,D\,\frac{\alpha}{1+\alpha}\,\frac{dw}{d\zeta}$$

$$+ \left[-\,[\xi + a]\,\frac{\alpha}{1-\alpha} + c\right]\left[R + \frac{d}{d\zeta}\,[\beta + k]\frac{dw}{d\zeta}\right]\frac{d\alpha}{d\zeta}$$

where

$$(28a) \qquad R = \frac{\Delta p\ h^2}{\mu\ \Delta x\ V(0)}$$

is the channel Reynolds number, and

$$(28b) \qquad D = \frac{S\ h^2}{\mu}$$

is the dimensionless drag per unit velocity.

The boundary conditions are

$$(29a) \qquad w(0) = 1$$

$$(29b) \qquad \frac{dw}{d\zeta}\,(0) = 0$$

$$(29c) \qquad w(1) = 0\ .$$

$$(29d) \qquad \int_0^1 \alpha(\zeta)\ d\zeta = \langle\alpha\rangle .$$

Let us seek a solution for D large. The outer solution assumes $\zeta = O(1)$. With D large, we must have β large, or $dw/d\zeta$ small, or α small. From eq. (26), we see that with $R = O(1)$, if $dw/d\zeta$ is small, β must be large. With $dw/d\zeta = D^{-q}g$, and $\beta = D^q\, b'$, we have $\alpha = \alpha_m + D^{-q}\,\alpha'$, and $b' = -\,(9/8)(a_m/\alpha')$. In order to obtain a balance in eq. (27), we must have $q = 1$.

Let us assume for the moment that the region with α small is near the wall, and that the region with $dw/d\zeta$ is in a region around the center of the channel, which we shall call the core. Then in the core we have $\alpha = \alpha_m$. Then the approximate particle concentration is given by

$$(30) \qquad \alpha \cong \begin{cases} \alpha_m\ , & 0 < \zeta < \zeta^* \\ 0\ , & \zeta^* < \zeta < 1\ , \end{cases}$$

where ζ^* is the location of the edge of the core. In the clear fluid region, the fluid velocity must satisfy

$$\frac{dw}{d\zeta} = - R\,\zeta \tag{31}$$

so that

$$w = -\frac{1}{2} R \left[\zeta^2 - 1\right] \tag{32}$$

At ζ^*, we have $w(\zeta^*) = -(R/2)(\zeta^{*2} - 1) = 1$. Thus, we see that $R = 2/(1 - \zeta^{*2})$. If $1 - \zeta^* = O(1)$, we see that $R = O(1)$. Since $\langle\alpha\rangle = \alpha_m \zeta^*$, we have

$$R = -\frac{2}{\left[\left[\frac{\langle\alpha\rangle}{\alpha_m}\right]^2 - 1\right]} \tag{33}$$

Note that R is not small when $\langle\alpha\rangle$ is near α_m .

The Transition Layer. We next show that a layer exists at $\zeta = \zeta^*$ where α makes a transition from α_m to 0, while $dw/d\zeta$ goes from 0 to $-R\,\zeta^*$. Note that finding the appropriate α forces $dw/d\zeta$ to have the proper behavior.

We let $\zeta = \zeta^* + D^{-p}\ \zeta'$. The right hand side of eq (26) becomes $- R\,\zeta^*$ to first approximation. Using this, we obtain a balance in eq (27) for $p = 1/2$, and $dw/d\zeta'$ can be eliminated to give

$$\frac{d^2\alpha}{d\zeta'^2} + f_1[\alpha]\left[\frac{d\alpha}{d\zeta'}\right]^2 + f_2[\alpha] = 0 \tag{34}$$

where

$$f_1[\alpha] = \frac{f''(\alpha)}{f'(\alpha)} + \frac{\left[-(\xi + a)\frac{\alpha}{1-\alpha} + c\right]}{\left[\xi + c - a\frac{\alpha}{1-\alpha}\right]}\,\frac{1}{2\alpha} \tag{35a}$$

$$f_2[\alpha] = \frac{1}{2\left[\xi + c - a\frac{\alpha}{1-\alpha}\right]}\,\frac{1}{1-\alpha}\,\frac{1}{1+[\beta - 1]\alpha}\,\frac{1}{f'(\alpha)} \tag{35b}$$

$$f(\alpha) = \frac{\beta + k}{1 + [\beta - 1]\alpha} \tag{35c}$$

It is not known whether the solutions to eq (34) exhibit the proper behavior. A numerical solution to equation (34) with $a = -1/5$, $c = -1/5$, and $\zeta = 1/4$ is shown in Fig 1.

When $\zeta^* = 1 - o(1)$, the clear-fluid layer no longer exists, and

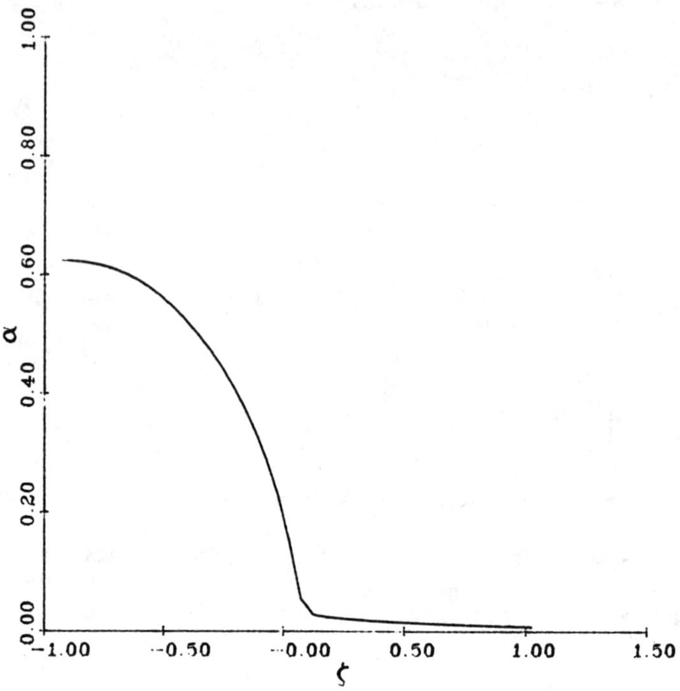

Figure 1. Distribution of α in the transition layer.

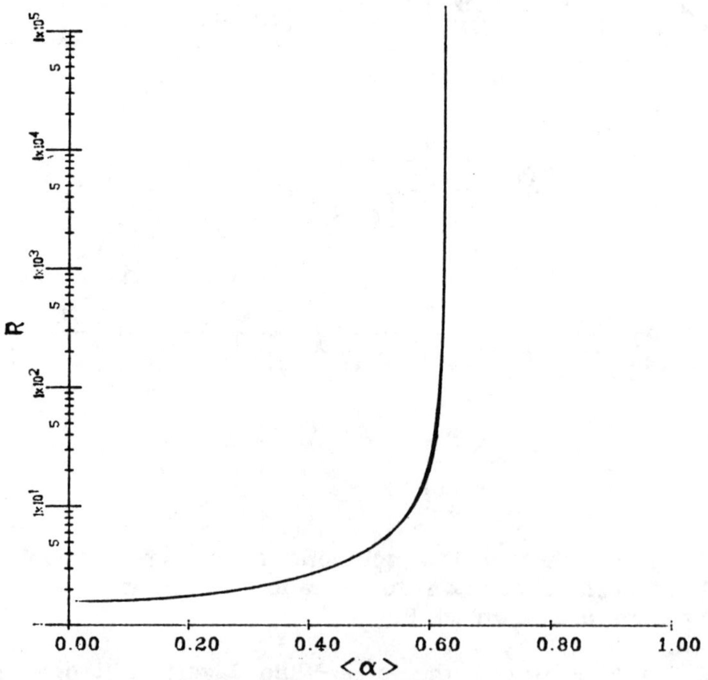

Figure 2. Reynolds number R versus <α>.

the wall lies in the transition layer. In this case, R need not be O(1). The analysis in this case is straight-forward, and is given in Drew (1986). Figure 2 is a plot of R versus $\langle\alpha\rangle$. For $\langle\alpha\rangle$ near α_m, the Figure uses the analysis alluded to above.

Fig. 2 can be interpreted in terms of the effective viscosity of the particle-fluid flow in Poiseuille flow. One sensible way to define the effective vicosity is to require that for a single-phase fluid of that viscosity, the same pressure drop gives the same centerline velocity. The value of R for a single-phase fluid is 2. Thus, the effective viscosity based on the centerline viscosity is given by

$$\frac{\mu_{eff}}{\mu} = R/2.$$

The centerline velocity is not a useful quantity for defining macroscopic parameters since it is not imposed on the flow through external means. Attempting to interpret effective viscosity in terms of the mass flux is equally unsatisfactory.

Conclusion. The equations of motion described in this paper show that the Poiseuille flow of a particle-fluid mixture in a channel results in a flow with a strong structure. The structure which occurs consists of a core of particles which are sufficiently "packed" so that they cannot shear, surrounded by a clear-fluid layer where all the shear occurs. The presence of this structure has several implications. First, if such a structure occurs in a flow, one must be careful in interpreting measurements of fundamental quantities such as mixture viscosity. The quantity measured may be strongly dependent on the structure. Second, this flow is not useful for measuring any of the quantities which cause the particle separation, even though one is large. The reason for this is that the flow becomes so degenerate. Finally, the presence of such a structure in this flow may allow the efficient filtration of such a mixture. Since no particles are near the walls, if fluid can be drawn through the walls without destroying the transition layer, then no particles will stick to the wall to impede the further flow of fluid through them.

The form of this structure is similar to many structures observed in particle-fluid systems, for example, the top of a sediment layer in a river, the "wall" of a bubble in a fluidized bed, and the wall of a spout in a spouted bed dryer. In all of these examples, the stability of the layer is important. Consider, for example, the top of a layer of sediment being overwashed with a fluid. If the fluid flows fast enough, the layer will become unstable and erosion will occur.

Acknowledgement. This work was supported by the National Science Foundation under Grant Number CPE83-14443 and the Mathematics Division of the U. S. Army Research Office under Contract DAAG29-85-G-0088.

References

[1] DREW, D. A. and LAHEY, R. T. Jr. The Virtual Mass and Lift Force on a Sphere in Rotating and Straining Inviscid Flow, to appear in International Journal of Multiphase Flow, 1986.

[2] STUHMILLER, J. The Influence of Interfacial Pressure on the Character of Two-Phase Flow Model Equations, International Journal of Multiphase Flow, 3(1977) pp. 551-560.

[3] ISHII, M. Thermo-fluid Dynamic Theory of Two-Phase Flow, Eyrolles, Paris, 1975.

[4] DREW, D. A. and LAHEY, R. T. Jr. Application of General Constitutive Principles to the Derivation of Multidimensional Two-Phase Flow Equations, International Journal of Multiphase Flow, 5 (1979) pp. 243-264.

[5] NIGMATULIN, R. I. Spatial Averaging in the Mechanics of Heterogeneous and Dispersed Systems, International Journal of Multiphase Flow, 5 (1979) pp. 353-385.

[6] EINSTEIN, A. Eine Neue Bestimmung der Molekuldimensionen, Ann. Phys. 19 (1906) pp. 289-306.

[7] FRANKEL, N. A. and ACRIVOS, A. On the Viscosity of a Concentrated Suspension of Solid Spheres, Chem. Eng. Sci. 22 (1967) pp. 847-853.

[8] GRAHAM, A. L. On the Viscosity of Suspensions of Solid Spheres, Applied Scientific Research 37 (1981) pp. 275-286.

Numerical Solution of the Three-Dimensional Stokes' Equations in the Presence of Suspended Particles

AARON L. FOGELSON*,† AND CHARLES S. PESKIN**,†

Abstract. A new fast numerical method for solving the three-dimensional Stokes' equations in the presence of discrete suspended particles is presented. The fluid dynamics equations are solved on a grid. The particles are represented by configurations of localized forces which are not constrained to move on the grid. These forces contribute to the force density term in the Stokes' equations. As a result, a single set of fluid dynamics equations holds at all points of the domain and there are no internal boundaries. Computational work increases only linearly with the number of particles, so large numbers (500-1000) of particles may be studied efficiently. Particle size, shape, and deformability may be prescribed. The method has been implemented to run on all CRAY computers: the implementation exploits the CRAY's vectorized arithmetic, and on machines with insufficient memory, it performs efficient disk I/O while storing most of the data on the disk. Applications of the method to sedimentation of one-,two-, and many-particle systems are described. Trajectories and settling speeds for two-particle sedimentation, and settling speed for multiparticle sedimentation from initial distributions on a cubic lattice or at random give good quantitative agreement with existing theories.

1. Introduction. We present a new fast numerical method for solving the three-dimensional Stokes' equations in the presence of discrete suspended particles. The fluid equations are solved on a computational grid and the particles are represented by configurations of localized forces which are not constrained to move on the grid. This method is an extension of the two-dimensional method for studying flows with immersed boundaries, which was introduced by Peskin [7] for studying blood flow in the heart, and which was applied by Fogelson [2] in an investigation of platelet aggregation during blood clotting. The salient feature of this immersed-boundary method is the replacement of material boundaries by suitable contributions to a force density term in the fluid dynamics equations. A single set of fluid dynamics equations then holds in the entire domain and there are no internal boundary conditions to treat. In the present context, this technique permits the study of particles of different sizes and

* Courant Institute of Mathematical Sciences, and Department of Mathematics, University of Utah, Salt Lake City, Utah 84112.

** Courant Institute of Mathematical Sciences, New York University, 251 Mercer Street, New York, New York 10012.

† This work was supported in part by the Multiphase Flow Project (under National Science Foundation grant DMS-8312229) at the Courant Institute of Mathematical Sciences, in part by the National Science Foundation under grant DMS-8502339, and in part by the Applied Mathematical Sciences Subprogram of the Office of Energy Research, U.S. Department of Energy under contract DE-AC03-76SF00098.

shapes as well as particles that deform. Since the computational work associated with the particles increases only linearly with the number of particles studied, the dynamics of large multiparticle systems, perhaps involving as many as 1000 particles, may be investigated. We expect that the method will be a powerful tool for studying many types of suspension flows. We present promising preliminary results from our application of the method to sedimentation problems. We also plan to incorporate the method into our ongoing study of platelet aggregation to extend that study to three dimensions.

2. Model System. We study the interaction of a viscous fluid with a collection of discrete model particles inside a three-dimensional cube B of side L. We assume that inertia is negligible and we describe the fluid motion by the Stokes' equations,

$$0 = -\nabla p + \mu \Delta \mathbf{u} + \mathbf{F}(\mathbf{x},t) \tag{1}$$

$$0 = \nabla \cdot \mathbf{u} \tag{2}$$

The force density $\mathbf{F}(\mathbf{x},t)$ will be constructed to have period L in each of the coordinate directions. We seek a solution $\mathbf{u}$, p which is similarly periodic. The Stokes' equations hold for *all* points $\mathbf{x}$ in the cube B.

Each model particle, say particle k, is constructed from a collection of points $\mathbf{x}_{kl}$, $l = 1,\ldots,M$, linked together by elastic springs. Each such point is assumed to have *no* inertial mass and to move at the local fluid velocity,

$$\frac{d\,\mathbf{x}_{kl}}{dt} = \mathbf{u}(\mathbf{x}_{kl}(t),t). \tag{3}$$

The points $\{ \mathbf{x}_{kl} \}$ which comprise the kth particle are linked to one another by a prescribed set of elastic links. For example, an octahedral particle can be constructed from seven points, six at the vertices and one at the center, with each vertex joined by elastic links to its four neighboring vertices and by a similar link to the center point. The links generate internal particle forces which act, through the mediation of the fluid, to keep the particle approximately rigid. It may be helpful to think of the link forces as approximations to the distribution of forces which a particle exerts on the adjacent fluid in order to obtain a constant fluid velocity at the particle's surface. For simplicity, we assume that each link acts as a linear spring. Thus, the resultant of the link forces which act on point $\mathbf{x}_{kl}$ is given by

$$\mathbf{f}_{kl}^{int} = \sum_{\mathbf{x}_{km} \in L_{kl}} S_{lm} \left(\| \mathbf{x}_{kl} - \mathbf{x}_{km} \| - r_{lm} \right) \frac{\mathbf{x}_{kl} - \mathbf{x}_{km}}{\| \mathbf{x}_{kl} - \mathbf{x}_{km} \|} \tag{4}$$

Here, L_{kl} is the set of points linked to point $\mathbf{x}_{kl}$, S_{lm} is the stiffness of the link joining points $\mathbf{x}_{kl}$ and $\mathbf{x}_{km}$, and r_{lm} is the resting length of this link. In addition to these internal link forces, application-dependent external forces, such as gravitational forces, as well as interparticle forces, such as electrical forces, may also act on the points of each particle. Both the internal and external forces influence particle motion only through the mediation of the fluid. They do this by contributing to the force density $\mathbf{F}(\mathbf{x},t)$ which appears in the Stokes' equations. More precisely, assuming that there are N particles each constructed from M points $\mathbf{x}_{kl}$, we define the force density by

$$\mathbf{F}(\mathbf{x},t) = \sum_{k=1}^{N} \sum_{l=1}^{M} \mathbf{f}_{kl}^{ext} \, \delta_s(\mathbf{x} - \mathbf{x}_{kl}(t)) + \sum_{k=1}^{N} \sum_{l=1}^{M} \mathbf{f}_{kl}^{int} \, \delta_s(\mathbf{x} - \mathbf{x}_{kl}(t)) - \mathbf{c} \tag{5}$$

Here, $\mathbf{f}_{kl}^{ext}$ is the applied external force on point $\mathbf{x}_{kl}$, $\mathbf{f}_{kl}^{int}$ is the resultant of the spring forces due to links joining point $\mathbf{x}_{kl}$ to other points in the kth particle, and $\delta_s(\mathbf{x})$ is a smooth function of integral 1 with support in a sphere of radius s. Thus, each force $\mathbf{f}_{kl}^{ext}$ or $\mathbf{f}_{kl}^{int}$ is "felt" by the fluid over a small volume surrounding the point $\mathbf{x}_{kl}$. This gives $\mathbf{x}_{kl}$ an effective radius of order s (see Fig. 1). We emphasize that the force density $\mathbf{F}$ is the only way that the fluid feels the presence of the particles. The vector $\mathbf{c}$ in Eq.(5) is a constant vector chosen so that

$$\int_B \mathbf{F}(\mathbf{x},t)\, d\mathbf{x} = 0. \tag{6}$$

This is a compatibility condition for the Stokes' equations in the periodic box B.

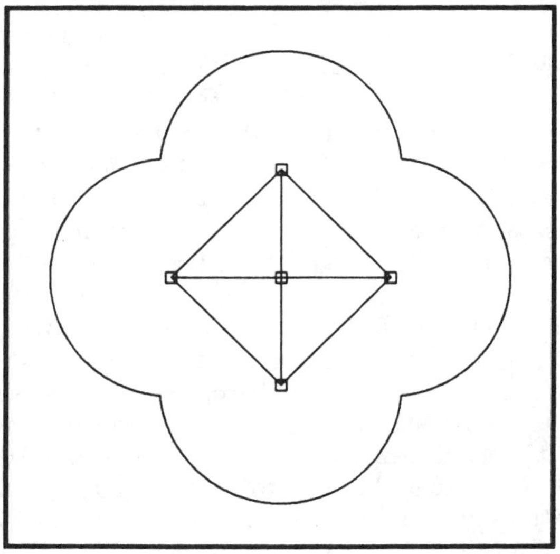

FIG. 1 *Cross-section of an octahedral particle. The boxes indicate points* $\mathbf{x}_{kl}$ *and are joined by line segments representing elastic links. The circular arcs enclose the region in which the particle makes a contribution to the force density* $\mathbf{F}(\mathbf{x},t)$.

To summarize the discussion thus far, we model physical particles by constructing each particle out of a set of points $\{ \mathbf{x}_{kl} \}$. These points are subject to external forces as well as to internal forces designed to maintain the particle size and shape. These forces all contribute to the force density $\mathbf{F}$ as described in Eq.(5); $\mathbf{F}$ determines the fluid motion by Eqs.(1)-(2); and the points $\{ \mathbf{x}_{kl} \}$ move at the local fluid velocity as described by Eq.(3). We reiterate that the main reason for representing particles by collections of localized forces is the simplicity of the resulting equations; a single set of fluid dynamics equations is valid everywhere in the domain, and there are no internal boundary conditions to treat.

3. Numerical Method. The numerical method for solving this system of equations involves approximating the fluid equations, Eqs.(1)-(2), at points of a computational grid, and tracking the particle points $\mathbf{x}_{kl}$ independently of this grid. The method is comprised of four pieces: 1) implicit calculation of the internal particle forces $\mathbf{f}_{kl}^{int}$; 2) definition of the force density $\mathbf{F}$ at points of the grid by "spreading" the internal and applied forces from the points $\mathbf{x}_{kl}$ to nearby grid points; 3) solution of finite-difference approximations to Eqs.(1)-(2) on the grid; and 4) interpolation of the fluid velocities from the grid to the points $\{ \mathbf{x}_{kl} \}$ in order to update these positions.

The internal particle forces are calculated implicitly to avoid numerical instabilities associated with explicit evaluation of these stiff forces. This calculation is performed by minimizing a nonlinear energy function.

Communication between the computational grid and the particle points $\{ \mathbf{x}_{kl} \}$ is involved both in "spreading" the particle forces to the grid and in interpolating the fluid velocities from the grid to the particle points. In both cases, the communication is achieved by means of a discrete version of the function $\delta_s(\mathbf{x})$ described above. The grid-particle communication, in a two-dimensional context, and the implicit force calculation are described in [2,7]; a description of the application of these techniques in the present context will be given in [4].

Discrete fluid dynamics equations are obtained by using finite-difference approximations to the differential operators appearing in Eqs.(1)-(2). The result is a discrete Poisson equation for p which involves differences on a staggered grid, and a discrete vector Poisson equation for $\mathbf{u}$ which involves differences on a tight grid. Despite this difference, essentially the same method of solution is used for both forms of discrete Poisson equation. This is the Fourier-Toeplitz method. It consists of performing a discrete Fourier transform in two coordinate directions on the nonhomogeneities; solving a set of periodic tridiagonal systems in the third coordinate direction to determine the Fourier coefficients of the solution $\mathbf{u}, p$; and computing the inverse Fourier transform to recover $\mathbf{u}, p$. This Stokes' equations solver will be described in more detail in [3].

The data storage requirements for the fluid dynamics calculation exceed the central memory storage capacity of computers even as large as the CRAY X-MP. For this reason, we have implemented an "out-of-core" version of the fluid-dynamics algorithm, in which, at any particular instant during the calculations, only a small subset of the data resides in central memory; the remainder of the data reside on the disk. Thus, data must be read from the disk to central memory and *vice versa*. Such Input/Output operations are slower than arithmetic computations, so we have implemented the numerical method in a way that overlaps steps of the algorithm that use the same data. We should point out that, with very minor programming changes, the numerical method can be (and has been) used on the CRAY-2, which has an enormous central memory, with all of the data residing in central memory.

We have designed the numerical method to exploit the fast vector registers available on the CRAY computers. This means that many of the arithmetic computations are, in effect, done in parallel.

It is important to measure the amount of computational work required to perform the calculations we have described. The work required to carry out the solution of the fluid dynamics equations is $NG\ log(NG)$ where NG is the number of grid points. The work required for the particle calculations is proportional to the number of particle points. This contrasts with other methods that involve direct particle-particle interactions and for which the work is proportional to the square of the number of particle points. This suggests that this method will be especially useful for calculations with large numbers of particles.

4. Results. We present simulations of one-, two-, and many-particle systems that involve the sedimentation of the particles under gravity. For all of these simulations, the external force is given by

$$\mathbf{f}_{kl}^{ext} = -g\,\mathbf{e}_3, \tag{7}$$

where $\mathbf{e}_3$ is a unit vector in the up direction. Fig. (2) is from a simulation of the settling of an octahedral particle. The picture depicts the situation after the particle has fallen approximately ten particle diameters and shows the *interpolated* velocity field near the particle in a vertical plane which bisects the octahedron. The frame of reference is one in which the average of the velocities of the left and right vertices is zero. The vertices of the octahedron are separated by distances that deviate by less than one percent from their initial values. The interpolated velocity field, which is plotted at points separated by 1/4 of the mesh space used in the fluid dynamics calculations,

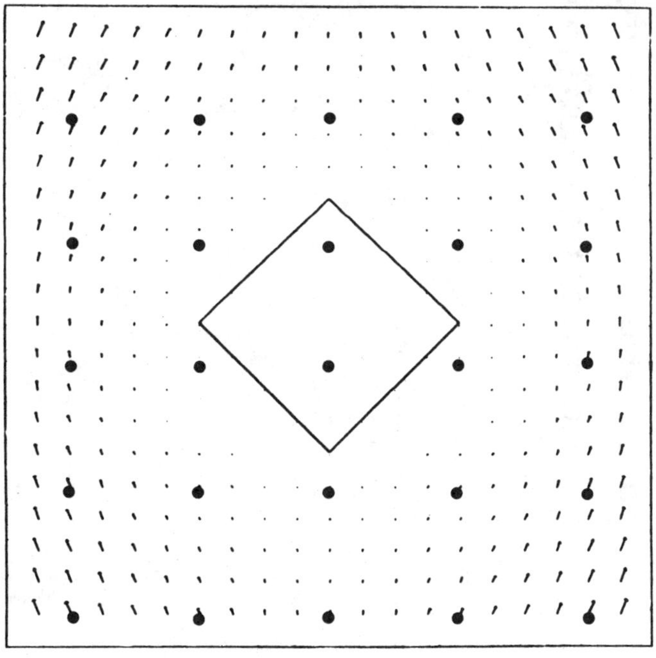

FIG. 2 *Vertical section through center of mass of octahedral particle settling under gravity. Arrows show interpolated velocity field in rest frame of particle. Circles indicate mesh points.*

is nearly zero for the fluid contained within the octahedron and in its immediate vicinity. This calculation shows the effectiveness of using a suitable configuration of point forces to cause nearby fluid to move as a rigid body.

Arguments based on symmetry and the linearity of the Stokes's equations can be used to show that two identical rigid spheres which are initially at the same height will settle along parallel vertical lines. We conducted a series of numerical experiments to see how well our method could reproduce this result. Fig. (3a) shows the settling of two particles each constructed from 21 points distributed at the vertices and center of a dodecahedron. The Stokes' equations were solved on a 64^3 grid. The two particles fall nearly in parallel; each particle's trajectory deviates from the vertical by an angle of about $0.2°$. We did a similar experiment with two octahedral particles (7 points per particle) and a 64^3 grid, and another with two "refined octahedra" (19 points per particle) each of which was constructed from an octahedral particle by inserting an additional point at the midpoint of each edge and linking together the new points on each face. This last calculation was performed using a 128^3 grid. In Fig. (3b), we plot the distance between the centers of mass of the two particles as a function of time for each of these experiments. The theoretical result to which these graphs should be compared is the horizontal line, $DISTANCE = 6$. We see that as the particles are made "more spherical" through the use of a larger number of points per particle, and also as the computational mesh is refined, the numerical results approach the theoretical result for rigid spheres. In all of these experiments, the nominal radius of each particle (i.e., the radius of the smallest circumscribing sphere) was 1.0, the initial interparticle distance was 6.0, and the imposed period in the calculation was 64.0.

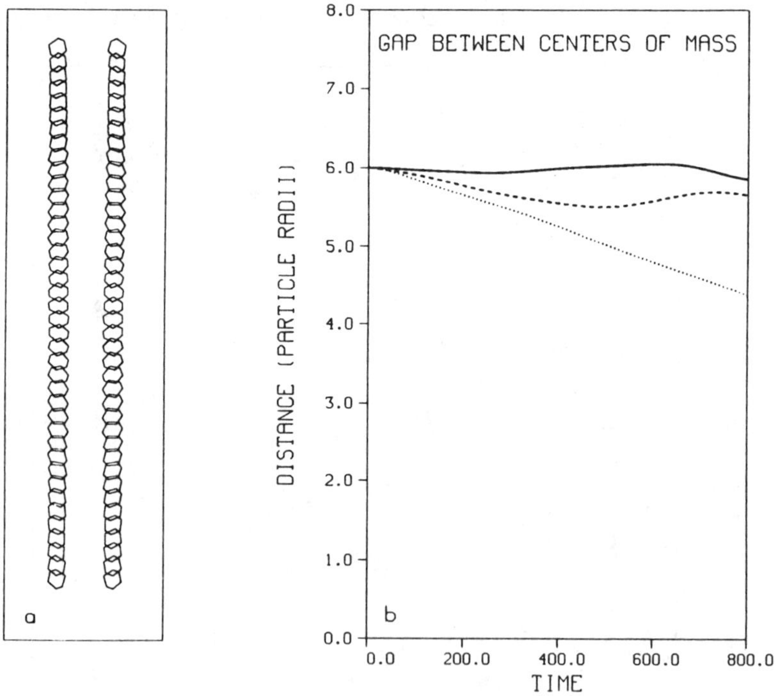

FIG. 3 *a) Sequence of vertical sections through centers of mass of two dodecahedral particles settling under gravity. Vertical extent of panel is one computational period L. b) Distance between centers of mass versus time for 2 octahedra (dotted curve), 2 dodecahedra (dashed curve), and 2 "refined octahedra" (solid curve) settling under gravity.*

We also calculated the ratio of the settling speed for two particles to that for a single particle, and compared this ratio to that predicted by a calculation using the method of reflections applied to two rigid spheres, as described, for example, in Happel and Brenner [6]. For each of the numerical experiments described above, the ratio was within 9% of that given by the method of reflections calculation; for the "refined octahedra", the ratios differed by less than 3%.

Two identical rigid spheres, which settle from different initial heights, move in parallel, each proceeding along a line that lies between the vertical line through it and the line through the two particles' centers of mass. Fig. 4 shows a numerical simulation of this phenomenon. The particles were modeled by dodecahedra and the calculation was done using a 64^3 grid. The angle of fall observed in this simulation was 4.5°, which compares favorably with a prediction of 5.0° based on a calculation using the method of reflections [6].

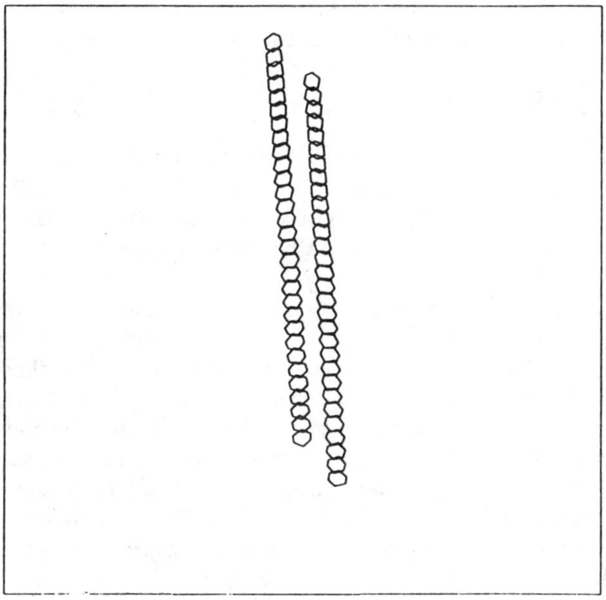

FIG. 4 *Sequence of vertical sections through centers of mass of two dodecahedral particles settling under gravity from different initial heights.*

We have conducted two types of multi-particle experiments each of which involved 125 particles. In the first of these experiments, octahedral particles were initially distributed on a cubic lattice which filled the periodic cube B used in our calculations. The observed average settling speed of the 125 particles was compared to a prediction based on Hasimoto's theory for the settling of a periodic cubic array of spheres [5]. For the situation we considered, Hasimoto's theory predicts a settling speed of approximately,

$$U_H = \frac{(1.0 - 1.76\ \beta^{1/3})\ g}{6\ \pi\ \mu\ a} \tag{8}$$

where a is the particle radius, g is the gravitational force on each sphere, and $\beta = 4/3\ \pi\ (a\ /l\ )^3$ is the volume fraction for particles spaced a distance l from one another. In order to use Eq.(8) to predict a settling speed for an array of particles, the particle radius must be known. The effective radius of one of our octahedral particles is larger than the radius of the sphere in which its vertices are inscribed. This is because each of the vertices has, in effect, a nonzero extent (recall Fig. 1) which augments the nominal radius of the octahedron. If we set U_H equal to the settling speed observed in our *single* octahedron experiments, and the interparticle spacing l equal to the imposed period L inherent in our calculations, then we can solve Eq.(8) for an effective radius, a_{eff}, of the octahedral particle. Determining a_{eff} in this way yields a volume fraction $\beta = .019$ for our 125-particle experiment. Thus, Hasimoto's theory predicts a settling speed of 0.00339 cm/sec for this experiment. This differs by just over 5% from our observed settling speed of 0.00357 cm/sec.

Our second multi-particle experiment involved 125 octahedral particles whose random initial positions were chosed from a uniform distribution without overlap. The observed settling speed early in the calculation (while the distribution was still random) was compared to Batchelor's prediction for the settling speed of a random array of spheres [1]. To first order, this prediction is

$$U_B = (1.0 - 6.55\,\beta)\,U_1 \tag{9}$$

where U_1 is the single particle settling speed. With volume fraction $\beta = .019$, as in the previous experiment, Eq.(9) predicts a settling speed which differs by 4% from the speed we observed. A similar experiment using dodecahedral particles yielded a settling speed which differed by 8% from Batchelor's prediction.

5. Concluding Remarks. We have presented a new fast numerical method for studying the dynamics of particles suspended in a very viscous fluid. The method's behavior compares favorably in a variety of tests with existing theories for sedimentation in two- and many-particle systems. Still, more sophisticated comparisons of the method with theory and experiment for a range of physical situations are needed to foster confidence in its predictions. Also, much work remains to be done to better characterize the influence of numerical parameters (e.g., mesh size, number and arrangement of points and links in a particle, stiffness of links) on the results produced by the method. Even so, we think it is not premature to say that we believe that the method will prove to be a powerful numerical tool for conducting multiparticle studies of a wide range of suspension flows. Among the special strengths of the numerical method are its relative efficiency in handling large numbers of particles, and its ability to treat particles of different sizes, shapes, and deformabilities. Further, the method is not limited to unbounded domains, as walls (e.g., those of a tube) can also be assembled out of arrays of the same kind of localized forces as those used to model the particles.

REFERENCES

[1] G.K. BATCHELOR, *Sedimentation in a dilute dispersion of spheres,* J. Fluid Mech., 52 (1972), pp. 245-268.

[2] A.L. FOGELSON, *A mathematical model and numerical method for studying platelet adhesion and aggregation during blood clotting,* J. Comput. Phys., 56 (1984), pp. 111-134..

[3] A.L. FOGELSON and C.S. PESKIN, *A fast out-of-core method for solving the three-dimensional Stokes' equations,* (in preparation).

[4] A.L. FOGELSON and C.S. PESKIN, *A numerical method for solving the three-dimensional Stokes' equations in the presence of suspended particles,* (in preparation).

[5] H. HASIMOTO, *On the periodic fundamental solutions of the Stokes equations and their application to viscous flow past a cubic array of spheres,* J. Fluid Mech., 5 (1959), pp. 317-328.

[6] J. HAPPEL and H. BRENNER, *Low Reynolds Number Hydrodynamics,* Martinus Nijhoff Publishers, Boston, 1983.

[7] C.S. PESKIN, *Numerical analysis of blood flow in the heart,* J. Comput. Phys., 25 (1977), pp. 220-252.

Interface Instabilities and Entrainment

JAMES GLIMM*

ABSTRACT

Jets, fingers, droplets, bubbles and vortices are commonly observed modes of fluid interface instability in the large amplitude, highly nonlinear regime. Entrainment of these modes provides a mechanism for the mixing of two fluids separated by a contact discontinuity. In this report we study the highly nonlinear (late time) regime of bubble amalgamation for a Rayleigh-Taylor unstable interface. The latter instability is a gravity or acceleration driven instability of an interface between two fluids of different densities. We utilize the statistical model developed by D. H. Sharp and J. A. Wheeler in 1961. In this model the dynamics is completely specified by the motion of a single bubble and the interaction (merger) of a pair of bubbles. The qualitative behavior of the model is discussed. Numerical solutions of the two fluid Euler equations for a gamma law gas are presented which show how to determine the parameters in the statistical model.

1. Introduction

Jets, fingers, droplets, bubbles and vortices are commonly observed modes of fluid interface instability in the large amplitude, highly nonlinear regime. Entrainment of these modes provides a mechanism for the mixing of two fluids separated by a contact discontinuity. The Rayleigh-Taylor instability is a gravity or acceleration driven instability between fluids of different densities. Its dominant modes are spikes (fingers) and bubbles. We study this problem for compressible fluids in the late time regime, and in the region adjacent to the undisturbed heavy fluid. This regime is dominated by the phenomena of bubble growth and merger. See [10] for a further discussion of the Rayleigh-Taylor problem and for additional references.

* Supported in part by the Applied Mathematical Sciences subprogram of the Office of Energy Research,U.S. Department of Energy, under contract DE-AC02-76ER03077.
Supported in part by the Army Research Office, grant DAAG29-85-K-0183.
Supported in part by the National Science Foundation, grant DMS-83-12229.
Courant Institute, NYU, New York, NY 10012 (Permanent address) and Los Alamos National Laboratory, Los Alamos, NM 87545.

The problem of fluid instabilities and mixing mechanisms is generic in multiphase flow. A generally valid program is to use a statistical model for the chaotic regime and to extract from it a new and simplified set of equations describing the macroscopic behavior, while numerical computations of the microscopic regime serve to set the parameters of the statistical model and to give an independent assessment of its validity.

Here we apply this point of view to the study of bubbles in the late time development of a Rayleigh-Taylor unstable interface. We present the initial steps in a program as outlined above. We use the Sharp-Wheeler statistical model [9, 10]. The further work on this program, including that discussed here, is a joint effort of the author and coworkers, and will be presented in full at a later time [4].

The main feature of the model is a description of the growth of bubbles due to the merger process. The main feature of the numerical program is a study of what we call the one and two body problems. The one body problem is the study of a single mode (spike, bubble, finger, jet, vortex, etc.), while the two body problem is a study of the interaction of two such modes. The aim of the statistical model is to capture the essence of a complex flow on the basis of one and two body interactions alone.

In the present report we describe the Sharp-Wheeler statistical model and the solutions of the one and two body problems as obtained by the front tracking method.

The front tracking method developed by the author and coworkers provides a high quality numerical method for the study of interface instabilities and mixing mechanisms. Previous papers have explained this method and have provided validation by comparison of front tracking solutions to laboratory experiment, analytic solutions and independently validated numerical solutions [2, 3, 5, 6, 7]. Internal consistency by convergence under mesh refinement has also been shown.

2. The Sharp-Wheeler Bubble Model

The main idea in the statistical model is to introduce a very reduced set of parameters ("idealized interfaces") and a simplified interface dynamics, which captures the essential features of the full flow in the mixing layer. For comparison purposes, we also need a map from arbitrary to idealized interfaces. It is then hoped that the stable statistical properties (average bubble size, etc.) will correspond to stable statistics of the full flow and that after adjustment of model parameters, the dynamics of such stable statistical properties will approximate the true dynamics of such stable properties of the full flow.

The first approximation we make in going from the full flow to the model is to assume that the interface position at time t is a single valued function $z = z(x,y,t)$ of the point x, y. Let us suppose that the heavy fluid is located in the region $z < z(x,y,t)$ below the interface and that the accelerating force is in the direction of the positive z axis.

Next we fix a set of bubble boundaries, which in mathematical language is a tiling of the surface $z = 0$ by polyhedra. In two space dimensions, which is the case we consider in this paper, the bubbles thus define a decomposition of the x axis into non-overlapping intervals.

Finally the model assumes that the fluid is piecewise constant, and constant within each bubble. Given a bubble decomposition (polyhedral tiling) of a general single valued interface $z = z(x,y,t)$, the principle of mass conservation again defines a corresponding interface within the model, i. e. piecewise constant on each bubble.

The kinematics of the model is now determined; we must specify its dynamics. There are two processes allowed in the dynamics of the model: vertical motion and merger. We assume that each bubble moves vertically with a velocity $\dot{z}$ which is independent of the bubble height and of the other bubbles. Thus $\dot{z}$ is a function of the bubble shape alone. In two dimensions, $\dot{z} = \dot{z}(r)$, is thus assumed to be a function of the bubble radius alone. On the basis of dimensional analysis,

$$-\dot{z} = \text{const.}\ (gr)^{\frac{1}{2}} ,$$

where const. is a function of the dimensionless model parameters.

The remaining aspect of the model bubble dynamics is merger. It is postulated that adjacent bubbles merge when one is sufficiently far ahead of the other. After merger, there is a single bubble, with cross section equal to the union of the cross sections of its two constituents and a height z determined by conservation of mass within this cross section. If the heights and radii of the two bubbles before merger are z_i and r_i, $i = 1, 2$, then we formulate the merger criterion as

$$z_2 - z_1 = m(r_1)$$

where $r_1 \leq r_2$ and $m(r)$ is an unknown function of r.

Thus the Sharp-Wheeler model is defined by two unknown functions, $\dot{z}(r)$ and $m(r)$, which express the essential features of the one and two body bubble problems respectively. These functions must then be set either by some external theory or experiments or by a numerical study of these one and two body problems in the full Euler equations. We have observed that

$$\dot{z}(r) = c_1 r^{\frac{1}{2}}$$

and we further postulate that

$$m(r) = c_2 r \,.$$

The Sharp-Wheeler model is now specified by two constants, and we next show that these constants can be eliminated by scaling, so that the model describes universal behavior. We consider the model in two dimensions in the x, z plane. Then the mean bubble size fixes the x units, c_2 fixes the relation of the z units to the x units, and c_1 fixes the time units. Moreover, the mean height fixes the origin of the z units. Thus the only essential parameters are the variances and higher moments characterizing the initial data of the bubble size and height distributions. Next we consider the model in an infinite strip, $x_0 \leq x \leq x_1$. The ratio of mean bubble size to interval width, $x_1 - x_0$, is a dimensionless measure of the distance of the initial data from the final equilibrium state of a single bubble. If the strip has a finite height, $z_0 \leq z \leq z_1$, then another dimensionless parameter is introduced which is a measure of the distance of the final data (at exit from the computational region) from the single bubble equilibrium state.

We have identified two renormalization group fixed points of a trivial (non-fractal) nature. There is an unstable fixed point consisting of uniform, small (infinitesimal) bubbles. There is a stable fixed point consisting of a single large (infinite, for the infinite width case) bubble. The dynamics starting in a neighborhood of the unstable fixed point moves to the stable fixed point. Analytic and numerical analyses of the Sharp-Wheeler model will be given separately [4], and comparison to a recently proposed model of Leith [8] will be discussed [4].

3. The Determination of Model Parameters

The compressible Rayleigh-Taylor problem depends on three dimensionless parameters: the density ratio $D = \dfrac{\rho_b}{\rho_a}$, where ρ_b is the density of the heavy gas just below the interface and ρ_a is the density of the light gas just above the interface, the polytropic gas constants γ_a and γ_b or other information to set the equation of state for the heavy and light fluids, and a Mach number M defining the ratio of a gravitational time scale to a sound speed time scale. We take $M^2 = \dfrac{g}{kc_b{}^2}$, where k is a wave number characteristic of the interface perturbation and c_b is the sound speed in the unperturbed heavy fluid. Observe that the mass of the heavy fluid per unit cross section is finite, even though the strip it is located in is semi-infinite, due to the exponential stratification of the fluid densities.

The initial growth at small time of both the spike and bubble is exponential. The for-

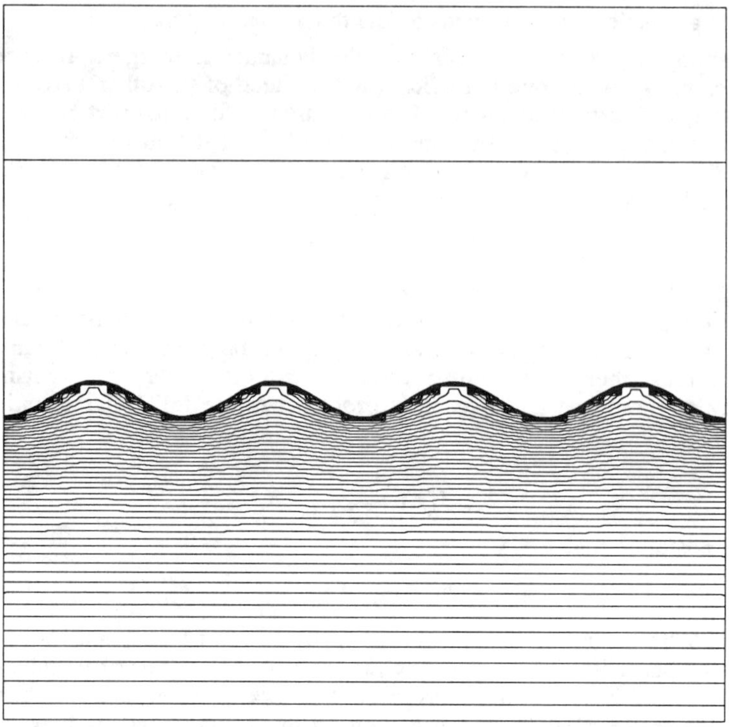

FIG. 1. Initial data for a Rayleigh-Taylor unstable interface. Density contours are shown.

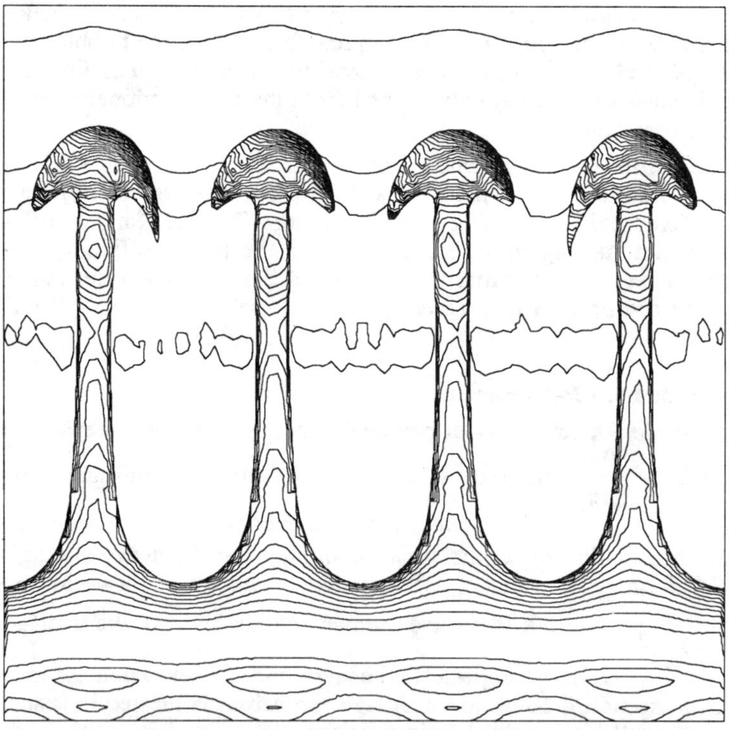

FIG. 2. Density contour plots in the late time development of a Rayleigh-Taylor unstable interface are shown. The parameters for this computation are as given in the text.

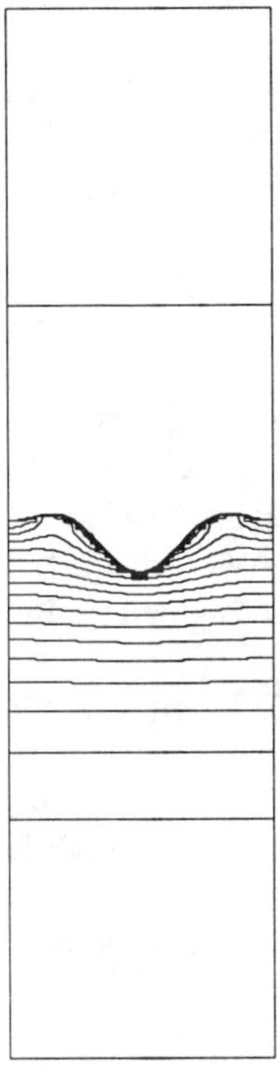

FIG. 3. The initial data for a bubble merge problem.

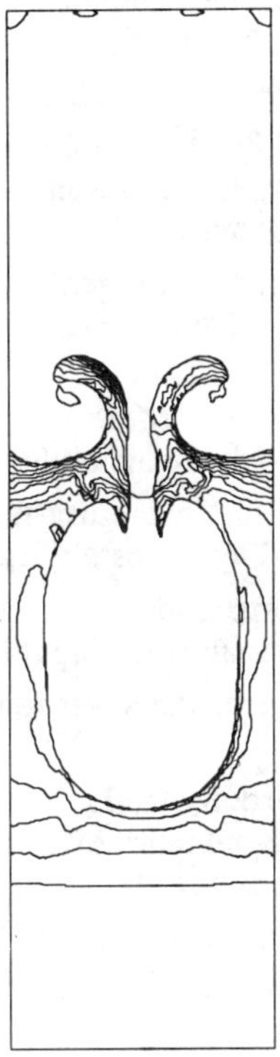

FIG. 4. A late time stage of bubble merger.

mulation of the linear theory, which governs the small time behavior was given by [1] for the infinite strip. These results are extended to apply to a finite strip with reflecting boundary conditions at top and bottom [4]. At late times the bubble and spike also approach an asymptotic constant velocity due to the form drag and the finite total mass supported above the bubble and spike. A possible intermediate time regime exists, which is free fall (spike) or rise (bubble). If this regime occurs for the bubble, bubble splitting is likely as the accelerating bubble interface is itself Rayleigh-Taylor unstable.

We illustrate this analysis by showing the numerical solutions of the full two fluid compressible Euler equations for a gamma law gas obtained by the front tracking method for typical one and two body bubble problems. In Fig. 1 and Fig. 2 we show the initial data and the late time development of a set of four equal bubbles and spikes. Density contours are shown, and as one can observe from the very minor asymmetries, all four fingers were computed. An 80×80 grid was used. The dimensionless parameters which characterize the run are $D = 10$, $M = 1.414$ and $\gamma_a = \gamma_b = 1.4$. In Figs. 3 and 4 we show the initial and late time density contours for a bubble capture process, with the same D and $\gamma_a = \gamma_b$ and an M similar to that of Fig. 1 and 2. (Because the perturbation is not a single mode, the bubble wave number is not clearly defined, and thus the value assigned to M is subject to interpretation.)

4. References

[1] I. B. Bernstein and D. L. Book, Effect of compressibility on the Rayleigh-Taylor instability, Phys. of Fluids, **26** (1983), pp. 453-458.

[2] I. L. Chern, J. Glimm, O. McBryan, B. Plohr and S. Yaniv, Front tracking for gas dynamics, J. Comp. Phys., **62** (1986), pp. 83-110.

[3] C. L. Gardner, Supersonic interface instabilities of accelerated surfaces and jets, Phys. of Fluids, **29** (1986), pp. 690-695.

[4] C. L. Gardner, J. Glimm, O. McBryan, R. Menikoff and D. H. Sharp, The dynamics of bubble growth for Rayleigh-Taylor unstable interfaces, In preparation.

[5] J. Glimm, Eli Isaacson, D. Marchesin and O. McBryan, Front tracking for hyperbolic systems, Adv. Appl. Math. **2** (1981), pp. 91-119.

[6] J. Glimm, C. Klingenberg, O. McBryan, B. Plohr, D. H. Sharp and S. Yaniv, Front tracking and two dimensional riemann problems, Adv. Appl. Math., **6** (1986), pp. 259-290.

[7] J. Glimm, O. McBryan, R. Menikoff and D. H. Sharp, Front tracking applied to Rayleigh-Taylor instability, SIAM J. Sci. Stat. Computing, **7** (1986), pp. 230-251.

[8] C. E. Leith, Instability notes II: A bubble growth model for final stage multiple scale Rayleigh-Taylor instability in two dimensions, Preprint, Mar. 1984.

[9] D. H. Sharp and J. A. Wheeler, Late Stage of Rayleigh-Taylor Instability, Institute for Defense Analyses, Unpublished Report, 1961.

[10] D. H. Sharp, An Overview of Rayleigh-Taylor Instability, Physica, **12d** (1984), pp. 3-18.

[11] C. P. Verdon, R. L. McCrory, R. L. Morse, G. R. Baker, D. I. Meiron and S. A. Orszag, Nonlinear effects of multifrequence hydrodynamic instabilities on ablatively accelerated thin shells, Phys. Fluids, **25** (1982), pp. 1653-1674.

[12] D. L. Youngs, Numerical simulation of turbulent mixing by Rayleigh-Taylor instability, Physica, **12d** (1984), pp. 32-44.

The Flow of Granular Materials and Aerated Granular Materials

R. JACKSON*

Abstract. Equations of motion for a non-cohesive granular material are presented, together with tentative constitutive relations for closure in the intermediate situation, where stress is generated by both frictional and collisional mechanisms. Corresponding boundary conditions are formulated, for both solid walls and free boundaries, and the influence of interstitial fluid is discussed briefly. Finally a solution of the complete equations is presented for the case of plane shear between horizontal plates.

Stresses and Equations of motion. The mechanical behavior of a flowing granular material results from forces exerted at points of contact between particles. For the purpose of formulating and solving problems whose shortest spatial dimension is much larger than the particle size, or spacing, we wish to replace this microscopic view of the assembly by a mechanically equivalent continuum, whose equations of motion can be formulated in terms of transmitted stress. The central problem is then to construct constitutive relations for the stresses in the continuum model which will mirror the behavior induced in the particle assembly by the contact forces.

Individual contacts between particles are sustained only for a limited time as the assembly moves and deforms,

---

*Department of Chemical Engineering, Princeton University, Princeton, NJ 08544. Part of this paper describes work supported by the U.S. Department of Energy, under grant No. DE-FG22-84PC70803 and by the National Science Foundation under grant No. CBT8504201.

and at each contact both normal and tangential forces are transmitted. When the assembly of particles is widely spaced and in rapid motion, individual contacts are brief and may be treated as "collisions", with transfer of normal and tangential momentum impulses. With close spacing and slow deformation, on the other hand, contacts persist for long periods, and the normal reaction forces and associated tangential frictional forces at these semi-permanent contacts dominate. As we shall see, theories of very different types have been developed to model these extreme cases, but unfortunately many situations of practical interest fall into an intermediate range, where contact times are neither very short, nor very long. An ad hoc constitutive model for this intermediate situation has recently been proposed (Johnson & Jackson [1]) and will be sketched here. It is based on the simple hypothesis that stresses due to collisions and to long term sustained contacts (referred to as collisional and frictional stress contributions, respectively) should be combined additively, each being calculated as though it were acting alone.

The bulk density ρ and velocity $\underline{u}$ of the continuum representing the particle assembly must satisfy a continuity equation and a momentum balance of the form

$$\frac{\partial \rho}{\partial t} + \nabla\cdot(\rho \underline{u}) = 0 \tag{1}$$

$$\rho \frac{D\underline{u}}{Dt} = \rho\, \underline{g} - \nabla\cdot(\underline{\underline{\sigma}}_c + \underline{\underline{\sigma}}_f) \tag{2}$$

where $\underline{g}$ is the specific gravity force vector, $\underline{\underline{\sigma}}_c$ and $\underline{\underline{\sigma}}_f$ are the collisional and frictional contributions to stress, referred to above, and D/Dt denotes the material derivative. It is customary to define stress in the compressive sense, since tensile stresses cannot be sustained by a noncohesive granular material.

The collisional contribution to stress is associated with a random component of motion of the particles, very similar to the thermal motion of molecules in a liquid, which is referred to as pseudo-thermal motion. Its associated kinetic energy is then the pseudo-thermal energy of the assembly, and with this it is possible to associate a grain temperature, T, in the same way as conventional temperature is associated with the energy of molecular random motion. If conditions in the assembly are spatially nonuniform, we can also identify a flux vector, $\underline{q}_{PT}$, of this pseudo thermal energy. However, the analogy with molecular motion is not quite complete since collisions between particles are dissipative, so the pseudo-thermal motion decays with time unless it is sustained by continuing deformation of the particle assembly. We denote by I the rate of dissipation of pseudo-thermal energy by collisions, per unit volume of the assembly. Then we can write a balance equation for pseudo-thermal energy, which takes the form

$$\frac{3}{2} \rho \frac{DT}{Dt} = -\nabla \cdot \underline{q}_{PT} - \underline{\underline{\sigma}}_c : \underline{\underline{\nabla u}} - I \tag{3}$$

The second term on the right hand side of this represents a source of pseudo-thermal energy associated with the rate of working of the collisional component of stress. Equation (3) is identical with the pseudo-thermal energy equation proposed by Jenkins and Savage [2] and by Haff [3] for the case in which there is no frictional contribution to stress. The pseudo-thermal energy dissipated in collisions is converted to true (molecular) internal energy of the solid material, so if the temperature of this material is of concern a fourth balance equation is needed, for the internal energy of molecular motion. I then becomes a source term in this equation. To close equations (1)-(3) constitutive relations are needed for $\underline{\underline{\sigma}}_f$, $\underline{\underline{\sigma}}_c$, $\underline{q}_{PT}$ and I, and these will now be discussed. The frictional stress is found from relations which are quite old and have their roots in soil mechanics and plasticity theory, while the collisional stress is deduced from a kinetic theory argument applied to the pseudo-thermal motion. These are quite disparate in nature, and will be described separately.

The Frictional Contribution to Stress. The frictional contribution to stress dominates at high densities and slow rates of deformation, and it is responsible for the initial yield behavior of the material. Granular materials exhibit a non-vanishing yield stress, and upon initial yield they may either dilate or compact as they shear, depending on the relative values of the mean stress and the bulk density. At low mean stress and high bulk density yield is accompanied by dilation and the material is said to be overconsolidated, while at high mean stress and low bulk density it is accompanied by compaction and is said to be underconsolidated. A number of models of initial yield have been developed for soils, which exhibit similar behavior. Here we describe a relatively early one due to the Cambridge school of soil mechanics (Schofeld and Wroth [4]) which is conceptually simple, though there are aspects of yield behavior which it probably does not reflect accurately. For simplicity we shall confine attention to two-dimensional motions, for which the model postulates the existence of a family of closed yield loci in the plane of the principal stresses (σ_1, σ_2), as sketched in Fig. 1. The loci are symmetric about the principal diagonal, and each is associated with a value of the bulk density of the material, the size of the locus increasing with increasing bulk density. For a state of stress represented by a point within the appropriate yield locus, a material of corresponding bulk density deforms elastically, while plastic yield occurs when the representative point lies on the yield locus.

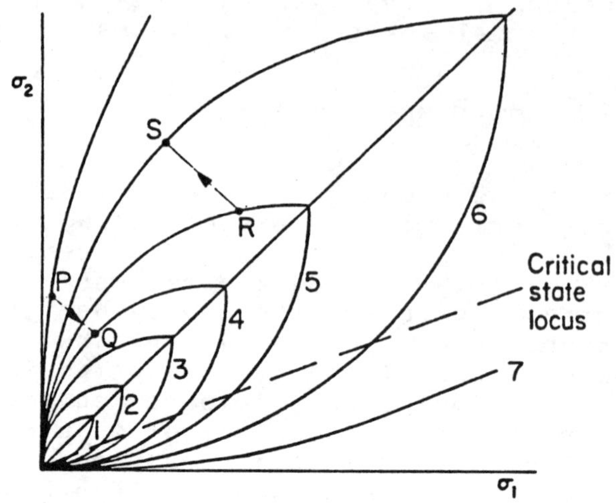

FIG. 1 Nest of yield loci in the plane of the principal stresses.

When yield occurs, the principal axes of the rate of deformation tensor are postulated to be parallel to the principal axes of stress, with the major principal rate of deformation axis aligned with the minor principal (compressive) stress axis. This is the so-called condition of coaxiality, or isotropy. Furthermore, the principal rates of deformation are postulated to be proportional to the corresponding components of the inward normal to the yield locus in principal stress space. This postulate, which is known as the plastic potential flow rule, is illustrated in Fig. 2, which shows the inward normals to a yield locus at three points. At point 1 the normal $\underline{n}_1$ has a large positive component and a small negative one, so one principal rate of deformation is large and positive, while the other is small and negative, and the overall result is a dilation of the material as it shears. Indeed, it follows from the plastic potential flow rule that the projection of the normal vector on the principal diagonal of the (σ_1,σ_2)-plane is proportional to the divergence of the velocity field, and for $\underline{n}_1$ this is seen to be positive. For yield at point 2, on the other hand, the normal $\underline{n}_2$ has a negative projection on the principal diagonal, so yield is accompanied by compaction. There are just two points, C and C', on the locus at which the normal is orthogonal to the principal diagonal, and here yield occurs without change in volume. These points are said to represent the <u>critical states</u> for material of the given bulk density. By joining the critical states of all the yield loci in Fig. 1 we obtain the <u>critical state locus</u>, represented by a broken line. Experiment indicates that critical state loci are straight lines through the origin, to a good approximation.

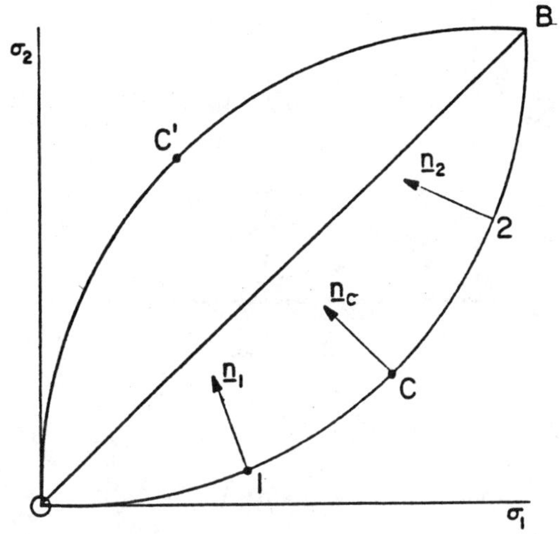

FIG. 2. Illustration of the plastic potential flow rule.

Now consider the effect of deforming a material from an initial state P, while keeping the mean stress σ $(=(\sigma_1 + \sigma_2)/2)$ constant. As the material deforms it dilates and yield occurs at a sequence of points along the trajectory PQ corresponding to constant mean stress, the shear stress required to maintain yield decreasing as deformation continues. When point Q is reached the material has dilated to a critical state, and further deformation occurs without change in volume. In contrast, if yield is initiated at a point such as R the path RS is followed, with compaction of the material until the critical point S is reached, after which deformation continues without change of volume.

The geometric picture presented in Fig. 1 can be mapped into the plane of shear stress and mean stress by defining the shear stress τ and the mean stress σ by

$$\tau = (\sigma_1 - \sigma_2)/2 , \qquad \sigma = (\sigma_1 + \sigma_2)/2$$

when the set of loci shown in Fig. 3 results. The maximum of each locus corresponds to the critical state, and the corresponding value of the mean stress σ will be denoted by σ_0. Then the set of loci can be described by a relation

$$\tau = f(\sigma, \sigma_0) \tag{4}$$

with σ_0 functionally related to the bulk density:

$$\sigma_0 = h(\rho) \tag{5}$$

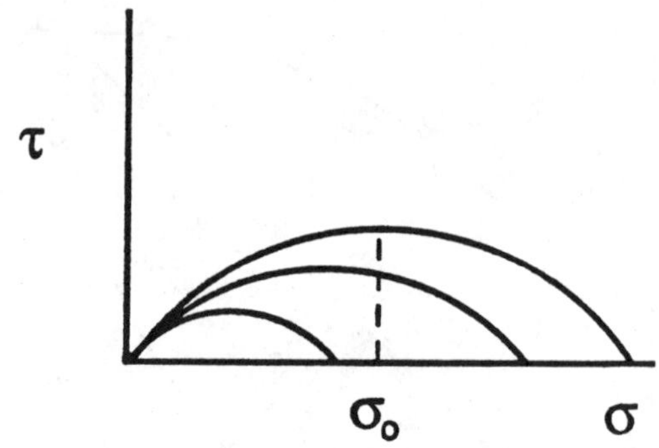

FIG. 3. Yield loci in the (σ,τ)-plane

$\sigma = \frac{1}{2}(\sigma_1 + \sigma_2)$, $\tau = \frac{1}{2}(\sigma_1 - \sigma_2)$

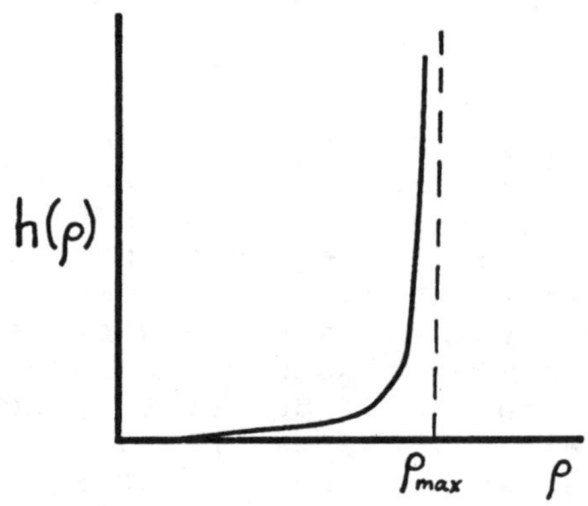

FIG. 4. Sketch of the function $h(\rho)$

The function $h(\rho)$ has the form sketched in Fig. 4, increasing rapidly and diverging to an asymptote as $\rho \to \rho_{max}$, corresponding to random close packing of the particles.

With this algebraic formulation of the yield loci the constitutive relations for the frictional component of stress can now be set out. The object is to express the stress tensor $\underline{\underline{\sigma}}_f$ in terms of the bulk density ρ and the rate of deformation tensor $\underline{\underline{D}}$ $(= \frac{1}{2}(\underline{\underline{\nabla u}} + \underline{\underline{\nabla u}}^T))$. The orientation of the principal stress axes is specified by the angle γ, measured clockwise from the x-axis of co-ordinates to the direction of the major principal stress. Then γ is determined by the isotropy condition, which can be written:

$$\cos 2\gamma \left(\frac{\partial u_x}{\partial y} + \frac{\partial u_y}{\partial x}\right) = \sin 2\gamma \left(\frac{\partial u_y}{\partial y} - \frac{\partial u_x}{\partial x}\right) \tag{6}$$

σ_o is related to ρ by (5) above then, knowing σ_o and γ, σ is found from the plastic potential flow rule, which takes the form

$$\frac{\partial \tau(\sigma,\sigma_o)}{\partial \sigma} \left(\frac{\partial u_y}{\partial y} - \frac{\partial u_x}{\partial x}\right) = \cos 2\gamma \left(\frac{\partial u_y}{\partial y} + \frac{\partial u_x}{\partial x}\right) \tag{7}$$

The shear stress τ then follows from equation (4), and the stress tensor is given by

$$\underline{\underline{\sigma}}_f = \begin{bmatrix} \sigma + \tau\cos 2\gamma & -\tau\sin 2\gamma \\ -\tau\sin 2\gamma & \sigma - \tau\cos 2\gamma \end{bmatrix} \tag{8}$$

There is no difficulty in generalizing these ideas to three dimensions, when the yield loci in the (σ_1,σ_2)-plane are replaced by yield surfaces in the three-dimensional $(\sigma_1,\sigma_2,\sigma_3)$-space, though information on the shape of the sections of these surfaces by planes normal to (1,1,1) is sparse.

Of course, a "theory" of this type contains unknown functions $f(\sigma,\sigma_o)$ and $h(\rho)$, whose forms would have to be determined by extensive experimental measurements. What we would like to do is determine the form of these functions from microscopic considerations, at the scale of the particles, leaving only one or two parameters to be measured experimentally. Not a great deal of progress has been made in this direction.

The Collisional Contribution to Stress. This has been the subject of an active investigation over the last few years. Early ideas of Bagnold [5] have been developed by Savage and Jeffrey [6], Ackerman and Shen [7], Jenkins and Savage [2], Haff [3], and Lun et al. [8]. These contributions are based more or less formally, on ideas from the kinetic theory of liquids, and they lead to constitutive relations for $\underline{\underline{\sigma}}_c$, $\underline{q}_{PT}$ and I of the following form:

$$\underline{\underline{\sigma}}_c = [f_1(\nu,T) + f_2(\nu,T)\nabla\cdot\underline{u}]\underline{\underline{I}} + f_3(\nu,T)\underline{\underline{S}} \tag{9}$$

$$\underline{q}_{PT} = -f_4(\nu,T)\nabla T - f_5(\nu,T)\nabla\nu \tag{10}$$

$$I = f_6(\nu,T) \tag{11}$$

where ν is the volume fraction solid (i.e., $\nu = \rho/\rho_s$, where ρ_s is the density of the solid material), while $\underline{\underline{S}}$ is the deviatoric part of the rate of deformation tensor, and I is the unit tensor.

$$(12) \qquad \underline{\underline{S}} = \underline{\underline{D}} - (\tfrac{1}{2}\, \mathrm{tr}\underline{\underline{D}})\underline{\underline{I}}$$

The forms of the functions f_1 - f_6 differ somewhat in the contributions of different authors, but those of Lun et al. [8] contain essentially no adjustable parameters, and are presented in Table 1.

With these relations the equations of motion are formally closed, though of very complex algebraic form. Some comments on the nature of these equations follow.

Comments on the Form of the Equations of Motion. The differential equations of motion are the continuity equation (1), the x- and y-components of the momentum balance (2), the pseudo-thermal energy balance (3), the isotropy condition (6), and the flow rule (7). Correspondingly there are six independent variables, namely u_x, u_y, ρ, σ, γ and T, since τ and σ_o are eliminated using the algebraic equations (4) and (5). The equations contain the first spatial derivatives of all the variables, the second spatial derivatives of u_x, u_y, T and ρ, and the first time derivatives of ρ, u_x, u_y, and T.

No solutions of the unsteady equations are known, and the complete steady state equations have been solved only for a simple example of plane shear (Johnson and Jackson, [1]). In addition to this solution, which will be discussed later, there have been attempts at solutions in the limiting cases where either frictional or collisional contributions to the stress predominate.

In the former case $\underline{\underline{\sigma}}_c$ is dropped from the equations, so it is not necessary to know the grain temperature T, and the pseudo-thermal energy equation can be omitted. This leaves five equations in the five unknowns u_x, u_y, ρ, σ, and γ, and these no longer contain any second derivatives with respect to the spatial co-ordinates. In particular, for steady motions, they can be shown to have five real characteristics, one of which is directed along the streamlines while the remaining four form two pairs of coincident characteristics, symmetrically disposed about the direction of the major principal stress axis and making angles $\pm\, \theta$ with it, where $\theta = \pi/4 - \tfrac{1}{2}\sin^{-1}(\partial\tau/\partial\sigma)$. The orientations of the characteristics are therefore determined by the position of the yield point on the (τ,σ)-loci in Figure 3. At the critical states $\partial\tau/\partial\sigma = 0$, so $\theta = \pi/4$, but on approaching the origin $\partial\tau/\partial\sigma \to 1$, and correspondingly $\theta \to 0$. Thus four of the characteristic directions collapse onto the major principal stress axis whenever the stress tends to zero.

TABLE 1

Functions Appearing in Collisional Constitutive Relations

$$f_1(\nu,T) = \rho T(1 + 4\eta\nu g_o)$$

$$f_2(\nu,T) = -\eta\mu_b$$

$$f_3(\nu,T) = -\left(\frac{2+\alpha}{3}\right)\left\{\frac{2\mu}{\eta(2-\eta)g_o}\left(1+\frac{8}{5}\eta\nu g_o\right)\left[1+\frac{8}{5}\eta(3\eta-2)\nu g_o\right] + \frac{6}{5}\mu_b\eta\right\}$$

$$f_4(\nu,T) = \frac{\lambda}{g_o}\left\{\left(1+\frac{12}{5}\eta\nu g_o\right)\left[1+\frac{12}{5}\eta^2(4\eta-3)\nu g_o\right] + \frac{64}{25\pi}(41-33\eta)(\eta\nu g_o)^2\right\}$$

$$f_5(\nu,T) = \frac{\lambda}{g_o}\left(1+\frac{12}{5}\eta\nu g_o\right)\frac{12}{5}\eta(2\eta-1)(\eta-1)\frac{d}{d\nu}(\nu^2 g_o)\frac{T}{\nu}$$

$$f_6(\nu,T) = \frac{48}{\sqrt{\pi}}\eta(1-\eta)\frac{\rho_p\nu^2}{d}g_o T^{3/2}$$

where

$$\eta = \frac{1}{2}(1+e_p), \quad \mu = \frac{5m(T/\pi)^{1/2}}{16d^2}, \quad \lambda = \frac{75m(T/\pi)^{1/2}}{8\eta(41-33\eta)d^2},$$

$$\mu_b = \frac{256\mu\nu^2 g_o}{5\pi}$$

and

$$g_o(\nu) = \frac{1}{1-(\nu/\nu_m)^{1/3}}$$

Note that the form of the yield loci is such that σ and τ tend to zero together, and this is important because the spatial derivatives of γ are multiplied by τ wherever they appear in the equations of motion. Consequently the equations are singular where the stress vanishes.

It is clear that surfaces of vanishing stress (for example, free surfaces of the granular material) are likely to be troublesome, since the equations of motion are singular on them and the number of independent characteristic directions is reduced by coincidence of pairs of characteristics. This has been a serious obstacle in the way of attempts to solve the equations for the important problem of flow of a granular material out of an orifice.

When collisional contributions to the stress dominate $\underline{\underline{\sigma}}_f$ can be dropped from the equations of motion, and the isotropy condition and flow rule are no longer needed. Explicit constitutive relations give $\underline{\underline{\sigma}}_c$ in terms of the density and elements of the rate of deformation tensor, and the continuity equation, the two componets of the momentum equation, and the pseudo-thermal energy equation then provide four equations for the four variables ρ, u_x, u_y, and T. These have been solved for plane shear between horizontal plates [9].

Boundary Conditions. Boundary conditions are needed in two situations: when a granular material is bounded by a solid containing wall, and when it is bounded by a free surface, unconstrained by external forces.

In the former case an appropriate condition is well known for the situation in which frictional stresses dominate. The particles in contact with the boundary then slide over it, impeded only by the forces of sliding friction at their points of contact with the solid wall. Coulomb's well known law of friction suggests that the ratio of tangential and normal stresses at the boundary will be a constant, denoted by $\tan\delta$, where δ is referred to as the angle of friction between the particles and the wall. Its value depends on the physical nature of the particles and the wall. At the opposite extreme, where collisional stresses dominate, Hui et al. [9] derived a boundary condition by simple kinetic theory arguments. Their approach can easily be generalized to the case where both stess contributors are significant, since it merely equates the tangential force per unit area exerted between the wall and the adjacent particles to the tangential stress within the material on a parallel plane, in the limit as that plane approaches the wall. The result is

$$\frac{\underline{u}_{s\ell}\cdot(\underline{\underline{\sigma}}_c+\underline{\underline{\sigma}}_f)\cdot\underline{n}}{|u_{s\ell}|}+\frac{\phi'\sqrt{3}\,\pi\rho\sqrt{T}\,|\underline{u}|_{s\ell}}{6\nu_m[1-(\nu/\nu_m)^{1/3}]}+N_f\tan\delta=0 \tag{13}$$

where $\underline{u}_{s\ell}$ denotes $\underline{u}-\underline{u}_{wall}$, the slip velocity at the wall, $\underline{n}$ is a unit vector normal to the wall directed into the granular material, N_f is the frictional component of the normal stress at the wall, ν_m is the volume fraction of solids at random close packing of the particles, and ϕ' is a parameter measuring the specularity of collisions between particles and the wall, whose value varies between zero for specular reflection and unity for diffuse scattering.

The first term in (13) is evaluated at the wall, and represents the limit of the tangential stress within the material on approaching the wall. The second term is the force exerted as a result of collisions between particles and the wall, and the third represents the frictional force due to particles sliding in contact with the wall. Clearly, a boundary condition of the form (13) permits a degree of slip between the granular material and the wall which depends on the roughness of the wall, but it is not at all similar to the familiar no-slip condition of fluid mechanics.

While (13) provides a mechanical boundary condition, it is also necessary to formulate a second condition describing the fate of the pseudo-thermal energy flux when it encounters a bounding wall. Within the solid material of the wall internal energy is simply the energy of molecular motion, and there is no analog of the pseudo-thermal energy of the granular material. Consequently there is no pseudo-thermal energy flux, and the flux of pseudo-thermal energy to the wall must match its net rate of dissipation at the wall. The boundary condition can, therefore, be derived by writing a pseudo-thermal energy balance about a control volume bounded by planes adjacent to the surface and on each side of it. The result is

$$(14) \qquad -\underline{u}_{s\ell}\cdot(\underline{S}_c\cdot\underline{n}) - \underline{n}\cdot\underline{q}_{PT} = D$$

where the right hand side is the rate of dissipation of pseudo-thermal energy, per unit area of the boundary, as a result of the inelasticity of particle-wall collisins. The second term on the left hand side is the flux of pseudo-thermal energy to the wall, and the first term gives the rate of generation of pseudo-thermal energy by the working of the collisional component of stress $\underline{S}_c$ exerted between the granular material and the wall. This term is positive and, when the slip velocity is large, it may dominate D. Then the wall serves as a source, rather than a sink, for pseudo-thermal energy. The rate of dissipation is given explicitly by

$$(15) \qquad D = \left[\frac{\pi\rho_s d^3 T(1-e_w^2)}{4}\right]\left[\frac{\sqrt{3T}}{d\{(\nu_m/\nu)^{1/3}-1\}}\right]\left[\frac{1}{d^2(\nu_m/\nu)^{2/3}}\right]$$

where e_w is the coefficient of restitution for particle-wall collisions. The first factor represents the loss of pseudo-thermal energy per collision, the second gives the numer of collisions per unit time between a particle and the wall, and the third represents the number of particles adjacent to unit area of the wall.

To complete the set of boundary conditions at a solid wall the mechanical and thermal conditions, (13) and (14), must be supplemented by the obvious kinematic condition $\underline{u}_{s\ell}\cdot\underline{n} = 0$.

There are situations of practical importance where these boundary conditions do not contain the whole story of interactions between the particles and the wall. For example, a number of practical devices use a solid surface vibrated normated to its own plane as a means of inducing flow of an otherwise recalcitrant granular material. If the frequency and amplitude of this vibration are comparable with those of the pseudo-thermal motion of the particles, then the vibration simulates a source of pseudo-thermal energy located in the plane of the wall, and this should be represented by an additional term in equation (14). However, the proper form for this term is presently unclear.

Boundary conditions at a free surface are, in principle, rather simpler than those at a solid wall. The kinematic condition is, of course, still applicable if the free surface is generated by streamlines. Furthermore, since there is no mechanism for dissipation or generation of pseudo-thermal energy, the flux $\underline{n}\cdot\underline{q}_{PT}$ must vanish at the surface. Finally, force balance requires that the total stress $\underline{\underline{\sigma}}$ should vanish at the surface.

These conditions are formally correct for a continous material, but in solving problems the last one tends to induce a singular behavior of the stress near the interfasce which is not physically realistic. The elements of the stress tensor, if forced to vanish on the free surface, drop rapidly to zero over a very thin layer adjacent to this surface. The thickness of the layer may be much less than the particle diameter, so it is physically meaningless and a mathematical nuisance, since it introduces an artificial element of "stiffness" into the solution of the boundary value problem. This difficulty can be avoided by recognizing that the material consists of grains of non-vanishing size, and writing the momentum balance for a layer adjacent to the surface with thickness equal to the grain diameter. This balance then serves as a boundary condition for the differential equations, replacing the condition of vanishing stress.

Effects due to an Interstitial Fluid. When the interstices of the particle assembly are occupied by a fluid forces are exerted between the two phases. If both particles and fluid are at rest the effect is simply a buoyancy force, but when there is relative motion other forces arise.

We must recognize two different causes for relative motion between a particle and its neighboring fluid:

(i) There is bulk streaming of the fluid phase relative to the particle phase, driven by a pressure gradient or a body force field. Then the local average velocities of the two phases, taken over a region containing many particles, differ from each other.

(ii) There is no bulk streaming (in other words, the local average velocities of fluid and particles are the same), but the fluid near a particle moves relative to the particle as a result of the relative motion of the particle and its neighbors generated by deformation of the particle assembly.

In case (i) the difference between the local average velocities of fluid and particles gives rise to drag forces, which are primarily responsible for important phenomena involving bulk relative movement of the two phases, such as fluidization, pneumatic transport, hindered settling and centrifugal separation. The forces arising from cause (ii) contribute to the rheological behavior of the assembly; for example, in dilute suspensions they are responsible for the well known modification of the fluid viscosity described by Einstein [10]. As the concentration of the particles increases the thickness of the fluid films separating adjacent particles decreases, shear rates in these films consequently increase, and the rheological behavior deviates more and more from that of the fluid alone. Finally, if the concentration is high, the particles are heavy, and the viscosity of the fluid is not too great, the intervening films are squeezed out when particles approach each other, and solid-solid contact occurs. The situation then approaches that discussed in detail above, modified to some extent by the buffering effect of the interstitial fluid.

Currently we have ideas about the rheology of relatively dilute suspensions and, as described in this paper, the rheology of granular materials with negligible interstitial fluid effects. However, the intermediate ground which provides the connection between the two is virtually unexplored.

Interaction effects due to bulk relative motion of the two phases are much easier to handle. Equations of motion which model these effects have been formulated by a number of authors (e.g. Anderson and Jackson [11]), and solutions have begun to provide a good deal of insight into the behavior of fluidized beds and particle transport devices using fluids, such as standpipes and pneumatic transport lines.

A Solution of the Equations of Motion. Plane Shear Between Horizontal Plates. At the present time a solution of the complete set of equations of motion has been constructed for only one non-trivial situation, namely fully-developed shearing motion between plane horizontal plates (Johnson & Jackson, [1]). The system analyzed is sketched in Fig. 5a. A total mass M_T of the granular material per unit area is loaded between the plates, a normal load N per unit area is applied, then the plates are set in relative motion by a shear force, whose value is S per unit area when steady motion has been established. The vertical separation of the plates is then H.

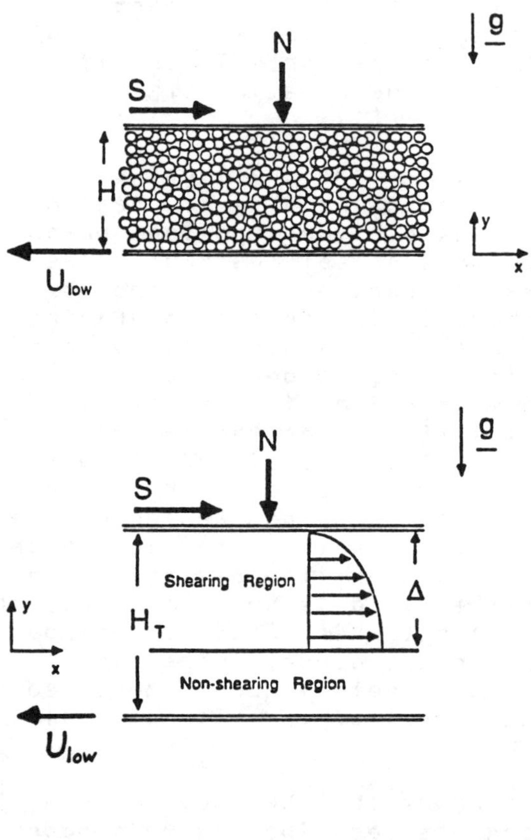

5a. Fully Shearing Experiment

5b. Partial Shearing Experiment

FIG. 5. Fully developed shear of a horizontal layer of granular material.

With this system two types of experiment can be performed:

(i) Constant load experiments.
The normal load N is fixed, the relative speed of the plates is varied, and the shear force S and plate separation H are measured as functions of N and $u_{\ell ow}$.

(ii) Constant density experiments.
The total depth H is fixed, $u_{\ell ow}$ is varied, and the shear force S and normal force N are measured as functions of H and $u_{\ell ow}$.

In both cases the upper plate is usually restrained from moving in the horizontal plane, while the lower one slides with speed u_{low} to generate the shear. The best available measurements are for experiments of the constant density type (Savage and Sayed [12]).

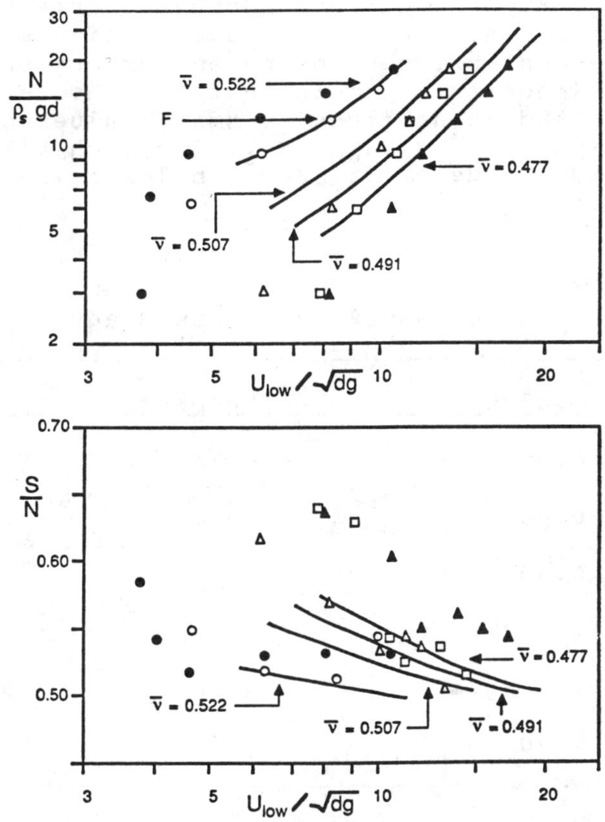

FIG. 6. Predictions of complete shearing analysis compared with constant $\bar{\nu}$ experiments of Savage and Sayed [12] using 0.0018 m diameter glass beads.

Experimental results: •, $\bar{\nu}$ = 0.522 (increasing load); o, $\bar{\nu}$ = 0.522 (decreasing load); Δ, $\bar{\nu}$ = 0.507; □, $\bar{\nu}$ = 0.491; ▲, $\bar{\nu}$ = 0.477.

The equations of motion were first applied to this problem with the assumption that the whole of the granular material between the plates is in fully developed shearing motion. Then it is everywhere in a critical state, and this greatly simplifies the analysis, which reduces to the numerical solution of a two point boundary value problem for a set of ordinary differential equations (for details see Johnson & Jackson [1]). Figure 6 shows comparisons of predicted values of N and S/N with results from constant density experiments of Savage and Sayed. The physical properties of the granular material (glass beads) and the values for parameters appearing in the equations of motion are listed in Table 2. All the parameters except α and Fr (see Table 1) are determinable by independent measurements, while the value is α is fixed arbitrarily and that of Fr is chosen to match theory and experiment at the single point labeled F in Fig. 6. The different curves correspond to different values of the overall average density $\bar{\nu}$, as indicated. The agreement between theory and experiment is fair, but it is important to note that the theoretical curves do not extend to arbitrarily small values of the shear rate, as measured by $u_{\ell ow}/\sqrt{dg}$. For each value of $\bar{\nu}$ there is a minimum value of $u_{\ell ow}/\sqrt{dg}$, below which solutions no longer exist.

TABLE 2

Properties of Sample of Glass Beads

Property	Value	Property	Value
d	0.0018m	δ	22.9°
e_p	0.80	ϕ	25°
e_w	0.50	ϕ'	0.60
F_r	3.65×10^{-32} kg/ms²	ϕ_y	25.2°
M_T	16.52 kg/m²	ρ_s	2980 kg/m³
α	1.60		

The reason for this behavior is revealed by the velocity profiles within the sample, shown in Fig. 7 for $\bar{\nu}$ = 0.522. Each curve intercepts the upper plate, y = 1, at a non-vanishing value of the velocity, which corresponds to the slip at that plate. Arrows on the horizontal axis also indicate the values of the lower plate speed corresponding to the curves, so the distance between the intercept of the curve with the axis y = 0 and the corresponding arrow is a measure of the slip velocity between the material and the lower plate. However, as the value of $u_{\ell ow}/\sqrt{dg}$ decreases, the velocity profile becomes progressively more convex in shape, and the slip at the lower plate decreases. In particular, when $u_{\ell ow}/\sqrt{dg}$ = 5.65, the profile becomes vertical when $y \to 0$, and there is no slip at the lower plate. In this condition the material is no longer shearing where it contacts the moving plate, and for still smaller values of $u_{\ell ow}/\sqrt{dg}$ the assumption that the granular material shears everywhere is no longer valid.

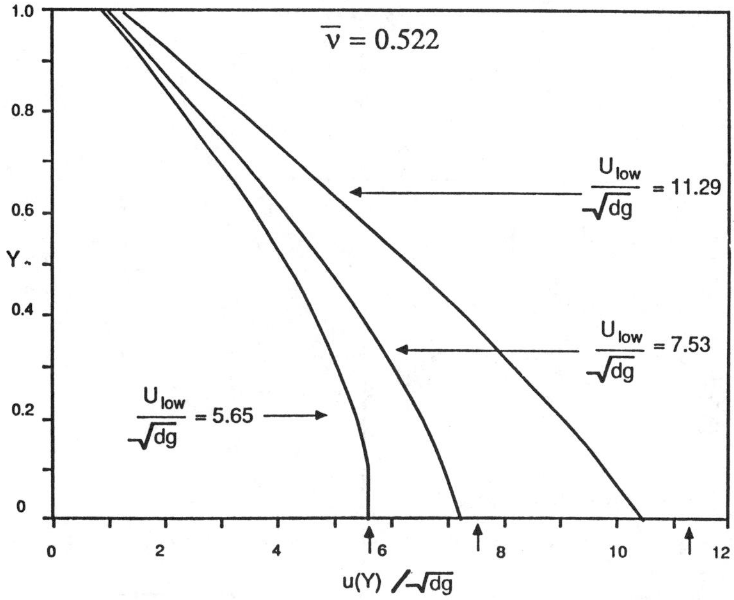

FIG. 7. Predicted velocity profiles with $\bar{\nu}$ = 0.522. Arrows indicate values of abscissa corresponding to lower plate velocity.

Because of this failure of the initial assumptions the problem was re-worked assuming the configuration of the granular material to be as sketched in Fig. 5b, where a layer of material in fully developed shearing motion lies above a second layer which remains at rest. The thicknesses and amounts of material in these two layers are to be determined as part of the problem. The stationary material is stressed below its yield point, so the nature of its packing before the experiment starts is important, since this determines its yield strength. Even for fully developed motion, it therefore follows that the solution is not determined unless the procedure for initial loading and preparation of the sample of granular material is fully specified. In our calculations we chose to assume that the material was initially compacted to a uniform and specified value of the bulk density. Calculations within the shearing layer resemble those leading to the results already presented, but they must now be supplemented by appropriate boundary conditions at the interface between the two layers in order to locate this interface. Details of the rather complicated procedure are given elsewhere (Johnson & Jackson, [1]).

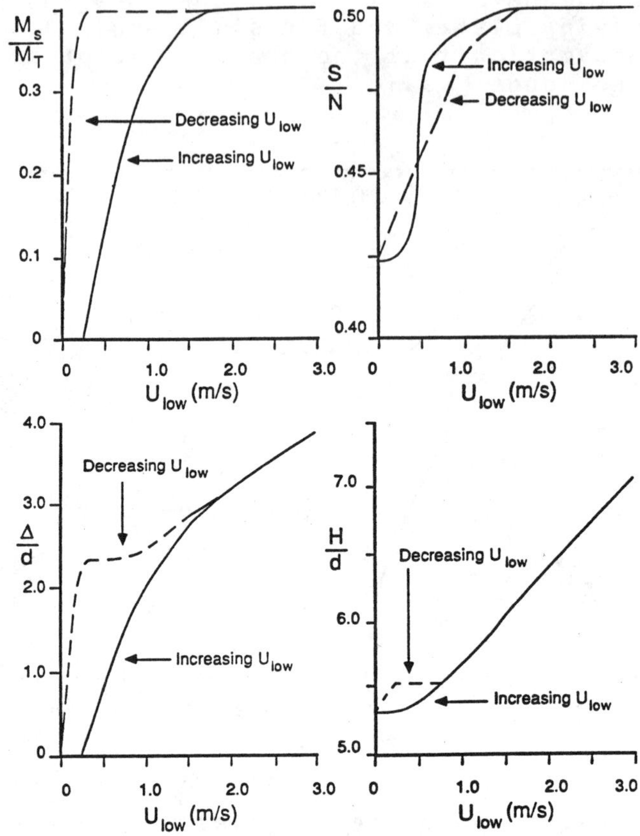

FIG. 8. Effect of $u_{\ell ow}$ on various properties, with constant normal load of 1000 kg/ms^2 on upper plate.

Results are presented in Figure 8, for the sample of glass beads whose properties are given in Table 2, subject to a constant load experiment with a normal load N = 1000 Kg/ms^2. The first graph shows the ratio M_S/M_T of shearing mass to total mass, that is, the fraction of the material which shears. This is seen to rise rapidly with $u_{\ell ow}$, after a short interval in which the material merely slips on the bounding plates without shearing. However, it quickly levels out to an asymptote at which only about 40% of the material is shearing. More than this cannot be induced to shear, no matter how rapidly the lower plate is moved. On decreasing $u_{\ell ow}$ from a high value, the curve for increasing $u_{\ell ow}$ is not retraced, but there is a substantial hysteresis effect, as indicated by the broken curve. The graph below presents similar information in a rather different form, showing the ratio of the thickness Δ of the shearing layer to the particle diameter d. This again increases with increasing $u_{\ell ow}$ and shows a hysteresis effect, but now there is no horizontal asymptote. The thickness of the shearing layer continues to increase with increasing overall shear rate, as a result of dilation of the deforming granular material, even though the mass in this layer remains constant. The remaining two graphs show the shear stress, as the ratio S/N, and the spacing H between the plates. The last graph therefore indicates the dependence of the overall degree of dilation of the sample on the overall shear rate.

Notice that the shear layer never exceeds three or four particle diameters in thickness. The limiting fraction of mass in the shear layer at high values of $u_{\ell ow}$ depends on the applied normal load. Thus, for N = 160 Kg/ms^2 it is found that only about 25% of the sample can be induced to shear, even at the highest overall shear rates, while the thicknes of the shear layer is two or three particle diameters. These results shed considerable doubt on the validity of a continuum picture of granular material flow, but they are in accord with qualitative observations that such materials tend to move in rigid blocks, separated by quite thin layers of rapidly shearing material. This tendency appears to be particularly marked in motions where the overall deformation rate is fairly slow.

While the theory used here is tentative, and available experimental observations are sparse, enough has been found to indicate that the motion of a granular material is much more complex than that of a fluid, and that its response to external stimuli will depend on its previous history of deformation and packing. If meaningful tests of theoretical predictions are to be made future experimenters will, therefore, have to standardize their procedures for preparing their samples prior to the experiment, as well as the procedure for establishing the shearing motion, and these will need to be documented in detail.

REFERENCES

[1] P. C. JOHNSON and R. JACKSON, Frictional-collisional constitutive relations for granular materials, with application to plane shear, to appear in J. Fluid Mech.

[2] J. T. JENKINS and S. B. SAVAGE, A theory for the rapid flow of identical, smooth, nearly elastic spherical particles, J. Fluid Mech., 130 (1983), pp. 187-202.

[3] P. K. HAFF, Grain flow as a fluid mechanical phenomenon, J. Fluid Mech., 134 (1983), pp. 401-430.

[4] A. N. SCHOFIELD and C. P. WROTH, Critical State Soil Mechanics, McGraw-Hill Publishing Company, 1968.

[5] R. A. BAGNOLD, Experiments on a gravity-free dispersion of large solid spheres in a Newtonian fluid under shear, Proc. Roy. Soc., A225 (1954), pp. 49-63.

[6] S. B. SAVAGE and D. J. JEFFREY, The stress tensor in a granular flow at high shear rates, J. Fluid Mech., 110 (1981), pp. 255-272.

[7] N. L. ACKERMAN and H. SHEN, Stresses in rapidly sheared fluid-solid mixtures, J. Eng. Mech., 108 (1982), pp. 95-113.

[8] C. K. K. LUN, S. B. SAVAGE, D. J. JEFFREY and N. CHEPURNIY, Kinetic theories for granular flow: inelastic particles in Couette flow and slightly inelastic particles in a general flow field, J. Fluid Mech., 140 (1984), pp. 223-256.

[9] K. HUI, P. K. HAFF, J. E. UNGAR and R. JACKSON, Boundary conditions for high shear grain flows, J. Fluid Mech., 145 (1984), pp. 223-233.

[10] A. EINSTEIN, A new determination of molecular dimensions, Ann. Phys., 19 (1906) pp. 289-306.

[11] T. B. ANDERSON and R. JACKSON, A fluid mechanical description of fluidized beds. Equations of motion, Ind. Eng. Chem. (Fundam.), 6 (1967) pp. 527-539.

[12] S. B. SAVAGE and M. SAYED, Stresses developed by dry cohesionless granular materials sheared in an annular shear cell, J. Fluid Mech., 142 (1984), pp. 391-430.

Nonlinear Mechanics of Fluidization of Spheres, Cylinders and Disks in Water

D. D. JOSEPH*, A. F. FORTES*, T. S. LUNDGREN* AND P. SINGH*

Abstract. Experiments on fluidization with water of spherical particles falling against gravity in columns of rectangular cross section are described. All of them are dominated by inertial effects associated with wakes. Two local mechanisms are involved: drafting and kissing and tumbling into stable cross stream arrays. Drafting, kissing and tumbling are rearrangement mechanisms in which one sphere is captured in the wake of the other. The kissing spheres are aligned with the stream. The streamwise alignment is massively unstable and the kissing spheres tumble into more stable cross stream pairs or doublets which can aggregate into larger relatively stable horizontal arrays. The stability of cross stream arrays in beds of spheres constrained to move in two dimensions is amazing. These arrays may even coalesce into aggregations of close packed spheres separated by regions of clear water. A somewhat weaker form of cooperative motion of cross stream arrays of rising spheres is found in beds of square cross section where the spheres may move freely in three dimensions. Horizontal arrays rise where drafting spheres fall because of greater drag.

Experiments using cylinders of different length to diameter ratios were also carried out. All objects float with their broad side perpendicular to the stream, broadside on. The broadside on position of particles is due to a turning couple at the front of bodies of the same nature as the couple which turns canoes broadside on in a stream and an explanation can be given from potential flow. Kissing spheres are a composite body equivalent to a long cylinder; they tumble because of the turning couple on the composite body. Broadside on position of single particles seems to be more stable than other positions.

* Department of Aerospace Engineering and Mechanics
University of Minnesota
Minneapolis, Minnesota 55455
This work was supported by the Fluid Mechanics branch of the National Science Foundation and by the Mathematics Division of ARO. The financial support from CNPq - Brazil given to one of us (Fortes) is gratefully appreciated.

Wakes behind blunt bodies, long cylinders and flat objects of various kinds are very strong. Single nonspherical particles oscillate around a mean horizontal position with a frequency that may be determined by vortex shedding. Other particles are sucked into wakes tending to suppress the oscillations. Clusters of particles, held together in the wake of larger particles, move as a group. Interesting particle clustering can be achieved by manipulation of wakes.

1. Introduction. In this paper we describe some experiments on the fluidization of spherical particles between two parallel glass plates and in a persplex column with a rectangular cross section. Some experiments between parallel plates were performed in beds of different width with spheres whose diameters were slightly smaller than the gap between the plates. We say that this fluidization of a single layer of particles is "two-dimensional". To our knowledge, the experiments here and in the papers of Volpicelli, Massimilla and Zenz [1966] and Garside and Al-Dibouni [1973] are the only ones using two-dimensional beds of a single layer of particles. Some of the observations which we shall make about the dynamics of fluidization in these beds can also be inferred by careful examination of photographs presented by Volpicelli, et al, though they did not make these inferences. They showed that fluidization in two-dimensional beds satisfies the fluidized bed correlations of Richardson and Zaki [1954] with a shifting of curves of correlation due to wall effects.

The fluid dynamics of two-dimensional fluidization of spheres is of intrinsic interest and is also useful as a diagnostic tool for three-dimensional fluidization. It is certain that there are some wall effects in two-dimensional beds. However, the dominant features of fluidization so clearly evident in the two-dimensional beds were also observed in three-dimensional beds at all Reynolds numbers for which fluidization was possible. The three-dimensional beds are the parallel plates with smaller spheres (two-layer beds) and columns with rectangular and square cross sections, using spheres of different densities and sizes.

Different flow regimes were observed in the experiments. All of them are dominated by inertial effects associated with wakes. A basic mechanism of fluidization involves pairwise coupling of spheres whose line of centers is parallel to the stream. We describe the dynamic scenario associated with this pairwise coupling as drafting, kissing and tumbling. The second sphere is drafted into the wake of the first sphere; they kiss, then tumble. The falling motion of kissing spheres whose centers are parallel to the stream is unstable to couples of the type which turn streamlined bodies broadside on. Neighboring spheres with centers aligned cross stream which appear to have the greatest stability are also a characteristic feature of the fluid dynamics of fluidized beds. The robust stability of the cross stream alignments are particularly dramatic in a "two-dimensional" bed of a single layer of particles. In this way drafting and kissing can lead to aggregation. The aggregates are closely packed arrays of spheres which can align in contact along vertical lines through their centers. Such aggregates are a characteristic feature of a two-dimensional fluidized bed, and they can take form as propagating or even as standing "shock waves" of aggregates separated by clear water. Propagation of stable cross stream arrays also occurs in the three-dimensional case. We see cases where the propagation occurs as a result of drafting, kissing and tumbling spheres collecting at the top and leaving the bottom of the aggregates.

The clear water regions between aggregates will be called voidage cracks. Voidage cracks, called "bubbles", are a characteristic feature of fluidization of beds of particles with air. Bubbles look much different than void cracks and the dynamics of fluidization with air must be hugely different than fluidization with water. The change of momentum in the liquid of liquid-solid systems is as important as the changes in the momentum of the particles, because their densities are not greatly different. The literature on fluidization is abundant with empirical and theoretical analysis of crack formation in fluidized beds. The latest works on the subject, such as those by El-Kaissy and Homsy [1976], Homsy, El-Kaissy and Didwania]1980] and Liu [1982 and 1983] hold that the crack formations can be traced to voidage instability waves. Though we are not yet committed to any theoretical model of a fluidized bed, our observations indicate that aggregation associated with drafting, kissing and tumbling can give rise to the propagation of voidage cracks. This type of propagation can be viewed as the shock waves of the effective density of spheres which were described in the works by Wallis [1962], Verloop and Heertjes [1970] and Fanucci, Ness and Yen [1979]. In the fluidization of a single layer of particles, suction in the wakes of spheres in relative motion lead to aggregation due to repeated capture of spheres and even to standing shock waves of closely packed spheres which percolate along vertical lines, as in Figure 1.

The topology of sphere aggregation in fluidization is a functional of the dynamics. Rowe [1961] and Rowe and Henwood [1961] studied drag forces on spheres in fixed aggregates. Unfortunately, they pushed their flow through the unnatural array of spheres shown in Figure 1 when U is turned through 90°. The hexagonal arrangements of similar spheres used by Richardson and Meikle [1964] are also misleading, because they artificially constrain the spheres.

Many investigators have reported the formation of natural clustering of particles in both gas and liquid-solid systems. As early as in the work of Wilhelm and Kwauk [1948], in the sedimentation experiments of Kaye and Boardman [1962], the formation of particle clusters have been reported. Experimental evidence of the importance of aggregates formed from drafting, kissing and tumbling was given by Happel and Pfeffer [1960]. Their study was restricted to free falling spheres along the axis of a cylindrical glass column of liquid with Reynolds numbers in the range of 0.3 and 0.7. Even at these low Reynolds numbers wake effects are important, even dominant. Happel and Pfeffer conclude their experimental study with the observation that

> "...it is possible the spheres suspended in random orientation may not maintain their positions relative to each other. Perhaps the formation of doublets (kissing spheres), and their corresponding higher velocities of fall, has been one of causes for the wide discrepancies in the presently available fluidization data."

Recent experimental work of Tsuji, Morikawa and Terashima [1982] on the hydrodynamic interaction between two spheres is relevant to our investigation. They used the pendulum method to measure the drag on the sphere and flow visualization. The pendulum method restrains the sphere to move like a pendulum at the end of a string. The method was extended to study flow interaction: a dummy sphere is set up in the front or rear of the test sphere, or a group of spheres connected by a rod in the transverse direction is hung by two strings. They used both streamwise and crosswise arrangements and determined the corresponding effects on the drag and vortex patterns. Their results agree with our findings on the existence of the suction in the wakes of spheres in relative motion even at low Reynolds numbers, when a single sphere does not shed vortices, and with our observation of increased drag on the cross stream arrays.

The compelling experimental demonstrations of wake dominated nonlinear mechanisms for fluidizing beds of particles ought to be addressed in the hydrodynamic modeling of fluidization and sedimentation.

2. Experimental parameters. The two-dimensional bed exhibited in Figure 2 was made of two parallel glass plates 0.285 in. apart with a 75 cm test section. The width was adjustable. The glass plates were framed in U-shaped beams, and changes of the bed thickness within ±0.007 in. were made possible with the use of adjustment screws so as to ensure an even distribution of the water flow throughout the bed. The working fluid was water supplied from a pressurized water tank through a needle valve. This valve gives fine control of the flow rate so that a fairly steady working pressure was available at the entrance of the bed. The maximum total throughput of the 5 HP centrifugal pump water supply system was approximately 50 ℓ/min., where ℓ stands for liter. An equalizing section consisting of a 6 in. layer of packed glass beads followed by a 2 in. layer of metal foam was placed upstream of the test section. Pressure taps attached to the rear plate of the bed were located right above the equalizing section and at the top of the test section. The water flowing from the top of the bed was collected in a trough open to the atmosphere. The flow rates were measured with a Micro Motion Mass Flow Meter, and the pressure drop across the entire bed was measured with piezometer tubes. The two-dimensional bed could be tilted at any angle from the vertical. This allows one to adjust the effective force of gravity.

The three-dimensional bed is exhibited in Figure 3. The rectangular cross sections used in this Perspex plate apparatus were 3 in. by 2 in. and 3 in. by 3 in. The square cross section, used for plastic beads, did not allow through flow of a magnitude sufficient to fluidize the heavier glass beads. The smaller cross section was used in the experiments on fluidization of glass beads. This bed had an equalizing section composed of a 4 in. layer of honeycomb aluminum followed by a 1 in. layer of metal foam.

Both beds could be illuminated from behind as well as from the front with 4 high intensity 650 watt quartz flood lamps. Illumination of the two-dimensional bed was made through an opaque Perspex plate graduated with a metric scale on its left side for measuring the bed height. The three-dimensional beds were illuminated from the front.

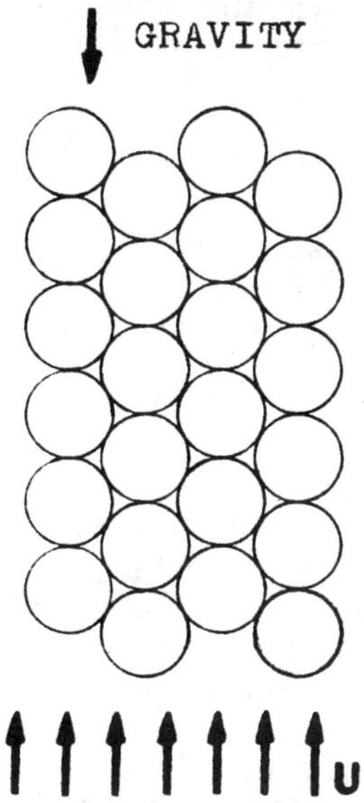

Figure 1. Close packed spheres contact on vertical lines.

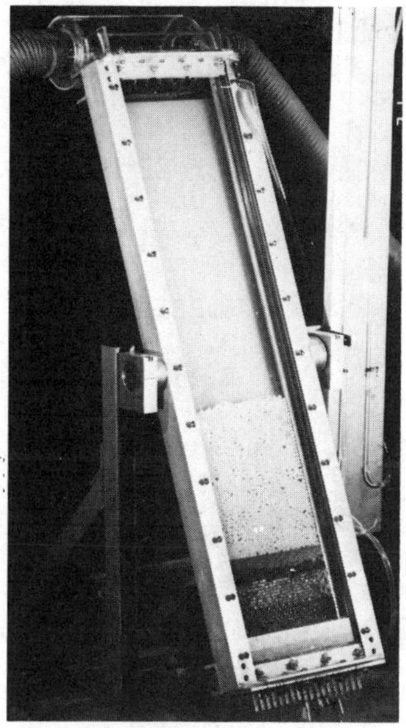

Figure 2. Two-dimensional apparatus shown in an inclined position.

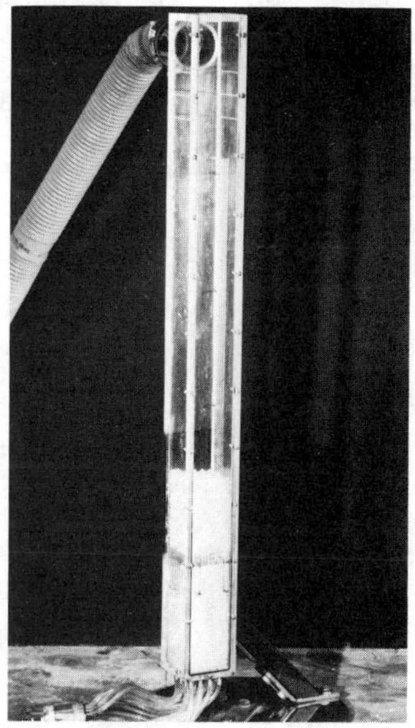

Figure 3. Three-dimensional apparatus.

(a) (b) (c) (d)

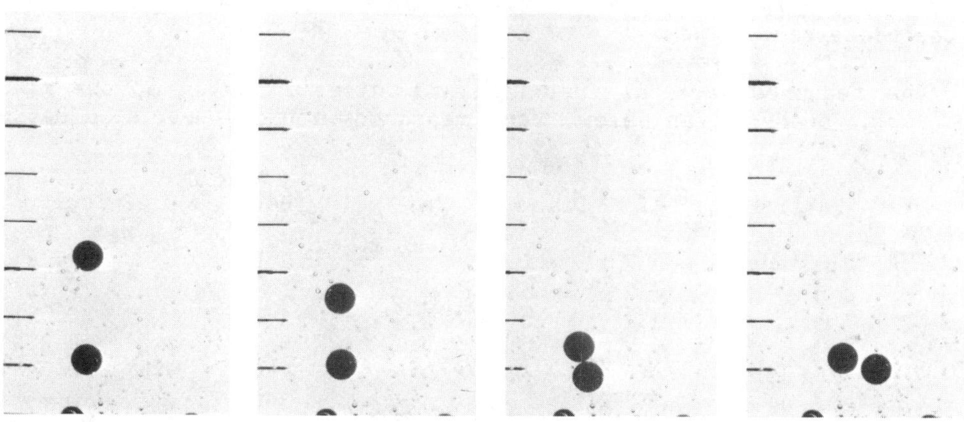

Figure 4. Drafting, kissing and tumbling of plastic spheres in a bed inclined 23° from the vertical, Re = 700. (a) The line of centers lies along the stream. (b) Both spheres are falling, but the second is falling faster than the first. (c) Kissing after drafting. (d) Tumbling after kissing.

Flow visualizations were recorded using still 35 mm photography and conventional video-cassette recording. Motion analysis was made possible with the help of a Spin Physics Motion Analysis System which incorporates a high speed video camera that can take up to 2000 frames per second. In order to enhance the contrast against a black background, the glass beads were etched in a solution of hydrofluoric acid. Some plastic beads were dyed in a solution of red oil dye and acetone in order to be better able to follow the motion of one single particle in the bed. The bed width was adjustable.

Average properties of the spheres are given below:

Particles	Diameter D, in.	Density ρ, g/cm³
Glass	0.231	2.462
	0.117	2.462
Plastic	0.251	1.119

In the two-dimensional bed, bed widths of 3 in., 6 in. and 8 in. were used. The flow rates used varied from 5.2 ℓ/min. to 21.5 ℓ/min. for fluidization with glass beads and from 1 ℓ/min. to 3.1 ℓ/min. for the plastic beads. In the three-dimensional bed, the maximum flow rate used was about 48 ℓ/min. for fluidization of glass spheres. Smaller flow rates were sufficient for the fluidization of plastic spheres.

The voidage ϕ is the fraction of water in the total volume occupied by the fluidized bed. The total volume is the box which is defined by the portion of the bed between the two planes perpendicular to the flow direction and tangent to the highest and lowest sphere in the bed. When using the smaller glass particles in the two-dimensional bed and for all the experiments in the three-dimensional bed, the void fractions were calculated from the volume of the particles determined by a careful measurement of the water added up to the least possible static bed height and from the known total volumes. The total volume changes as the bed expands.

The measured vertical free-fall velocities in water of the glass and plastic beads, given below, were measured within a precision better than 1%:

Particles	Diameter, in.	Bed	Velocity, m/s
Glass	0.231	2-D	0.341
	0.117		0.337
		3-D	0.551
			0.372
Plastic	0.250	2-D	0.094
		3-D	0.154

The Reynolds number is $Re = DU_s/\nu$ where D is the diameter of the sphere, ν is the kinematic viscosity and U_s is the superficial velocity of the fluid based on the net volume flux of water across a plane perpendicular to the flow direction. The Reynolds numbers of our experiments were always greater than 270, making our flow regimes definitely non-Stokesian. Potential flow analysis of the dynamics of fluidization is equally deficient, because the dynamics is controlled by suctions in the wakes of spheres in relative motion. For example, analysis shows that two spheres in steady potential flow will attract one another when they move perpendicular to their line of centers and they will repel one another when they move parallel to their line of centers

(Lamb [1938], p. 191). In fact, these predictions do not apply to flows with wakes. It seems probable that many features of the flows we observed could be explained by a proper introduction of vortical regions (wakes) into otherwise potential flow. In our experiments, spheres moving broadside on do not attract and spheres following along their line of centers do, a result already known even for weak inertial effects, as the Oseen's equations show (Oseen [1927], pp. 199 ff.).

3. Analysis of drafting, kissing and tumbling. Fundamental features of the fluid dynamics may be observed by fluidizing two spheres. Figures 4a and 4b show two plastic spheres suspended in a rising stream in the two-dimensional bed. The bed was inclined 23° 35' from the vertical. This inclination introduced a gravity force component acting on the spheres normal to the walls of the channel, which has increased the stabilizing effects of the wall friction. We will show later that this wall friction does not significantly change the particle interactions. The Reynolds number was 730. In Figure 5a the same plastic spheres were suspended in a vertical channel. The Reynolds number was slightly smaller than in the inclined channel. A single sphere is in a force equilibrium between the net weight of the sphere and the drag on it. If the two spheres do not interact, they are nearly in equilibrium and drift slowly from place to place. One sphere always drifts into the wake of the other. The second sphere which is now on the line of centers parallel to the stream begins to accelerate, first slowly, then rapidly, to the falling first sphere. This can occur when the centers of spheres are even five or six diameters apart. This is the drafting part of the drafting, kissing and tumbling scenario as shown in Figures 4a, 4b and 5a. The second sphere is very rapidly sucked into contact with the first, then kissed. This is shown in Figures 4c and 5b. The falling motion of contacting spheres aligned in the direction of motion is very unstable. As soon as they kiss, they tumble and are thrown apart as shown in Figures 4d and 5c. A slight angular displacement of the line of centers of two contacting spheres will induce an unsymmetric wake which evidently gives rise to a strong destabilizing couple. Drafting, kissing and tumbling can occur repeatedly and with great frequency. Moreover, since the wake of spheres is determined by the direction of streaming, the events of this scenario are controlled locally and also occur in the three-dimensional bed, as can be seen in Figures 6a, 6b and 6c. The still picture camera sequence of Figures 7a, 7b and 7c taken 1/3 seconds apart, shows the drafing, kissing and tumbling occurring with the dyed pair of plastic spheres on the right side of the pictues, in the 3 in. by 3 in. cross section channel.

The stabilizing effect of the wall friction accounts for the higher Re in the inclined channel, but the evidence shows that the mechanisms of drafting, kissing and tumbling are affected neither by inclination nor by wall effects. Generally speaking, the dynamics in the regions near the wall is not inertially dominated. Thus, the particle interactions near the wall could not be controlled by inertial suction, which contradicts blatantly the experimentally observed drafting of one sphere in the wake of another. Walls can have an important quantitative effect on frictional drag without altering the qualitative balance with opposing forces associated with inertial suction.

(a) (b) (c)

Figure 5. Drafting (a), kissing (b) and tumbling (c) in the vertical two-dimensional bed. The mechanisms are not affected by the inclination of the bed.

(a) (b) (c)

Figure 6. Drafting (a), kissing (b) and tumbling (c) in the three-dimensional bed.

Figure 7. The interaction of two dyed plastic spheres in the three-dimensional apparatus. Rising cross stream arrays of spheres are also apparent. Re = 850. Lubrication layers of clear fluid at the side walls are evident.

Since it appeared impossible to maintain a water flow of absolute uniformity over the cross sectional area of the three-dimensional bed, the increased level of the fluid turbulence reduced the distance of interaction of the spheres. In fact, the shrinking of the wake region behind bluff bodies with increasing Reynolds number has been reported in the physical explanations of Gerrard [1966], as well as in the experimental work of Achenbach [1974]. Turbulence does not overcome the capture phenomenon, though it does reduce the distance of interaction.

In Figures 8a and 8b we have exhibited one more feature of fluidization seen in two-dimensional flows with many particles in purer form. The total number of spheres in Figure 8a is eight; eight spheres can span the 2 in. width of the channel with some play. The picture shows seven spheres standing stationary in the upward stream of water. The eighth sphere is stationary, standing on the equalizing section. Figure 8b shows the same configuration in the vertical two-dimensional bed. They are stable, for a time, in the configuration in which their lines of centers are perpendicular to the stream. Since they rise, they are being accelerated in their wakes, pulled along by the suction. We call attention to the fact that spheres aligned cross stream do not touch, but they are evidently spaced to produce stability. There are many different stable spacings. These cross stream lines of spheres can be found in the many photographs of the two-dimensional beds of spheres shown in Section 5, especially Figures 15a and 15b.

To explore the stability of the cross stream alignments, we decided to cement three spheres together. The cemented spheres are massively stable when their line of centers is perpendicular to the stream. In a striking display of this basic feature of fluidization, Figures 9a and 9b show that the stability of the cross stream arrays are indeed unaffected by the inclination of the walls. The increased drag experienced by the spheres in this configuration (Tsuji, et al [1982]) testifies to the fact that the cemented triplet and the free spheres rose abreast only when spaced in stable configurations. The factors which enter into the breakdown of these configurations are not perfectly understood. However, it was obvious that interparticle collisions associated with drafting, kissing and tumbling and fluid turbulence play an important role. This breakdown may be more or less identified with the propagation of banded aggregates of rising cross stream arrays of spheres which gain spheres from the tumbling of kissing spheres drafted into the aggregate from the top and lose spheres from the collapse of the aggregate from the bottom. This regulating mechanism for propagation of aggregates is evident in Figures 7 through 14.

4. Experimental results for the fluidization of many spheres. Figures 10a, 10b and 10c show the onset and propagation of void cracks in a channel 8 in. wide using glass beads in the inclined channel. They are separated by time intervals of about 1/3 second. The cracking of closely packed structures is called void cracking and can be viewed as a density perturbation or even as a shock wave of density (Wallis [1962], Verloop, et al [1970] and Fanucci, et al [1979]). The average void fraction in these figures is 0.53 and $R = 1080$. Under these conditions, the bed is nearly stationary and nearly packed. Essentially, the

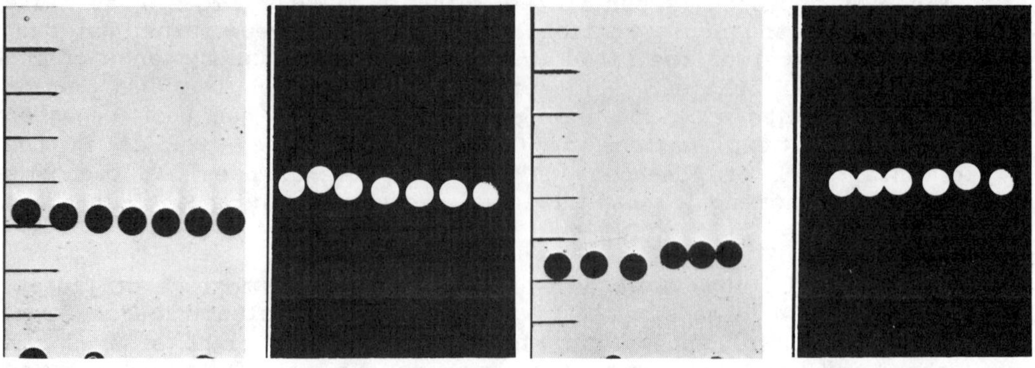

(a) (b) (a) (b)

Figure 8. Rising stable cross stream array of spheres in the two-dimensional bed. Different spacings are stable. (a) Seven distinct spheres in the inclined channel. (b) Seven distinct spheres in the vertical channel.

Figure 9. A different stable cross stream array. (a) Three distinct and cemented spheres in the inclined channel. (b) Three distinct and three cemented spheres in the vertical channel.

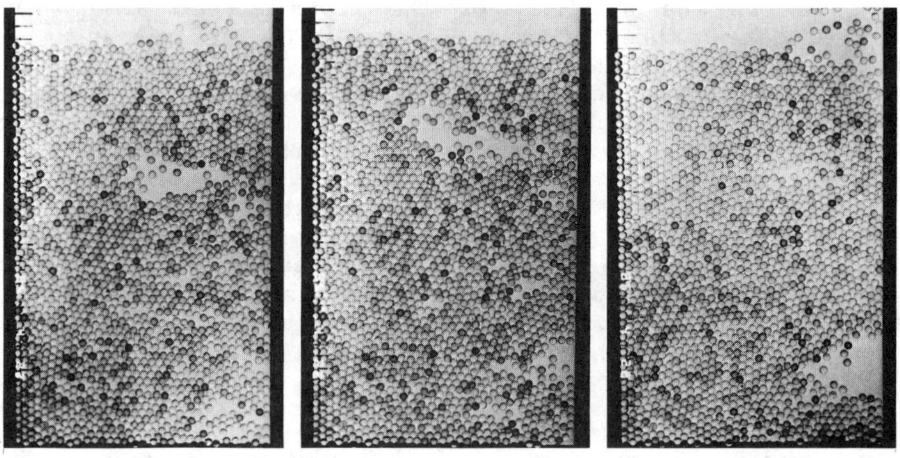

(a) (b) (c)

Figure 10. Fluidization of glass beads at 0.53 average voidage and Re = 1080 in the two-dimensional inclined bed. The channel is 8 inches wide. Readers should focus on the motion of voids. This motion occurs as a withdrawal of spheres at the void roof and the collection of spheres at the void floor.

same structure can be seen in the vertical channel, although not with the same stability. We might expect that the pressure drop across the stationary closely packed bed is related to the superficial velocity by the well-known formula for the pressure drop-velocity relation in the state of incipient fluidization, like the Ergun equation (Richardson [1971]).

The effects of inertia of the liquid in flow through a porous media appear as a drag proportional to the square of the velocity. The idea is that this drag is due to the "dead water" region behind particles (Joseph, Nield and Papanicolau [1982] derive the quadratic drag (in the equation below) from this idea). The same idea seems to operate in fluidized beds, i.e., the contributions of the inertial hydrodynamic interaction between particles to the bulk properties of the flow can be expressed as

$$\mathbf{f} = g(\phi)\ |\mathbf{u}|\mathbf{u}\ ,$$

where $\mathbf{u}$ is the local fluid velocity relative to the particle and $g(\phi)$ is an unknown function of the voidage ϕ. Although a discussion of constitutive relations is beyond the scope of the present work, the quadratic drag term suggested here is physically consistent with the observed phenomena.

Packed beds of fluidized spheres are different from porous material because the drag forces may move the spheres. This type of local density perturbation usually takes form as void cracking of the packed structure. The upper surface of the crack moves upward, widening the crack. This type of cracking occurred only in the two-dimensional beds. The cracks are always perpendicular to the bulk flow direction, more or less, and never parallel to the flow. The flow across the roof of the crack thus creates a local pressure gradient that drives the packed structure above the crack upwards. The relief of this pressure gradient in the final bursting of the ascending void crack in the upper layer of the bed shown in Figure 10c is indicative of its presence as a driving force. The existence of this local pressure gradient is consistent with the more general observation of Harrison and Davidson [1963, 1971] that for a fluidized bed of given voidage the pressure drop at a given velocity is less than for a randomly packed fixed bed at the same velocity and voidage.

Another mechanism for the propagation of cracks is the continuous dislodging of spheres from the void roof. This occurred whenever there was a local defect in the otherwise stable lattice of the aggregate. The rising velocities of the void roof which propagates by depletion of spheres from the roof were always smaller than the velocity of single particles in free fall and the velocity of the flow across the crack was never sufficient to keep the dislodged particles fluidized.

Figures 11a and 11b show fluidization of glass beads in the 3 in. wide inclined channel with Re = 1250. The presence of large voids, the big clear water regions, is not well described by specifying a void fraction. The slugging flow of packed spheres displaces regions of clear water. The same cracks or shocks observed in the wider channel

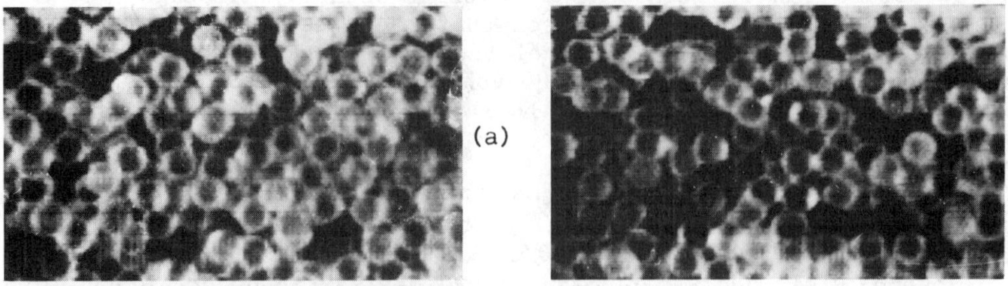

(a) (b)

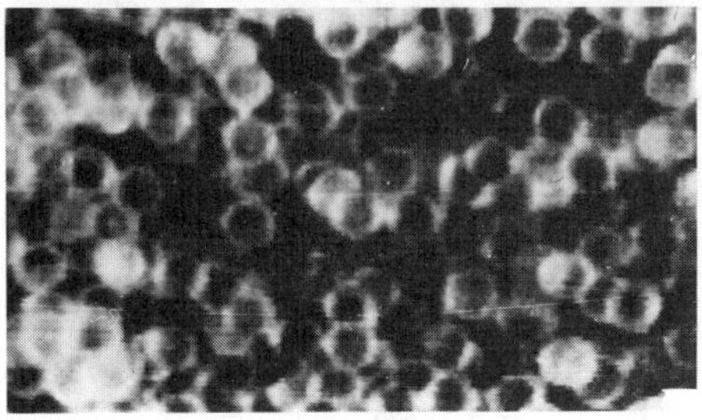

(c)

Figure 12. Close-up of the high speed video camera sequence of fluidization of the 0.117 inch diameter glass spheres in the three-dimensional bed.
Re = 690, voidage = 0.90

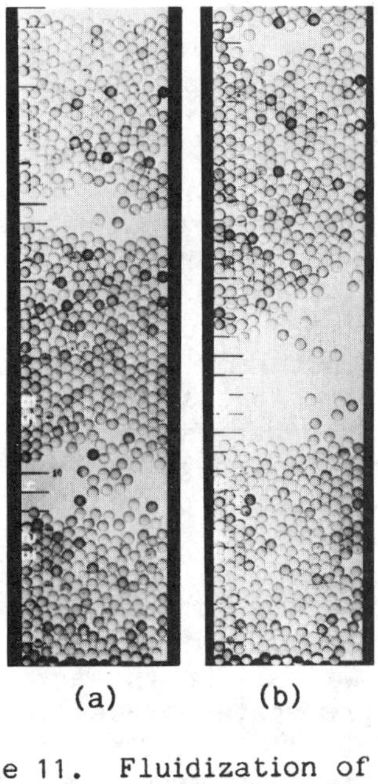

(a) (b)

Figure 11. Fluidization of glass beads at Re = 1250 in a 3-inch wide two-dimensional inclined bed.

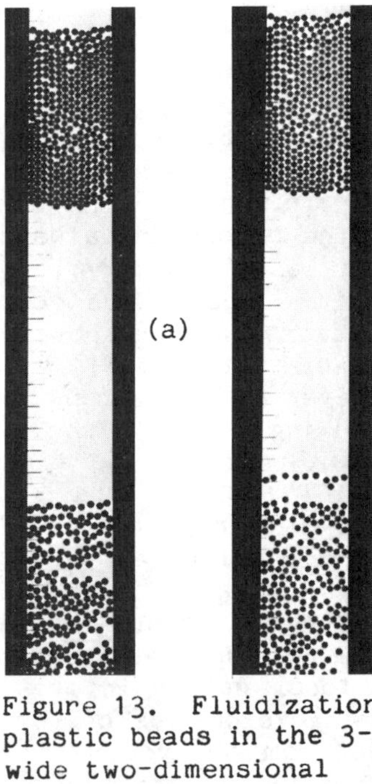

(a) (b)

Figure 13. Fluidization of plastic beads in the 3-inch wide two-dimensional inclined bed at Re = 290. The spheres at the top of the bed are stable and close packed with contacting vertical lines of centers, as in Figure 1. The bottom configuration with stable cross stream row is nearly stationary.

now span the whole channel. Also, the expansion mechanism, as described before, is revealed here in the same way. The different sequences taken with the high speed video camera shown in Figures 12a, 12b and 12c, separated by time intervals of 0.05 seconds, reveal the distinctive features referrred to thus far. First, the characteristic topological structure of closely packed spheres whose line of contact lies along the stream, as is vividly depicted in the aggregates at the top of Figure 13a and 13b, manifests itself in the preferential way the spheres collide with one another, i.e., the top sphere collides with the bottom one following in its wake. This topological feature is compatible with the dynamics associated with drafting and kissing. The lateral collisions can be attributed to the eddying motion of the liquid, since there is no possible lateral attraction between two spheres, as pointed out earlier. Secondly, even for these highly turbulent flows, the uprising motion is entirely determined by cross stream arrays of spheres. The fluid is pushed upward through the channels around the vertical line of centers of contacting spheres. There is a wake above each sphere, providing the glue. Isolated particles drift into these stream channels and are dragged into their own wakes, gluing more spheres to form the aggregates.

The remarkable line of particles at the top of the bottom part of the beds shown in Figures 13a and 13b was a characteristic feature of the flow in both the vertical and inclined channels.

Further increase of the Reynolds number does not change the basic mechanisms, and although the eddying motion sets up the lateral collisions of the particles, the expanded bed reveals essentially the same structure.

5. Nonspherical particles float broadside on. We did some fluidization experiments with nonspherical bodies, long cylinders with rounded and square ends, disks and a raft of glued long cylinders. No matter what, these bodies float broadside on. This is explained by the classical potential flows analysis of translating motions of a solid in a liquid (Thompson and Tait [1879], Lamb [1945, Section 124], Milne-Thompson [1960]). For any body there are three directions in which permanent translations are possible, but only one of these three, with the broadside perpendicular to the stream, is stable. The position of the stagnation points in any perturbed equilibrium is such as to turn the body broadside on, as in Figure 14.

The flows in our experiments are not potential flows but potential flow is probably a good approximation for the flow on the forward side of the body. Behind the body there is essentially a dead water region of low pressure whose main effect is to produce a drag. Evidently the turning couples on the forward side are large enough to turn the body broadside on. Many examples of the broadside on positioning of moving bodies have been put in evidence. For example, Milne-Thompson [1960] writes that

> "Thus a ship has to be kept on her course by the helmsman, an elongated airship requires similar attention. A sailing ship will not sail permanently before the wind with the helm lashed, but tends to set itself at right angles to the

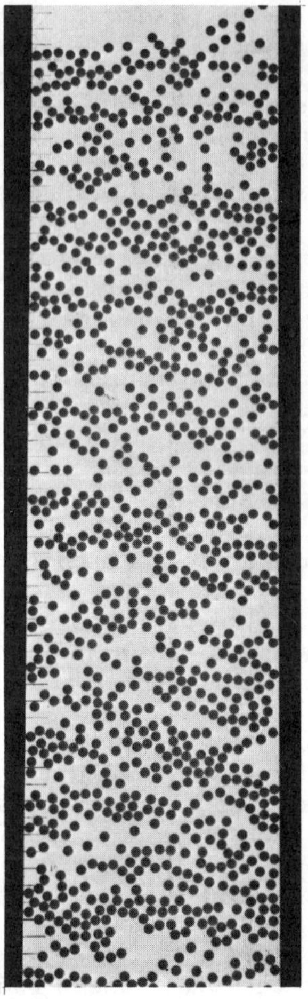

Figure 15. Fully expanded fluidized bed of plastic beads Re = 300, voidage = 0.88. Most spheres line up in horizontal arrays. Vertical pairs are drafting.

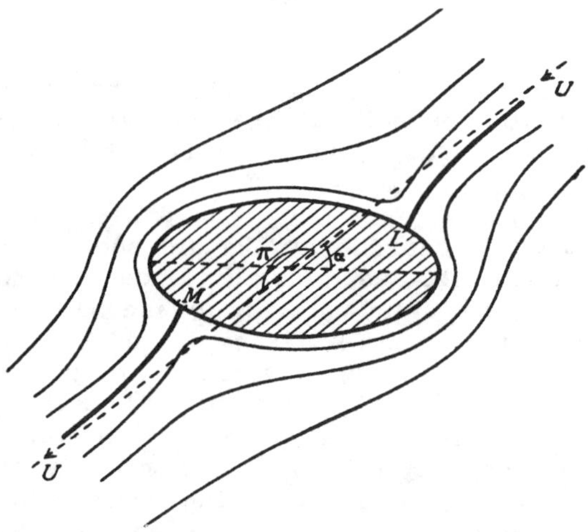

Figure 14. Potential flow solution for uniform flow past a slender body. The high pressure at the points of stagnation give rise to a couple causing the body to turn broadside on. In real flows this mechanism operates only on the front side, with a wake on the back.

wind. A body sinking in liquid tends to sink with its longest dimension horizontal."

6. Arrays of spherical particles tend to float broadside on. It seems to us that broadside on orientations of distinct spheres are more stable than other orientations (see Figures 7 through 13, 15).

We are aware that this sweeping statement is an interpretation of our experiences and not an established fact. Relatively greater stability of the cross stream doublets can be supported by an analysis of kissing and tumbling of the type shown in Figure 4, 5 and 6. Drafting is obviously a phenomenon which is dominated by wakes, and not by potential flow. When two spheres kiss, they instantaneously form a single body, equivalent to cemented spheres nearly aligned with the vertical plus the attached wakes at the back end and between the spheres. The front part of this body might be thought to be in potential flow. The upward force at the forward stagnation point of the composite body will give rise to a turning couple causing the kissing spheres to tumble. It is obviously impossible for the tumbling composite body to align itself broadside on without first separating the kissing spheres, and the separated spheres tend to align broadside on. The result of the tendency in the expanded bed shown in Figure 15, and in other figures, is a stratified structure in which at any moment most spheres are aligned in cross stream doublets and larger cross stream arrays.

7. Fluidized cylinders. Experiments were carried out with two different kinds of cylinders, round ends with length/diameter = 1.0 in./0.255 in. and square ends with length/diameter = 0.504 in./0.255 in. The diameter of these cylinders was slightly smaller than the gap of 0.285 inch between the parallel plates of the narrow gap apparatus. The motion of cylinders in the narrow gap apparatus is constrained to two dimensions. The specific gravity of the cylinders was 1.2. The Reynolds numbers for these experiments was in the range of 500 to 800. A single long particle was never in equilibrium under its weight and drag no matter whether it was fluidized in the two- or three-dimensional bed or even if it was sedimenting. The single cylinders would always oscillate around a mean horizontal position in a rocking mode which we think is associated with vortex shedding. Infrequently a single cylinder would rock momentarily into a vertical position and suddenly drop several inches before being turned to the horizontal. This shows that the major fluidizing force on long particles is a pressure drag due to wakes.

There are very strong wakes behind fluidized cylinders. Other particles which are sucked into the wakes seem to suppress unsteadiness. Some interesting figures of equilibrium consisting of long cylinders with round ends glued together by wake forces are shown in Figure 16a and 16b. These curious configurations were stable and stationary for minutes. A similar, but slowly changing configuration of long particles is shown in Figure 16c. In Figure 17 we have exhibited two examples of shorter cylinders with squared ends in the two-dimensional bed. Wake interactions are very evident in these figures and take form as piles of cylinders in contact along their long sides, and in a characteristic inverted T formation in which the top cylinder stands end on in the wake of the bottom cylinder. There are a few

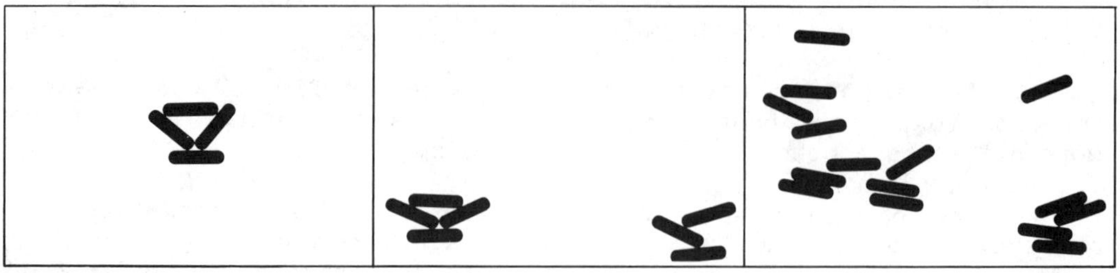

(a) (b) (c)

Figure 16. Fluidized cylinders in the vertical two-dimensional bed, Re = 650. The flow is up, against gravity. In (a) and (b) the cylinders are stationary. In (c), some of the cylinders are in motion.

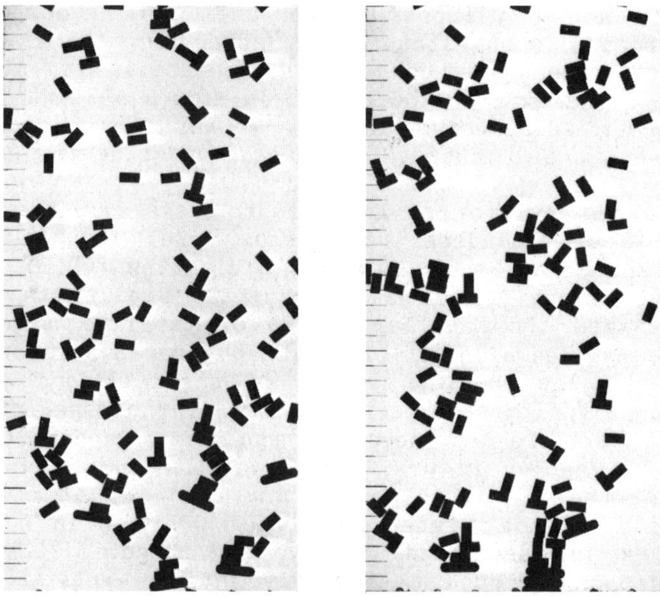

Figure 17. Fluidization of cylinders in the two-dimensional vertical bed. Most of the cylinders have squared ends. They are together in wakes and touch either on their long sides or edgewise in an inverted T. Short cylinders stack on long cylinders as a natural arrangement induced by the flow.

longer cylinders with round ends in the beds. These long particles collect shorter ones in their wakes creating a kind of fluidized bed architecture.

A different kind of inverted T is shown in Figure 18. In this figure one sees an inverted T formed from three spheres. The two on the bottom are glued and the one on top is held tenaciously in place by strong suction in the wake.

It is not possible to obtain all of the structured configurations which occur in the two-dimensional beds when the cylinders are not constrained to move in two-dimensions. As in two-dimensions, the dominant orientation of the cylinders is broadside on. There are very strong wake interactions in three-, as well as two-, dimensional beds. The cylinders cluster in horizontal arrays separated by clear water. Clear water regions are also evident at the side walls as in the case of the plastic spheres shown in Figure 7.

8. Fluidized disks. We made disks from the same stock as the cylinders of diameter d = 0.255 inch. The length/diameter ratio of these disks is approximately one half. In Figure 20 we have exhibited a photograph of the fluidized disks and some cylinders in the two-dimensional bed. The disks draft, kiss and tumble in much the same way as spheres. Most disks float stably broadside on in an oscillating rocking mode but disks which float edgeside on do appear and can be stable for seconds. Disks and small cylinders collect in the wakes of large cylinders. Some disks stand on the cylinders in edgeside on position, forming the characteristic inverted T shown in Figure 17. Inverted T's can occasionally be seen in three-dimensional beds. An example is exhibited at the top of Figure 21a. There are very strong wake effects in three dimensions; drafting, kissing and tumbling are commmon and tend to produce the clusters shown in Figure 21b and 21c.

9. Stacking experiments in two and three dimensions. We built a raft by glueing five long cylinders together. We fluidized the raft in the three-dimensional bed. It floated broadside on, on the average, whilst rocking in the characterisitc oscillation which we suppose is associated with vortex shedding. We matched the drag on the raft with the drag on a small cylinder length/diameter = 0.25 in./.0255 in. and fluidized many such cylinders with the raft. The cylinders are sucked into the wake of the raft like debris behind a fast moving truck, as shown in Figure 22. The composite body, raft plus entrained cylinders, is very stable, the oscillations are entirely suppressed.

Conclusions. The following results are suggested by the experiments discussed in this paper.

1. The dynamics of beds of particles fluidized by water at moderate and high Reynolds numbers is dominated by local mechanisms associated with wakes.

2. The results of Happel and Pfeffer [1960] indicate that even at Reynolds numbers as small as 1/2 the fluidization of spheres is strongly influenced by nonlinear effects of wakes and can not be described by models based in Stokes equations.

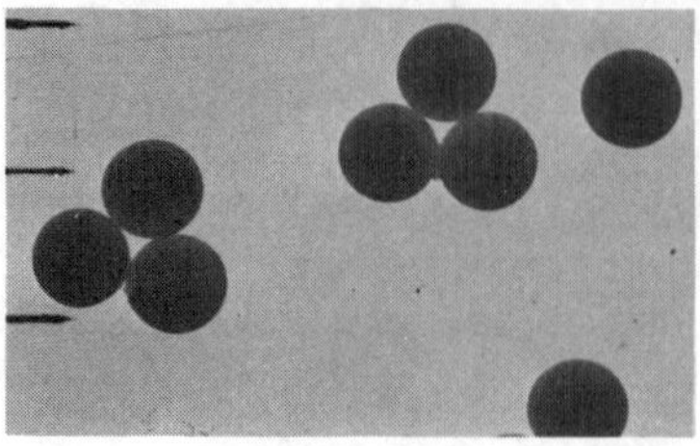

Figure 18. Fluidized spheres in a two dimensional vertical bed. There are three triplets. The bottom pair of the three is glued. The top sphere is nested tenaciously in place by the wake.

Figure 20. Fluidized disks and cylinders in the vertical two-dimensional bed.

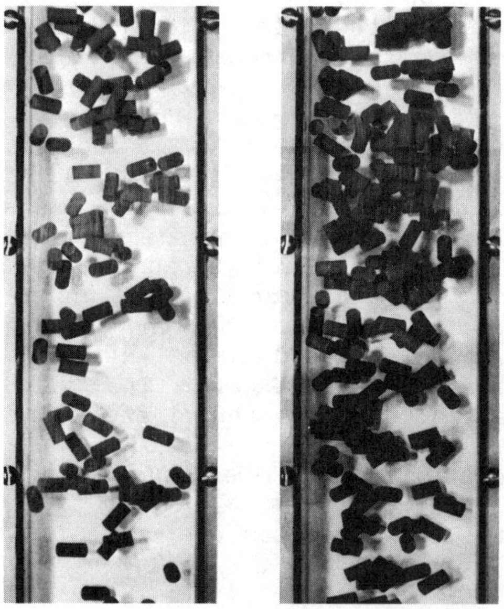

Figure 19. Fluidization of cylinders in the three-dimensional bed. These are the same cylinders as in Figure 17. The wake stacking and inverted T configurations are unstable in three-dimensions. The dominant orientation of the cylinders is broadside on. There are horizontal clusters separated by clear water and clear water is on the side walls (cf. Figure 17).

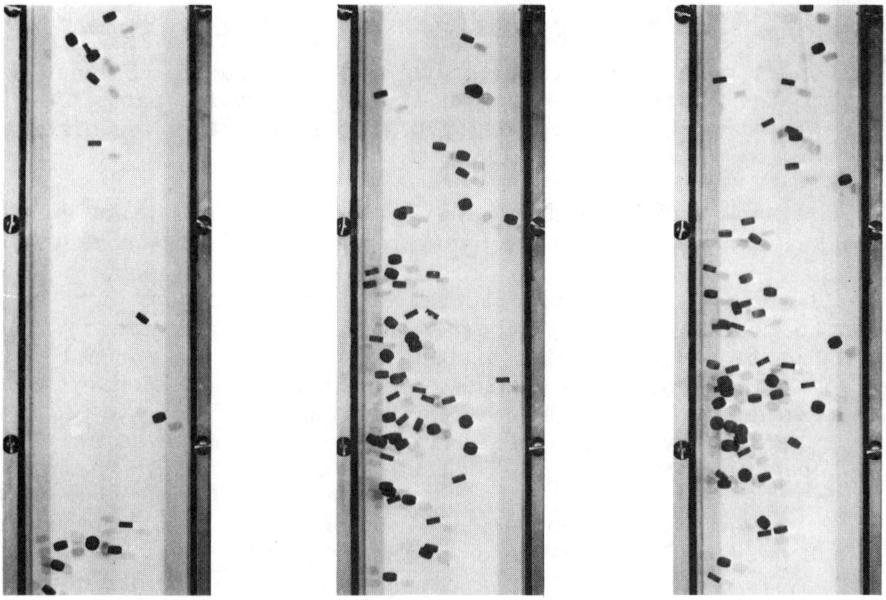

(a) (b) (c)

Figure 21. Fluidized disks in the three-dimensional bed. Strong wake effects are evident in the inverted T shown in (a) and in the clusters shown in (b) and (c).

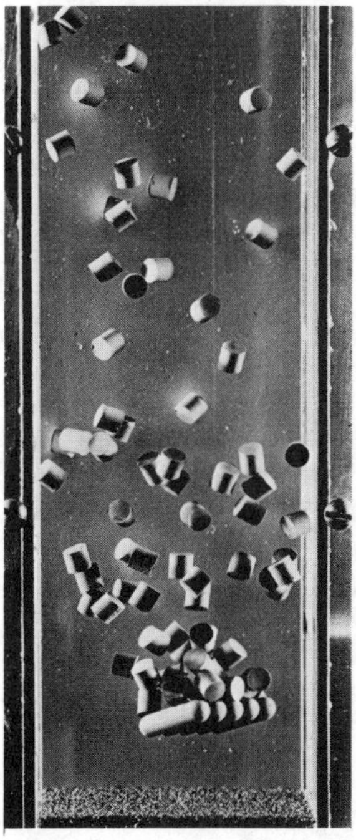

Figure 22. Short cylinders are sucked into the wake of the raft.

3. Drafting, kissing and tumbling appears to be the major rearrangement mechanism in the fluidization of beds of spheres with water.

4. Long particles and long arrays of single particles line up broadside onto the stream. The instability of streamwise pairs of kissing particles and the relative stability of crosswise pairs (doublets) seem to play a major role in aggregation.

5. Drafting particles fall and cross stream arrays of particles rise in a stream in which single particles are neutrally buoyant.

6. The stability of cross-stream arrays in beds of spheres constrained to move in two dimensions is amazing. Single horizontal lines of rising particles, even with different spacings, are robustly stable in the vertical and inclined channel. Single stable stationary lines of particles in horizontal arrays and stable aggregates of close packed spheres separated by regions of clear water were observed only in the inclined channel.

7. A weaker form of cooperative motion of cross stream arrays of rising spheres is found in beds of square cross section where the spheres may move freely in three dimensions. In a fluidized bed some particles must go up and some must come down. The going up and the coming down are very different. Falling spheres move fast. They accelerate in the wakes of other falling spheres. Rising arrays are generated from falling spheres by tumbling into cross stream doublets after drafting and kissing. Cross stream arrays rise because they have a greater drag.

8. It was not possible to fluidize a single long particle in a stationary position, even in the two-dimensional bed. Single long particles always oscillate in a rocking motion about the broadside on position. Single sedimenting long particles also oscillate in a rocking motion. The oscillation is possibly produced by vortex shedding.

9. The oscillation of single long particles tends to be suppressed by the presence of other particles. This suppression takes form by interacting wakes of cooperating particles. The effect is greatly enhanced in beds where the motion of the particles is confined to two dimensions. In these cases it was possible to fluidize stationary structures glued together by wakes.

10. Fluidized disks always align broadside on. Single disks oscillate in a rocking motion around the preferred alignments. Many disks draft, kiss and tumble like fluidized spheres.

11. Big particles always suck small paticles into their wake.

12. There was a lubrication layer of clear water at the walls of the three-dimensional beds of particles of all shapes.

Acknowledgement. A portion of this work is taken from the forthcoming Ph.D. thesis, "Nonlinear Mechanics of Fluidization of Beds of Spherical Particles", by Antonio Fortes and the forthcoming paper by Fortes, Joseph and Lundgren to appear in J. Fluid Mech.

REFERENCES

[1] E. ACHENBACH, Vortex shedding from spheres, J. Fluid Mech., **62**, Part 2 (1974), pp. 209-221.

[2] M. M. EL-KAISSY and G. M. HOMSY, Instability waves and the origin of bubbles in fluidized beds, Int. J. Multiphase Flow, **2** (1976), pp. 379-395.

[3] J. B. FANUCCI, N. NESS and R. H. YEN, On the formation of bubbles in gas-particulate fluidized beds, J. Fluid. Mech., **94**, Part 2 (1979), pp. 353-367.

[4] J. GARSIDE, and M. AL-DIBOUNI, Behavior of liquid fluidized beds containing a wide size distribution of solids, Fluidization and Its Applications, Toulouse, 1973, pp. 53-62.

[5] J. H. GERRARD, The mechanics of the formation region of vortices behind bluff bodies J. Fluid Mech., **25**, Part 2 (1966), pp. 401-413.

[6] J. HAPPEL and R. PFEFFER, The motion of two spheres following each other in a viscous fluid, A. I. Ch. E. Journal, **6**, 1 (1960), pp. 129-133.

[7] D. HARRISON and J. DAVIDSON, Fluidized particles, Cambridge University Press, 1963, pp. 15-19.

[8] D. HARRISON and J. DAVIDSON, Fluidization, Academic Press, 1971, pp. 47-50.

[9] G. M. HOMSY, M. M. EL-KAISSY and A. DIDWANIA, Instability waves and the origin of bubbles in fluidized bed, II, Int. J. Multiphase Flow, **6** (1980), pp. 305-318.

[10] D. D. JOSEPH, D. A. NIELD and G. PAPANICOLAU, Nonlinear equation governing flow in a saturated porous medium, Water Res. J., **18**, 4 (1982) pp. 1049-1052; Erratum 19 (1983), pp. 591.

[11] B. H. KAYE and R. P. BOARDMAN, Cluster formation in dilute suspensions, Symp. Interact. Fluids and Particles, London (1962). p. 17.

[12] H. LAMB, Hydrodynamics, Dover Publications, 1945, p. 191.

[13] J. T. C. LIU, Notes on a wave-hierarchy interpretation of fluidized bed instabilities, Proc. Roy. Soc., London, **A 380** (1982), pp. 229-239.

[14] J. T. C. LIU, Nonlinear instabilities in fluidized beds, Mechanics of Granular Materials: New Models and Constitutive Relations, J. T. JENKINS and M. SATAKE, eds., 1983, pp. 357-364.

[15] L. M. MILNE-THOMPSON, Theoretical Hydrodynamics, 4th Ed. Macmillan, 1960, p. 530.

[16] C. W. OSEEN, Hydrodynamik, Akademishe Verlagsgesellschaft M.B.H., 1927, pp. 199 ff.

[17] J. F. RICHARDSON, Incipient Fluidization and Particulate Systems, In Fluidization, J. F. DAVIDSON and D. HARRISON, eds., Academic Press, 1971, pp. 50-51.

[18] J. F. RICHARDSON and R. A. MEIKLE, Sedimentation and fluidization, Part IV: Drag force on individual particles in an assemblage, Trans. Inst. Chem. Eng., **39** (1964), pp. 357-362.

[19] J. F. RICHARDSON and W. ZAKI, Sedimentation and fluidization, Part 1, Trans. Inst. Chem. Eng., **32** (1954), pp. 35-53.

[20] P. N. ROWE, Drag forces in a hydraulic model of a fluidized bed, Part II, Trans. Inst. Chem. Eng., **39** (1961), p. 175.

[21] P. N. ROWE and G. A. HENWOOD, Drag forces in a hydraulic model of a fluidized bed, Part I, Trans. Inst. Chem. Eng., **39** (1961), p. 43.

[22] W. THOMPSON and P. G. TAIT, Natural Philosophy, 2nd Ed., Cambridge, 1879.

[23] Y. TSUJI, Y. MORIKAWA and K. TERASHIMA, Fluid-dynamic interaction between two spheres Int. J. Multiphase Flow, **8**, 1 (1982), pp. 71-82.

[24] J. VERLOOP and P.M. HEERTJES, Shock waves as a criterion of the transition from homogeneous to heterogeneous fluidization, Chem. Eng. Sci., **25** (1970), p. 825.

[25] G. VOLPICELLI, L. MASSIMILLA and F. S. ZENZ, Nonhomogeneities in solid-liquid fluidization, Chem. Eng. Symp. Series, **62**, 67 (1966), p. 42.

[26] G. B. WALLIS, A simplified one-dimensional representation of two-component vertical flow and its applications to batch sedimentation, Symp. Interact. Fluids and Particles, London 9 (1962), p. 17.

[27] R. H. WILHELM and M. KWAUK, Fluidization of solid particles, Chem. Eng. Progress, **44**, 3 (1948) p. 201.

The Effective Viscosity of a Mixture of Two Stokes Fluids

ROBERT V. KOHN* AND ROBERT LIPTON**

Abstract

We are concerned with homogenization of the Stokes system, and with the averaging of equations for two-phase flow. In the former direction, section one reports on a new, optimal lower bound for the energy dissipation rate of a mixture of two Stokes fluids. In the latter direction, section two explores the relevance of homogenization to the derivation of averaged equations. Linking the two is the (highly speculative) "maximum lubrication hypothesis," that the effective viscosity of a mixture should saturate the lower bound. Surface tension effects are ignored in the first two parts, but discussed in section three along with other directions for future work.

1. An optimal lower bound on the energy dissipation rate.

Much attention has recently been devoted to seeking optimal bounds for the effective moduli of composite materials. The search has been motivated in large part by applications to structural optimization [1,17,18,23,24,28-30]. Areas of recent progress include both conductivity [15,21,22,38] and plate theory [7]. The work reported here is discussed in the context of Stokes flow; but it applies equally to mixtures of two incompressible, elastic solids, for which the equations are identical.

*Courant Institute of Mathematical Sciences, 251 Mercer Street, New York, NY. 10012. Partial support from NSF grant DMS-8312229, ONR grant N00014-83-0536, and the Sloan Foundation is gratefully acknowledged.

** Mathematical Sciences Institute, Caldwell Hall, Cornell University, Ithaca, NY. 14853. Presently supported by the U.S. Army Research Office through the Mathematical Sciences Institute.

We are interested in immiscible, incompressible, viscous fluids, and in regimes that can be modeled by the stationary Stokes system. By a "mixture" we mean an arrangement of two such fluids in a container. The composition is assumed to vary on a length scale ε, small with respect to the overall length scale of the container. It can be characterized by specifying the viscosity at each point,

$$a_\varepsilon(x) = \alpha \text{ or } \beta \ , \tag{1.1}$$

where $0 < \alpha < \beta$ are the viscosities of the two components. We shall ignore the effect of surface tension, so the fluid velocity u_ε solves

$$-\mathrm{div}(a_\varepsilon(x)D[u_\varepsilon] - p_\varepsilon I) = f \ , \ \nabla\cdot u_\varepsilon = 0 \ , \tag{1.2}$$

in which $D[u_\varepsilon] = (1/2)(\nabla u_\varepsilon + \nabla u_\varepsilon^T)$ is the strain rate. The condition at fluid interfaces, implicit in (1.2), is that both u_ε and the normal stress $(a_\varepsilon D[u_\varepsilon] - p_\varepsilon I)\cdot n_\varepsilon$ are continuous. (One might reasonably question this as a model for real fluids, since surface tension profoundly influences the fine-scale structure of a mixture [12]. See section 3 for discussion of this point.)

Our interest is in the behavior of u_ε when the length scale ε of the microstructure is small. The spatially periodic case is discussed in [31]. A more general mathematical setting is provided by the theory of H-convergence [6,27,37], also called strong G-convergence [35,36,41]. (Though the Stokes system is not explicitly addressed in these references, it can easily be treated by the methods they develop.) This theory deals with <u>sequences</u> of coefficients, corresponding conceptually to letting the length scale ε tend to zero. After passage (if necessary) to a subsequence, it characterizes the limiting velocity $u_0 = \lim u_\varepsilon$ as the solution of a new, <u>effective equation</u>

$$-\mathrm{div}(AD[u_0] - p_0 I) = f \ , \quad \nabla\cdot u_0 = 0 \ . \tag{1.3}$$

The <u>effective viscosity</u> A is a fourth order tensor, relating the weak limits (intuitively, local averages) of the strain rate and the deviatoric stress:

$$\sum_{k,\ell} A_{ijk\ell} D_{k\ell}[u_0] = \lim_{\varepsilon\to 0} a_\varepsilon D_{ij}[u_\varepsilon] \ . \tag{1.4}$$

It determines a quadratic form on trace-free, symmetric matrices, which represents the energy dissipation rate of the mixture:

$$\langle AD[u_0],D[u_0]\rangle = \lim_{\varepsilon\to 0} \langle a_\varepsilon D[u_\varepsilon],D[u_\varepsilon]\rangle \ . \tag{1.5}$$

The effective viscosity A depends only on the microstructure, <u>i.e.</u> on the sequence $\{a_\varepsilon\}$, not on the external force f or the boundary conditions.

We are interested in bounds for A in terms of the <u>proportions</u>

$\theta, 1 - \theta$ of the original two fluids. In this general setting the proportion θ is defined by taking the weak limit of a_ε itself:

$$\text{wk}\lim_{\varepsilon \to 0} a_\varepsilon = \theta\alpha + (1 - \theta)\beta . \tag{1.6}$$

Both the effective viscosity A and the proportion θ may vary with x, though for a macroscopically homogeneous mixture both will be constant. If the mixture were isotropic then A would be multiplication by a scalar effective viscosity, but we are explicitly interested in the anisotropic case.

The ultimate goal in the search for optimal bounds is to specify which tensors A can arise as the effective viscosity of some microstructure $\{a_\varepsilon\}$ with specified proportions θ and $1 - \theta$ of the original components. The analogous scalar problem has been solved completely [21,22,38], but for elasticity or the Stokes system a full characterization seems as yet beyond reach. It is therefore natural to seek bounds just for the energy dissipation rate $\langle A\xi,\xi\rangle$. The analogous question for Kirchhoff plate theory was settled in [7]. (In the context of incompressible elasticity, such bounds and the methods of [18] allow one to solve optimal design problems for minimum or maximum compliance under a fixed load.)

It is well-known that the energy dissipation rate is bounded above by the arithmetic mean and below by the harmonic mean of the component viscosities:

$$(\alpha^{-1}\theta + \beta^{-1}(1 - \theta))^{-1}|\xi|^2 \leq \langle A\xi,\xi\rangle \leq (\alpha\theta + \beta(1 - \theta))|\xi|^2 \tag{1.7}$$

for any trace-free, symmetric matrix ξ. These were proved by Paul [33], using a somewhat different mathematical model for a composite material, and they are easily established in the context of H-convergence, too [6]. It is also well-known that (1.7) can be improved if A is assumed to be isotropic [8-10]. Here, however, we are interested in whether improvement is possible without further restriction on A. Briefly, the answer is no for the upper bound, and yes for the lower bound. The complete story is as follows:

A) The upper bound of (1.7) is optimal in any dimension $n \geq 2$. Indeed, given α, β, θ, and ξ, equality is achieved by a laminar microstructure of rank one, using any eigenvector of ξ as the layer normal. For this microstructure the deviatoric stress is a multiple of ξ: $A\xi = (\theta\alpha + (1 - \theta)\beta)\xi$.

B) The optimal lower bound in dimension $n \geq 2$ is

$$\langle A\xi,\xi\rangle \geq \alpha|\xi|^2 + (1 - \theta)f(\alpha,\beta,\theta,\xi) , \tag{1.8}$$

where f is defined variationally by

$$f = \max_{\mu} 2\langle\mu,\xi\rangle - (\beta - \alpha)^{-1}|\mu|^2 - \frac{\theta}{2\alpha} \max_{1 \leq i,j \leq n} \{(\mu_i - \mu_j)^2\} . \tag{1.9}$$

Here μ ranges over $n\times n$, symmetric, trace-free matrices, and $\{\mu_i\}_{i=1}^n$ are the eigenvalues of μ. For any α,β,θ, and ξ, equality is achieved in (1.8) by a suitably chosen laminar microstructure of rank at most $n - 1$. For this microstructure the deviatoric stress is simultaneously diagonal with ξ, and it has the form

$$A\xi = \alpha\xi + (1 - \theta)\mu^* \tag{1.10}$$

where μ^* is the maximizer of (1.9).

C) Our new lower bound (1.8) agrees with the familiar one (1.7) precisely when ξ has rank two. Thus the harmonic mean bound is optimal in dimension 2 and when ξ has rank 2 (planar shear flow), but suboptimal in all other cases. If ξ has rank 2, with nontrivial eigenvectors e_1 and e_2, then the lower bound is achieved by a rank-one laminar microstructure with $(e_1 \pm e_2)/\sqrt{2}$ as the layer normal, for which (1.10) becomes $A\xi = [\alpha^{-1}\theta + \beta^{-1}(1 - \theta)]^{-1}\xi$.

D) In dimension $n = 3$, the optimal bound (1.8) may be expressed explicitly in terms of the eigenvalues $\xi_1 \geq \xi_2 \geq \xi_3$ of ξ. There are three cases, depending on the signs of

$$A = \frac{3(\beta-\alpha)\theta}{2\alpha}\xi_2 - (\xi_3 - \xi_2)$$

$$B = \frac{-3(\beta-\alpha)\theta}{2\alpha}\xi_2 + (\xi_1 - \xi_2) .$$

We have

$$\alpha|\xi|^2+(1-\theta)f = \begin{cases} [\alpha^{-1}\theta+\beta^{-1}(1-\theta)]^{-1}|\xi|^2 + \dfrac{3\theta(1-\theta)(\beta-\alpha)^2\xi_2^2}{2[\theta\beta+(1-\theta)\alpha]} & \text{if } A \geq 0,\ B \geq 0 \\[2ex] \alpha|\xi|^2 + \dfrac{6(1-\theta)(\beta-\alpha)\alpha\xi_3^2}{4\alpha+3\theta(\beta-\alpha)} & \text{if } A \geq 0 ,\ B < 0 \\[2ex] \alpha|\xi|^2 + \dfrac{6(1-\theta)(\beta-\alpha)\alpha\xi_1^2}{4\alpha+3\theta(\beta-\alpha)} & \text{if } A < 0 ,\ B \geq 0 . \end{cases}$$

In the first case the bound is achieved by a rank-one laminate with layer normal $(e_1 \pm e_3)/\sqrt{2}$, where e_i is the eigenvector associated to ξ_i. The other two cases require rank-two laminates, using $(e_2 \pm e_3)/\sqrt{2}$ or $(e_1 \pm e_2)\sqrt{2}$ respectively as the second layer normal.

The full analysis appears in [14,20], and will not be repeated here. There are three main steps:

1) Reduction to the spatially periodic case.
2) Proof of the bound.
3) Proof of attainability.

The first of these is explained in [13]. Physically, such a reudction is almost obvious: given a microstructure that achieves some effective behavior A, take a cubic "sample" and extend it periodically. As the ratio between the scale of the microstructure and the size of the sample tends to zero one obtains a family of periodic composites -- with increasingly complicated unit cells -- whose effective behavior tends to A. This idea has been used implicitly by many authors, *e.g.* [10,15,26]. Its justification in the context of H-convergence rests heavily on the local character of optimal bounds, *c.f.* [2].

Our proof of the new lower bound uses the Hashin-Shtrikman variational principle. First introduced in 1962, it is easily adapted to the spatially periodic context [15,26]. The same method was applied in [8,9] to bound the effective viscosity of a macroscopically isotropic mixture. We get a different result precisely because we make no hypothesis on the symmetry of the composite. Other applications of the Hashin-Shtrikman variational principle to anisotropic composites are given in [15,16].

The third step, attainability, is based entirely on the use of laminar composites. A "laminar composite of rank one" is obtained by arranging the two initial components with specified proportions in layers normal to a particular direction; a "laminar composite of rank two" is obtained by a similar arrangement of two laminar composites of rank one; and so forth. This construction yields a rather large family of microstructures whose effective behavior can be computed exactly. It has been used by many authors to show the optimality of various bounds [6,7,21-25,38]. Our analysis dpends heavily on a convenient, recursive approach developed in [6,38].

It is instructive to consider the example of a shear flow in the plane. According to (A) and (C) above, the familiar bounds (1.7) are achieved in this case by rank-one lamination. In fact, each bound can be achieved in two different ways, as shown in figures 1 and 2.

Figure 1. Laminar microstructures achieving the upper bound when the overall velocity is $u_0 = (\lambda x_2, 0)$. In both cases the deviatoric stress is $AD[u_0] = [\theta\alpha+(1-\theta)\beta]^{-1}\begin{pmatrix} 0 & \lambda/2 \\ \lambda/2 & 0 \end{pmatrix}$.

Figure 2: Laminar microstructures achieving the lower bound when the overall velocity is $u_0 = (\lambda x_2, 0)$. In both cases the deviatoric stress is $AD[u_0] = (\theta\alpha^{-1}+(1-\theta)\beta^{-1})^{-1}\begin{pmatrix}0 & \lambda/2\\ \lambda/2 & 0\end{pmatrix}$.

We note that only one of the four geometries, the one with horizontal layers, is preserved by the flow.

For a three dimensional flow with constant shear rate ξ, the picture is similar for achieving either the upper bound or the first regime of the lower bound ($A \geq 0$, $B \geq 0$) when layers of rank one suffice. For the other two regimes of the lower bound layers of rank two are required. There are two choices for each layer direction, giving four constructions in all. Each looks roughly as in figure 3.

Figure 3. Laminar microstructure of rank 2 achieving the lower bound in dimension three.

2. Homogenization and averaged equations for two-phase flow.

The definition of "effective viscosity" used in section one is more or less the standard one [4,8,9,12,31,34,39,40]. Because it is based on the stationary Stokes equations, variational principles and the theory of homogenization can be applied to prove rigorous results such as bounds. Two fundamental issues remain unresolved, however. One is the appropriateness of starting from the stationary equations: even if the overall flow is stationary, in many cases its fine-scale structure will be rapidly varying. The second is that the fine-scale structure of the mixture -- on which the effective viscosity depends -- is itself somehow determined by the flow. These are profound and difficult questions, to which there can be no easy response.

There is an alternative, more phenomenological approach, involving the introduction of averaged equations [3]. Its close analogy with homogenization has been noted before, e.g. [19], but has in our opinion received insufficient attention. We shall therefore present an averaging procedure for the Navier-Stokes equations, in the context of two incompressible fluids with equal density, which parallels as closely as possible the ideas of H-convergence. Our treatment is purely formal, posing many questions and offering no conclusive answers. We hope, however, that some of these questions will prove amenable to mathematical analysis in the future.

We begin with the dynamical analogue of (1.2):

$$(u_\varepsilon)_t + u_\varepsilon \cdot \nabla u_\varepsilon - \nabla \cdot (a_\varepsilon D[u_\varepsilon] - p_\varepsilon I) = f \tag{2.1a}$$

$$\nabla \cdot u_\varepsilon = 0 \tag{2.1b}$$

$$(a_\varepsilon)_t + \nabla \cdot (u_\varepsilon a_\varepsilon) = 0 \ . \tag{2.1c}$$

The parameter ε is intended once again to mark the length scale of the mixture. The viscosity $a_\varepsilon(x,t)$ still takes only the two values α and β, but now according to (2.1c) it is transported by the flow. Density -- assumed to be the same for both fluids -- has been set equal to one in (2.1a). All equations are to be understood in the weak sense; we are again ignoring surface tension, so the interface condition implicit in (2.1a) is continuity of the velocity u_ε and the normal stress $(a_\varepsilon D[u_\varepsilon] - p_\varepsilon I) \cdot n_\varepsilon$ where the two fluids meet.

The procedure for averaging these equations must necessarily be formal, as no existence theorem is known. One approach, developed in [19], is to use the method of multiple scales. Here we shall use the notions of weak and strong convergence instead. Consider a family of solutions $\{u_\varepsilon, p_\varepsilon, a_\varepsilon\}$, for which the length scale ε of the fine structure tends to zero. We suppose that they converge to limits $\bar{u}$, $\bar{p}$, $\bar{a}$ as $\varepsilon \to 0$:

$$\begin{aligned} & u_\varepsilon \to \bar{u} && \text{strongly} \\ & \nabla u_\varepsilon \to \nabla \bar{u} \ , \ (u_\varepsilon)_t \to \bar{u}_t && \text{weakly} \\ & a_\varepsilon \to \bar{a} \ , \ p_\varepsilon \to \bar{p} && \text{weakly} \end{aligned} \tag{2.2}$$

The spaces are left unspecified out of ignorance; presumably, a good set of choices would be parallel to the existence theory for weak solutions of the (single-fluid) Navier-Stokes equations, see e.g. [42]. Applying the rule that $g_\varepsilon h_\varepsilon \to \bar{g} \cdot \bar{h}$ weakly if $g_\varepsilon \to \bar{g}$ weakly and $h_\varepsilon \to \bar{h}$ strongly, and recalling that differentiation preserves weak convergence, (2.2) allows passage to the limit in each term of (2.1a-c) except $a_\varepsilon D[u_\varepsilon]$. Denoting the limiting deviatoric stress by

$$\bar{\sigma} = \lim_{\varepsilon \to 0} a_\varepsilon D[u_\varepsilon] \ ,$$

we obtain the averaged equations

$$\bar{u}_t + \bar{u}\cdot\nabla\bar{u} - \nabla\cdot(\bar{\sigma} - \bar{p}I) = f$$

(2.3) $$\nabla\cdot\bar{u} = 0$$

$$\bar{a}_t + \nabla\cdot(\bar{a}\,\bar{u}) = 0 .$$

Notice that $\bar{a}$ is the "average" viscosity, i.e.

(2.4) $$\bar{a}(x,t) = \alpha\theta(x,t) + \beta(1 - \theta(x,t)) ,$$

where θ is the proportion of fluid α (more precisely, the weak limit of the characteristic function of the set it occupies).

To close the averaged system we need a constitutive law for $\bar{\sigma}$ in terms of $\bar{u}$ and $\bar{a}$. By analogy with (1.4), it is natural to expect a linear law

(2.5) $$\bar{\sigma}_{ij} = \sum_{k,\ell} B_{ijk\ell}\, D_{k\ell}[\bar{u}] .$$

The fourth order tensor B should again be called an "effective viscosity", as it relates the overall deviatoric stress to the overall strain rate. Its relation to our earlier notion of effective viscosity is not obvious, since a_ε was a function of space alone in (1.2) while it is transported by the flow in (2.1). Our guess, however, is that the time dependence does not affect the averaging. More formally, we offer the

Conjecture: The tensor $B(x,t)$ of (2.5) is, for almost every t, the H-limit of $a_\varepsilon(\cdot,t)$ as $\varepsilon \to 0$ with t fixed.

This is consistent with the two-scale asymptotic analysis of [19]. Further evidence is supplied by an analogous periodic homogenization problem for the heat equation: if $b(y,\tau)$ is a periodic function of $y \in \mathbb{R}^n$ and $\tau \in \mathbb{R}$, and if $0 < \alpha < 2$, then the averaging of

(2.6) $$u_{\varepsilon t} - \nabla\cdot(b(x/\varepsilon, t/\varepsilon^\alpha)\nabla u_\varepsilon) = f$$

is properly done first with respect to space, then with respect to time [43]. Since our a_ε is transported by the flow, the length scales of its space and time variation should be comparable -- suggesting the analogy of (2.6) with $\alpha = 1$. We have not proposed a further averaging with respect to time, because we expect the (spatially averaged) effective viscosity to vary slowly in time.

If our conjecture holds, then the bounds established for the stationary problem -- summarized in section one -- carry over to this dynamical context. Of course, these bounds do not determine the effective viscosity; they merely restrict the range of possibilities. There is a long tradition, however, of using bounds as "rules of thumb" for the effective moduli of composites. Moreover, certain two-fluid systems have been observed to prefer configurations that give enhanced lubrication [5,11]. We therefore propose, as a constitutive relation, that the effective viscosity should saturate the optimal lower bound. With the further assumption that $\xi = D[\overline{u}]$ and $A\xi = \overline{\sigma}$ are simultaneously diagonal, it follows by elementary linear algebra that (1.10) must hold. Our idea thus leads to an unambiguous constitutive law:

> Maximum Lubrication Hypothesis: $\overline{\sigma}$ is determined from the strain rate $D[\overline{u}]$ and the proportion θ of fluid α by $\overline{\sigma} = \alpha D[\overline{u}] + (1 - \theta)\mu^*$, where μ^* maximizes (1.9) with $\xi = D[\overline{u}]$.

We hasten to add that this proposal has not yet been checked against experimental data.

3. Discussion.

Our approach has the fundamental limitation of omitting surface tension. In fact, surface tension effects profoundly influence the microstructure geometry: in a mixture of two real fluids, one will typically break up into spherical droplets of roughly uniform size surrounded by the other. A thoughtful analysis has been given by Keller, Rubenfeld, and Molyneux, who propose that the droplet boundaries be restricted to move rigidly [12]. They modify the variational principles for Stokes flow to account for this restriction, and use them to bound the effective viscosity of an isotropic mixture. For spherical droplets of viscosity β and concentration $c = 1 - \theta$ suspended in a fluid of viscosity α, their upper and lower bounds agree to first order in c and match the results obtained formally by Einstein [4] ($\beta = \infty$) and Taylor [39] ($0 < \beta < \infty$): the (scalar) effective viscosity is

$$\frac{a_{eff}}{\alpha} = 1 + \frac{2+5\beta/\alpha}{2(1+\beta/\alpha)}\, c + \dots \quad . \tag{3.1}$$

This is naturally different from the result obtained by ignoring surface tension. That calculation, analogous to ours (sketched in section one) but assuming isotropy, was done by Hashin [8,9]. He obtained upper and lower bounds for any value of $c = 1 - \theta$, and his bounds are now known to be optimal [6,25,32]. Specialized to $c \approx 0$ for comparison with (3.1), Hashin's result is

$$1 + c\,\frac{(\frac{\beta}{\alpha}-1)}{1+\frac{2}{5}(\frac{\beta}{\alpha}-1)} + \dots \le \frac{a_{eff}}{\alpha} \le 1 + c(\frac{\beta}{\alpha}-1)[1 - \frac{2}{5}(1-\frac{\alpha}{\beta})] + \dots \quad . \tag{3.2}$$

Notice that when $\beta = \alpha$ (3.2) gives $a_{eff} = \alpha$, but (3.1) gives $a_{eff} = \alpha(1 + \frac{7}{4}c)$. Indeed, when $\beta = \alpha$ and surface tension is ignored the two fluids are indistinguishable and the mixture is just another fluid with viscosity α. Surface tension restricts the possible fluid motions to those that preserve the shape of the droplets, thus increasing the effective viscosity. The approach of [12] is clearly superior as a model of real fluids; we consider it a challenge for the future to derive bounds analogous to (1.8) from this more sophisticated viewpoint.

A second challenge is to achieve a better understanding of constitutive relations. Our proposed "maximum lubrication hypothesis" can hardly be correct in all situations, and is in any event without theoretical foundation. There is perhaps an analogy to hyperbolic conservation laws, for which weak solutions are not unique; a selection mechanism is provided by perturbing the equations to include a vanishingly small viscosity. The true constitutive law of a mixture is, we presume, similarly determined by "small effects" -- including surface tension -- that were omitted from (2.1). An analysis based on perturbing the Navier-Stokes equations seems far beyond reach; but perhaps something could be done for a suitably chosen model problem.

Acknowledgement: We gratefully acknowledge fruitful discussions with D.D. Joseph on many aspects of this work.

References

[1] J.-L. Armand, K.A. Lurie, and A.V. Cherkaev, Optimal control theory and structural design, in Optimum Structure Design, Vol. 2, R.H. Gallagher et al.. eds., J. Wiley and Sons, 1983.

[2] E. Cabib and G. Dal Maso, On a class of optimum problems in structural design, preprint, 1985.

[3] D. Drew, Mathematical modeling of two-phase flow, Ann. Rev. Fluid Mech. 15 (1983), pp. 261-291.

[4] A. Einstein, Eine neue bestimmung der moleküldimensionen, Ann. Phys., 19 (1906) pp. 289-306 and 34 (1911) pp. 591-592.

[5] A.E. Everage, Jr., Theory of stratified bicomponent flow of polymer melts I: equilibrium Newtonian tube flow, Trans. Soc. Rheology 17 (1973) pp. 629-646.

[6] G. Francfort and F. Murat, Homogenization and optimal bounds in linear elasticity, Arch. Rat. Mech. Anal., to appear.

[7] L.V. Gibiansky and A.V. Cherkaev, Design of composite plates of extremal rigidity, A.F. Ioffe Physico-Technical Institute Preprint no. 914, 1984.

[8] Z. Hashin, Bounds for viscosity coefficients of fluid mixtures by variational methods, Proc. Int'l. Symp. on Second-Order Effects in Elasticity, Plasticity, and Fluid Dynamics, Reiner and Abir, eds., Pergamon Press, New York, 1964, pp. 434-449.

[9] Z. Hashin, Viscosity of rigid particle suspensions, Contributions to Mechanics, Reiner Anniversary Volume, Pergamon Press, New York, 1967, pp. 347-359.

[10] Z. Hashin and S. Shtrikman, A variational approach to the theory

of the elastic behavior of multiphase materials, J. Mech. Phys. Solids 11 (1963) pp. 127-140.

[11] D. Joseph, K. Nguyen, and G. Beavers, Nonuniqueness and stability of the configuration of flow of immiscible fluids with different viscosities, J. Fluid Mech. 141 (1984) pp. 319-345.

[12] J.B. Keller, L.A. Rubenfeld, and J.E. Molyneux, Extremum principles for slow viscous flows with applications to suspensions, J. Fluid Mech., 30 (1967), pp. 97-125.

[13] R.V. Kohn and G. Dal Maso, in preparation.

[14] R.V. Kohn and R. Lipton, in preparation.

[15] R. Kohn and G. Milton, On bounding the effective conductivity of anisotropic composites, in Homogenization and Effective Moduli of Materials and Media, J. Ericksen et al. eds., Springer Verlag, 1986.

[16] R.V. Kohn and G. Milton, Bounds for anisotropic composites by variational principles, in preparation.

[17] R. Kohn and G. Strang, Structural design optimization, homogenization, and relaxation of variational problems, in Macroscopic Properties of Disordered Media, R. Burridge et al. eds., Lecture Notes in Physics 154, Springer-Verlag, 1982, pp. 131-147.

[18] R. Kohn and G. Strang, Optimal design and relaxation of variational problems, Comm. Pure Appl. Math. 39 (1986) pp. 113-137, 139-182, 353-377.

[19] T. Levy and E. Sanchez-Palencia, Suspension of solid particles in a Newtonian fluid, J. Non-Newtonian Fluid Mech. 13 (1983) pp. 63-78.

[20] R. Lipton, An Optimal Lower Bound on the Energy Dissipation Rate for Homogenized Stokes Flow, Ph.D. Thesis, New York University, 1986.

[21] K.A. Lurie and A.V. Cherkaev, Exact estimates of conductivity of composites formed by two materials taken in prescribed proportion, Proc. Royal Soc. Edinburgh 99A (1984) pp. 71-87.

[22] K.A. Lurie and A.V. Cherkaev, Exact estimates of the conductivity of a binary mixture of isotropic components Proc. Roy. Soc. Edinburgh A, to appear.

[23] K.A. Lurie, A.V. Cherkaev, and A.V. Fedorov, Regularization of optimal design problems for bars and plates, J. Opt. Th. Appl. 37 (1982) pp. 499-522 and 523-543.

[24] K.A. Lurie, A.V. Cherkaev, and A.V. Fedorov, On the existence of solutions to some problems of optimal design for bars and plates, J. Opt. Th. Appl. 42 (1984) pp. 247-281.

[25] G. Milton, Modelling the properties of composites by luminates, in Homogenization and Effective Moduli of Materials and Media, J. Ericksen et al. eds., Springer Verlag, 1986.

[26] G. Milton and N. Phan-Thien, New bounds on the effective elastic moduli of two-component materials, Proc. Roy. Soc. Lon. A, 380 (1982) pp. 305-331.

[27] F. Murat, H-convergence, Seminaire d'Analyse Fonctionelle et Numerique, Univ. d'Alger, 1977/8, mimeographed notes.

[28] F. Murat, Control in coefficients, in Systems and Control Encyclopedia: Theory, Technology, Applications, Pergamon Press, Oxford, 1985.

[29] F. Murat and L. Tartar, Calcul des variations et homogénéisation, in Les Methods de l'Homogénéisation: Theorie et Applications en Physique, Coll. de la Dir. des Etudes et Recherche de Electricite de France, Eyrolles, Paris, 1985, pp. 319-370.

[30] F. Murat and L. Tartar, Optimality conditions and homogenization, in Nonlinear Variational Problems, A. Marino et al. eds., Research Notes in Math. No. 127, Pitman Press, 1985, pp. 1-8.

[31] G. Nguetseng, Etude asymptotique du comportement macroscopique d'un melange de deux fluides visquex, J. de Mec. Theor. et Appl. 1 (1982), pp. 951-961.

[32] A. Norris, A differential scheme for the effective moduli of composites, Mech. of Materials, (1985).

[33] B. Paul, Prediction of elastic constants of multiphase materials, Trans. A.I.M.E. 218 (1960), pp. 36-41.

[34] S. Prager, Diffusion and viscous flow in concentrated suspensions, Physica 29 (1963) pp. 129-139.

[35] L. Simon, On G-convergence of elliptic operators, Indiana Univ. Math. J., 28 (1979), pp. 587-594.

[36] S. Spagnolo, Sulla convergenza di soluzioni di equazioni paraboliche ed ellitiche, Ann. Scuola Norm. Sup. Pisa 22 (1968), pp. 577-597.

[37] L. Tartar, Cours Peccot, College de France, 1977.

[38] L. Tartar, Estimations fines de coefficients homogénéisés, in Ennio De Giorgi's Colloquium, P. Krée ed., Research Notes in Math. No. 125, Pitman, Boston, 1985, pp. 168-187.

[39] G.I. Taylor, The viscosity of a fluid containing small drops of another fluid, Proc. Roy. Soc. A138 (1932) pp. 41-48.

[40] H.L. Weissberg and S. Prager, Viscosity of concentrated suspensions, Trans. Soc. Rheology 9 (1965).

[41] V. Zhikov, S. Kozlov, O. Oleinik, and K. Ngoan, Averaging and G-convergence of differential operators, Russian Math Surveys, 34 (1979) pp. 69-147.

[42] R. Temam, Navier-Stokes Equations: Theory and Numerical Analysis, North Holland, Amsterdam and New York, 1977.

[43] A. Bensoussan, J.-L. Lions, and G. Papanicolaou, Asymptotic Analysis for Periodic Structures, North-Holland, Amsterdam and New York, 1978.

The Proof that Laminates Generate All Possible Effective Conductivity Functions of Two-Dimensional, Two-Phase Materials

GRAEME W. MILTON*

Abstract. The effective conductivity tensor of any two-dimensional, two-isotropic-phase composite is known to have certain general analytic properties as a function of the component conductivities. It is established that every matrix-valued rational function compatible with these known analytic properties is the effective conductivity function of some sequentially layered laminate material and of some hierarchical coated ellipse assemblage. Thus there are no hidden, undiscovered, general analytic properties common to all two-dimensional conductivity functions of two-phase composites. In higher dimensions the sequentially laminated materials and the ellipsoid assemblages do not suffice to generate all possible conductivity functions, but they do generate all isotropic conductivity functions compatible with a generalized Keller duality relation.

1. Introduction. In my talk I discussed a new field equation recursion method that yields continued fraction representations and hence bounds for the effective conductivity tensor A of multicomponent composites [1]. By contrast the present paper will focus on whether the analytic properties of A as a function of the component conductivities are at present completely known for two dimensional composites comprised of two isotropic components with conductivities a_1 and a_2.

Such effective conductivity functions $A_*(a_1, a_2)$ take values in the space of 2×2 symmetric matrices and are well known [2-6] to be

(i) homogeneous, satisfying

$$A_*(\lambda a_1, \lambda a_2) = \lambda A_*(a_1, a_2) \tag{1.1}$$

for all λ;

(ii) normalized, to the extent that

$$A_*(1,1) = I \ , \tag{1.2}$$

* Department of Physics 405-47, California Institute of Technology, Pasadena, CA 91125 U.S.A.

where $\boldsymbol{I}$ is the identity matrix;

(iii) dissipative, which by definition implies

$$\operatorname{Re}\,[\boldsymbol{e}\,\boldsymbol{A}_*(a_1,a_2)\,\boldsymbol{e}^T] > 0 \quad \text{whenever} \quad \operatorname{Re} a_1 > 0 \quad \text{and} \quad \operatorname{Re} a_2 > 0 \tag{1.3}$$

for all real vectors $\boldsymbol{e}$ (physically implying that the composite dissipates energy whenever both components dissipate energy); and

(iv) self-dual, in the sense that

$$\boldsymbol{A}_*(a_1,a_2) = [\boldsymbol{R}^T \boldsymbol{A}_*(1/a_1, 1/a_2)\,\boldsymbol{R}]^{-1}, \tag{1.4}$$

where

$$\boldsymbol{R} = \begin{bmatrix} 0 & 1 \\ -1 & 0 \end{bmatrix} \tag{1.5}$$

is the 90° rotation matrix with transpose $\boldsymbol{R}^T$.

Here the converse will be established: for every 2×2 symmetric-matrix-valued rational function $\boldsymbol{A}_*(a_1,a_2)$ satisfying these four properties there exists a two-dimensional composite with effective conductivity function $\boldsymbol{A}_*(a_1,a_2)$. In this sense the analytic properties are completely known: one cannot deduce additional properties without placing restrictions on the underlying microgeometry.

The main result of the paper is that *every* rational function within this set of all possible effective conductivity functions is realized within the class of *sequentially layered laminate materials*. Such laminate materials are constructed in stages. At stage zero either component 1 or component 2 is taken as the laminate. More generally at the jth stage the laminate constructed at the previous $(j-1)$th stage is sliced in a specified j-dependent direction on a length scale much larger than existing inhomogeneities and layered with slices of component 1 (or component 2) in prescribed j-dependent proportions. The choice of the component to be layered at each stage is j-dependent and completely arbitrary. This procedure is repeated for j=1 up to j=r to produce a laminate of rank r. These and other types of laminates with inhomogeneities on widely separated length scales have been utilized in a variety of contexts: see for example [7] and the references cited there. They were first introduced by Bruggeman [8] as a basis for generating approximations for the effective elastic moduli of isotropic polycrystals.

Previously it was recognized [7,9,10] that any sample of composite, built from two isotropic components with conductivities a_1 and a_2 and placed between two parallel superconducting plates at fixed uniform potentials, could be replaced by a multiscaled laminate without altering the dependence of the total current on a_1 and a_2. This observation coupled with the duality relation (1.4) implies [7] that the class of effective conductivity functions generated by two-dimensional laminates includes all rational functions $\boldsymbol{A}_*(a_1,a_2)$ that have the same eigenvectors for all a_1 and a_2, and that satisfy (1.1)-(1.4). Here we include more general functions with eigenvectors that vary as a_1 and a_2 are changed.

The outline of the paper is as follows. In Section 2 a characterization is given of the set of all possible conductivity functions of d-dimensional sequentially layered laminates of finite rank. These and only these functions can be represented in the form of a certain type of continued fraction incorporating as coefficients a set of positive semidefinite matrices $\boldsymbol{M}_i$, $i = 1,2,\ldots,k$ of unit trace and a set of constants θ_i, $i = 1,2,\ldots,k$, that lie in the interval (0,1). In Section 3 we prove how an arbitrary rational function satisfying (1.1)-(1.4) can be represented in the required form of continued fraction with a set of matrices $\boldsymbol{M}_i$ and parameters θ_i corresponding to some two-dimensional sequentially layered laminate. A prescription is given for determining these matrices and parameters from the given analytic function. This completes the proof of the main result of the paper. In Section 4 another class of geometry is described that generates all possible effective conductivity functions in two dimensions, namely hierarchical assemblages of coated ellipses. We also investigate what family of isotropic conductivity functions are generated by sequentially layered laminates in higher dimensions. Such functions are found to satisfy a duality relationship that generalizes Keller's two-dimensional result [4] but which is not valid for arbitrary composite geometries unless $d=2$.

2. The Conductivity Functions of Sequentially Layered Laminates. Tartar and Murat [11] have established that by taking an anisotropic material of conductivity tensor $\boldsymbol{A}$ and successively laminating it in various directions with an isotropic material of conductivity a in a space of arbitrary dimension d one obtains a composite having effective conductivity $\boldsymbol{A}'$ satisfying

$$\theta\,(\boldsymbol{A}'-a\,\boldsymbol{I})^{-1} = (\boldsymbol{A}-a\,\boldsymbol{I})^{-1} + (1-\theta)\,a^{-1}\boldsymbol{M}\,, \tag{2.1}$$

where θ is the proportion of the starting material $\boldsymbol{A}$ in the final composite and $\boldsymbol{M}$ is a symmetric semipositive definite matrix with the property that

$$\mathrm{Tr}\,\boldsymbol{M} = 1\,, \tag{2.2}$$

and which depends on the geometry of the laminate and not on the material properties of the components. Conversely, subsequent to work of Lurie and Cherkaev [12] on the two-dimensional problem, Tartar and Murat [11] proved there exists such a laminate with conductivity $\boldsymbol{A}'$ given by the formula (2.1) for every matrix $\boldsymbol{M}$ with these properties and for every parameter $\theta\in(0,1)$. Given a matrix $\boldsymbol{M}$ or rank $c\le d$ it suffices to take c orthogonal layerings, each in the direction of the eigenvectors of $\boldsymbol{M}$. The amount layered in each direction is taken proportional to the associated eigenvalue of $\boldsymbol{M}$ and adjusted by a common factor until the materials $\boldsymbol{A}$ and a occupy volume fractions θ and $1-\theta$ in the final composite.

Now suppose we start with component 1, which has conductivity tensor

$$\boldsymbol{A}_0 = a_1\boldsymbol{I}\,, \tag{2.3}$$

and layer it in various directions with component 2 to obtain a laminate with effective conductivity tensor $\boldsymbol{A}_1$ which in turn we layer in various directions with component 1 to obtain a laminate effective conductivity tensor $\boldsymbol{A}_2$, and continue this process until we obtain a sequentially layered laminate of effective conductivity $\boldsymbol{A}_k$. From (2.1) the resulting hierarchy of effective tensors are related via the pair of recursion formulae,

$$\theta_{2n+1}(\boldsymbol{A}_{2n+1}-a_2\boldsymbol{I})^{-1} = (\boldsymbol{A}_{2n}-a_2\boldsymbol{I})^{-1} + (1-\theta_{2n+1})\,a_2^{-1}\boldsymbol{M}_{2n+1}\,, \tag{2.4}$$

$$\theta_{2n+2}(\boldsymbol{A}_{2n+2}-a_1\boldsymbol{I})^{-1} = (\boldsymbol{A}_{2n+1}-a_1\boldsymbol{I})^{-1} + (1-\theta_{2n+2})\,a_1^{-1}\boldsymbol{M}_{2n+2}\,, \tag{2.5}$$

where the θ_i, $i=1,2,\ldots,k$, are a sequence of real numbers in the range

$$0<\theta_i<1\,, \tag{2.6}$$

and the $\boldsymbol{M}_i$, $i=1,2,\ldots,k$, are a sequence of positive semidefinite real matrices satisfying

$$\mathrm{Tr}\,\boldsymbol{M}_i = 1 \quad \text{for all } i\,. \tag{2.7}$$

The numbers θ_i and the matrices $\boldsymbol{M}_i$ can be regarded as a set of geometric parameters that characterize the relevant features of the laminate geometry.

By rewriting the recursion relations in the form,

$$\boldsymbol{A}_{2n+1} = a_2\boldsymbol{I} + \theta_{2n+1}\,[(1-\theta_{2n+1})\,a_2^{-1}\boldsymbol{M}_{2n+1} - (a_2\boldsymbol{I}-\boldsymbol{A}_{2n})^{-1}]^{-1}\,, \tag{2.8}$$

(2.9) $$A_{2n+2}=a_1 I+\theta_{2n+2}\,[(1-\theta_{2n+2})\,a_1^{-1}\,M_{2n+2}-(a_1 I-A_{2n+1})^{-1}]^{-1},$$

making the substitution $A_0=a_1 I$, and successively eliminating the tensors A_1, A_2, up to A_{k-1} it is clear that we generate a continued fraction expansion for the effective tensor A_k of any sequentially layered laminate that has component 1 as the starting material. Conversely any analytic function $A_k(a_1,a_2)$ that can be represented in this form is an effective conductivity function of a sequentially layered laminate. The matrices M_i and parameters θ_i effectively serve as a set of instructions for constructing the laminate.

Thus it remains to show that any rational function $A_*(a_1,a_2)$ satisfying (1.1)-(1.4) is expressible as continued fraction either of the form implied by (2.3)-(2.5) or of the complimentary form obtained by interchanging a_1 and a_2 in the above formulae.

3. Representation of an Arbitrary Rational Function with the Required Analytic Properties. Suppose we are given any 2×2 matrix valued rational function $A_*(a_1,a_2)$ satisfying (1.1)-(1.4). By homogeneity $a_2^{-1}A_*(a_1,a_2)$ only depends on the ratio a_1/a_2. Therefore let us define the degree m of the rational function $A_*(a_1,a_2)$ to be the number of negative real (non-zero and finite) values of the ratio a_1/a_2 at which $a_2^{-1}A_*(a_1,a_2)$ has exactly one infinite eigenvalue plus twice the number of such ratios at which $a_2^{-1}A_*(a_1,a_2)$ has two infinite eigenvalues. This definition ensures m is independent of the labelling of the variables. Given

(3.1) $$A_*(a_1,a_2)\equiv A_m(a_1,a_2),$$

our aim is to construct a hierarchy of functions $A_m(a_1,a_2), A_p(a_1,a_2), A_q(a_1,a_2),\dots A_0(a_1,a_2)$ of successively lower degree, $m>p>q>\cdots>0$, that have the required analytic properties and which are related via recursion relations of the form (2.3)-(2.5).

To understand the physical idea behind the ensuing analysis, suppose we were given a rational function $A_m(a_1,a_2)$ and wanted to find an associated sequentially layered laminate. Assuming component 1 was the last component layered we should get an indication of how much of this component was layered and in what directions by setting $a_2=0$ (or $a_1=\infty$): for instance if $A_m(a_1,0)$ has only one non-zero eigenvalue then at the last stage component 1 must have been layered parallel to the direction of the associated eigenvector. By subtracting from $A_m(a_1,a_2)$ the contribution from this layering (in effect removing the last layering from the laminate) one would expect to get a rational function $A_p(a_1,a_2)$ of reduced degree, still satisfying (1.1)-(1.4). By repeating this procedure and where necessary setting $a_1=0$ (or $a_2=\infty$), we would eventually obtain information about the entire sequence of layerings from which a laminate with conductivity function $A_m(a_1,a_2)$ could be constructed.

From the work of Bergman [2] and from the form of the recursion relations (2.4) and (2.5) it seems appropriate to introduce the matrix valued functions,

(3.2) $$S_m^{(2)}(s)=[I-a_2^{-1}A_m(a_1,a_2)]^{-1},$$

(3.3) $$S_m^{(1)}(t)=[I-a_1^{-1}A_m(a_1,a_2)]^{-1},$$

and the variables

(3.4) $$s=(1-a_1/a_2)^{-1};\qquad t=(1-a_2/a_1)^{-1}.$$

These are related via the identities

(3.5) $$S_m^{(2)}(s)=s\,[I-(1-s)\,[S_m^{(1)}(1-s)]^{-1}\,]^{-1},$$

(3.6) $$S_m^{(1)}(t)=t\,[I-(1-t)\,[S_m^{(1)}(1-t)]^{-1}\,]^{-1},$$

$$s+t=1, \tag{3.7}$$

obtained by eliminating $A_m(a_1,a_2)$ and the ratio a_1/a_2 from the above formulae. Now following Bergman [2] let us study the analytic properties of $S_m^{(2)}(s)$. First we may as well set $a_2=1$ and study the properties of $A_m(a_1,1)$. The dissipation property (1.3) of $A_m(a_1,a_2)$ and the homogeneity of this function implies

$$\text{Re}\,[e^{i\gamma}A_m(a_1,1)]>0 \quad \text{when} \quad \text{Re}(e^{i\gamma}a_1)>0 \quad \text{and} \quad \text{Re}(e^{i\gamma})>0. \tag{3.8}$$

When a_1 is real and positive we can take $\gamma=\pi/2-\varepsilon$ and $\gamma=-\pi/2+\varepsilon$ where ε is infinitesimal and thereby deduce that $A_m(a_1,1)$ is real and positive definite. When a_1 has positive (or negative) imaginary part we can take $\gamma=-\pi/2+\varepsilon$ (or $\gamma=\pi/2-\varepsilon$) and deduce that $\text{Im}\,A_m(a_1,1)$ is positive (negative) semidefinite. Hence any sign change in the eigenvalues of $\text{Im}\,A_m(a_1,1)$ can only occur when a_1, and hence when $1/a_1$, moves across the negative real axis.

Next we establish that $A_m(a_1,1)$ is a strictly monotonic increasing function of a_1 along the real axis except at those points that are poles of the function. In the reciprocal $1/a_1$ plane a pole of $A_m(a_1,1)$ can be defined as a real or complex value of $1/a_1$ at which $\text{Re}\,A_m(a_1,1)$ or $\text{Im}\,A_m(a_1,1)$ has at least one infinite eigenvalue. Since some of the eigenvalues of $\text{Im}\,A_m$ must change sign in the vicinity of such poles, the poles must be located on the negative real $1/a_1$ axis. They must be simple poles since the eigenvalues of $\text{Im}\,A_m$ change sign several times around higher order poles. Furthermore the residues must be real positive semidefinite matrices to ensure the eigenvalues of $\text{Im}\,A_m$ have the correct sign on each side of the axis.

Consequently $A_m(a_1,1)$ has the representation

$$A_m(a_1,1)=B_0+a_1B_{h+1}+\sum_{i=1}^{h}\frac{B_i}{b_i+1/a_1}, \tag{3.9}$$

where the constants $b_i, i=1,2,\ldots,h$ (where $h\leq m$) are real and positive and the matrices B_j, $j=0,1,2,\ldots,h+1$ are positive semidefinite. (The matrix B_0, although not a residue in this representation, is positive semidefinite because $A_m(a_1,1)$ is positive definite for all $a_1>0$, including infinitesimally small a_1.) The parameters are related via the identity

$$I=B_0+B_{h+1}+\sum_{i=1}^{h}\frac{B_i}{1+b_i}, \tag{3.10}$$

which follows from the normalization (1.2).

Except at the poles of $A_m(a_1,1)$ the derivative

$$\frac{dA_m}{da_1}=B_{h+1}+\sum_{i=1}^{h}\frac{B_i}{(a_1b_i+1)^2} \tag{3.11}$$

is a finite positive semidefinite matrix and has a zero eigenvalue only when the residues B_i, $i=1,2,\ldots,h+1$, share a common eigenvector with zero eigenvalue. From (3.9) and (3.10) such an eigenvector would be an eigenvector of $A_m(a_1,1)$ having unit eigenvalue for all a_1 and (1.4) implies the other eigenvalue of $A_m(a_1,1)$ would likewise be unity for all a_1. Let us ignore this trivial case which evidently only occurs when $A_m(a_1,1)=I$. Then (3.11) is a positive definite matrix and $A_m(a_1,1)$ is strictly monotonic increasing along the real a_1 axis, except at the poles. On the positive real axis this and the normalization (1.2) implies the positive eigenvalues of $A_m(a_1,1)$ take values less than unity when $a_1<1$ and values greater than unity when $a_1>1$.

From this analysis and by manipulation of (1.4) we conclude that $S_m^{(2)}(s)$ has the properties

(3.12) $$\operatorname{Im} S_m^{(2)}(s) = 0, \quad S_m^{(2)}(s) \geq I \quad \text{for all} \quad s \geq 1;$$

$$\operatorname{Im} S_m^{(2)}(s) = 0, \quad S_m^{(2)}(s) \leq 0 \quad \text{for all} \quad s \leq 0,$$

(3.13) $$\operatorname{Im} S_m^{(2)}(s) > 0 \quad \text{whenever} \quad \operatorname{Im} s > 0;$$

$$\operatorname{Im} S_m^{(2)}(s) < 0 \quad \text{whenever} \quad \operatorname{Im} s < 0,$$

(3.14) $$S_m^{(2)}(s) + R^T S_m^{(2)}(1-s) R = I,$$

where R is the 90° rotation matrix (1.5). Conversely, any rational function $S_m^{(2)}(s)$ satisfying (3.12)-(3.14) generates via (3.2) and (3.4) a function $A_m(a_1, 1)$ that can be represented in the form (3.9) for some set of positive constants b_i and positive semidefinite matrices B_i. Subsequently from (1.1) it follows that $A_m(1, a_2)$ has the representation

(3.15) $$A_m(1, a_2) = B_0 a_2 + B_{h+1} + \sum_{i=1}^{h} \frac{B_i / b_i}{1/b_i + 1/a_2},$$

which clearly has the same form as the representation (3.9). After excluding the trivial case that occurs when $A_m(1, a_2) = I$, (3.15) together with (3.10) and (3.14) implies the function $S_m^{(1)}(t)$ has the properties

(3.16) $$\operatorname{Im} S_m^{(1)}(t) = 0, \quad S_m^{(1)}(t) \geq I \quad \text{for all} \quad t \geq 1;$$

$$\operatorname{Im} S_m^{(1)}(t) = 0, \quad S_m^{(1)}(t) \leq 0 \quad \text{for all} \quad t \leq 0,$$

(3.17) $$\operatorname{Im} S_m^{(1)}(t) > 0 \quad \text{whenever} \quad \operatorname{Im} t > 0;$$

$$\operatorname{Im} S_m^{(1)}(t) < 0 \quad \text{whenever} \quad \operatorname{Im} t < 0,$$

(3.18) $$S_m^{(1)}(t) + R^T S_m^{(1)}(1-t) R = I,$$

directly analogous to the properties (3.12)-(3.14) of $S_m^{(2)}(s)$.

When $S_m^{(1)}(0) = 0$ (3.18) implies $S_m^{(1)}(1) = I$ and by taking the limit $s \to 0$ in the formula (3.5) we deduce that

(3.19) $$\left[S_m^{(2)}(0) \right]^{-1} = I - \frac{dS_m^{(1)}(t)}{dt} \Big|_{t=1}.$$

Since $S_m^{(1)}(1) = I$, the point $t=1$ is clearly not a pole of $S_m^{(1)}(t)$ and therefore the derivative in (3.19) is finite: $S_m^{(2)}(0)$ is negative definite. Similarly $S_m^{(1)}(0)$ is negative definite if $S_m^{(2)}(0) = 0$.

Thus the matrices $S_m^{(1)}(0)$ and $S_m^{(2)}(0)$ are never simultaneously zero. So let us suppose $S_m^{(2)}$ is non-zero, by relabelling the components if necessary. Then the matrix

(3.20) $$M_m \equiv S_m^{(2)}(0) \,/\, \operatorname{Tr} S_m^{(2)}(0)$$

is well defined and real, satisfying

(3.21) $$\operatorname{Tr} M_m = 1; \quad M_m \geq 0.$$

Now the form of the recursion relation (2.4) suggests we should introduce the function

(3.22) $$S_p^{(2)}(s) \equiv \theta \, S_m^{(2)}(s) + (1-\theta) \, M_m,$$

and study its analytic properties as a function of s for fixed θ in the range

(3.23) $$1 \geq \theta \geq [1 - \operatorname{Tr} S_m^{(2)}(0)]^{-1}.$$

From (3.13), (3.14) and the identity

(3.24) $$M_m + R^T M_m R = I \,,$$

which holds because M_m is a two-dimensional matrix with unit trace, it follows that

(3.25) $$\mathrm{Im}[S_p^{(2)}(s)] > 0 \quad \text{whenever} \quad \mathrm{Im}\, s > 0;$$
$$\mathrm{Im}[S_p^{(2)}(s)] < 0 \quad \text{whenever} \quad \mathrm{Im}\, s < 0,$$

(3.26) $$S_p^{(2)}(s) + R^T S_p^{(2)}(s) R = I \,,$$

and (3.12) clearly implies

(3.27) $$\mathrm{Im}\, S_p^{(2)}(s) = 0 \quad \text{for all} \quad s \geq 1 \text{ or } s \leq 0 .$$

Since $A_m(a_1,1)$ is a monotonic increasing function of a_1 along the positive real axis both $S_m^{(2)}(s)$ and $S_p^{(2)}(s)$ are likewise monotonic increasing functions of s in the intervals $s \geq 1$ and $s \leq 0$. So to establish

(3.28) $$S_p^{(2)}(s) \geq 1 \quad \text{for all} \quad s \geq 1;$$
$$S_p^{(2)}(s) \leq 0 \quad \text{for all} \quad s \leq 0,$$

it suffices to prove

(3.29) $$S_p^{(2)}(0) \leq 0; \quad S_p^{(2)}(1) \geq I \,.$$

The first inequality is easily verified from (3.23) and the relation

(3.30) $$S_p^{(2)}(0) = S_m^{(2)}(0) \left[\frac{\theta\,(\mathrm{Tr}\, S_m^{(2)}(0) - 1) + 1}{\mathrm{Tr}\, S_m^{(2)}(0)} \right]$$

implied by (3.22) and (3.20), while the second inequality follows from the first and the identity (3.26).

In summary the function $S_p^{(2)}(s)$ and hence the auxiliary function $S_p^{(1)}(t)$, defined via a relation analogous to (3.6), have the same analytic properties as $S_m^{(2)}(s)$ and $S_m^{(2)}(t)$. Now consider the degree of the associated function $A_p(a_1,a_2)$. The poles of $A_p(a_1,1)$ occur where $S_p^{(2)}(s)$ has at least one zero eigenvalue. Equivalently, from (3.22) and assuming M_m is non-singular, these poles occur at points s in the interval [0,1) where the matrix

(3.31) $$G(s) \equiv M_m^{-1/2}\, S_m^{(2)}(s)\, M_m^{-1/2}$$

has at least one eigenvalue equal to $1-1/\theta$. In practice we could plot the two eigenvalues $g_1(s), g_2(s)$ of $G(s)$ as function of s in the (g,s) plane, and mark the points of intersection with the line, denoted as L_θ, at $g = 1-1/\theta$. The degree of $A_p(a_1,a_2)$ by definition is the number of times the curve $g_1(s)$ intersects L_θ in the open interval (0,1) plus the number of times $g_2(s)$ intersects L_θ in the same interval. Now, from the monotonicity of $S_m^{(2)}(s)$ it follows that $g_1(s)$ and $g_2(s)$ are monotonic increasing functions of s except at their poles. Furthermore, since M_m and hence $S_m^{(2)}(0)$ have been assumed to be non-singular, (3.2) implies that the points $g_1(0) = g_2(0) = \mathrm{Tr}\, S_m^{(2)}(0)$ lie below L_θ until θ is decreased to the lower limit

(3.32) $$\theta_m \equiv [1 - \mathrm{Tr}\, S_m^{(2)}(0)]^{-1}$$

at which both $g_1(s)$ and $g_2(s)$ intersect L_θ at $s = 0$.

Therefore the degree of $A_p(a_1,a_2)$ is preserved, and thus equals m, for $\theta < \theta_m$. Once we set $\theta = \theta_m$ the two infinite eigenvalues of $A_p(a_1,1)$ at $a_1 = \infty$ no longer contribute to the degree count: the resulting degree of $A_p(a_1,a_2)$ is $p = m-2$.

When M_m is singular the poles of $A_p(a_1, 1)$ occur at points where

(3.33) $$S_m^{(2)}(s) - M_m + \theta^{-1} M_m$$

has at least one zero eigenvalue. Since $S_m^{(2)}(s) - M_m$ and $\theta^{-1} M_m$ are monotonic increasing functions of respectively s and $1/\theta$ the poles of $A_p(a_1, 1)$ move continuously towards $s = 0$, i.e. $a_1 = \infty$, as θ is decreased towards θ_m. The degree of $A_m(a_1, a_2)$ is preserved for $\theta < \theta_m$ and once we set $\theta = \theta_m$ it is reduced to $p = m-1$.

Irrespective of whether M_m is singular or not singular we are left with a function $A_p(a_1, a_2)$ of reduced degree. Furthermore $S_p^{(2)}(s)$ has the required analytic properties (3.25)-(3.28) and in addition satisfies the identity

(3.34) $$S_p^{(2)}(0) = 0,$$

which implies $S_p^{(1)}(0)$ is negative definite. Hence the matrix

(3.35) $$M_p \equiv S_p^{(1)}(0) / \mathrm{Tr}\, S_p^{(1)}(0)$$

is positive definite with unit trace and for

(3.36) $$\theta_p \equiv [1 - \mathrm{Tr}\, S_p^{(1)}(0)]^{-1}$$

the function

(3.37) $$S_q^{(1)}(t) \equiv \theta_p\, S_p^{(1)}(t) + (1 - \theta_p) M_p ,$$

has the required analytic properties and generates a function $A_q(a_1, a_2)$ of degree $q = p-2$. Furthermore since $\mathrm{Tr}\, S_p^{(1)}(0) < 0$ the parameter θ_p, like θ_m, lies in the open interval (0,1).

By continuing this procedure, we must eventually obtain a function

(3.38) $$A_0(a_1, a_2) = a_1 I \quad \text{or} \quad A_0(a_1, a_2) = a_2 I$$

of zero degree. From (3.22) and (3.37) the resulting hierarchy of functions $A_0(a_1, a_2)$, $A_2(a_1, a_2)$ $A_4(a_1, a_2)$, ..., $A_m(a_1, a_2)$ are clearly related by recursion relations of the form (2.4) and (2.5).

Thus we have established $A_m(a_1, a_2)$ is the effective conductivity function of a sequentially layered laminate material. Note it suffices to consider laminate materials associated with sequences of *positive definite* matrices M_i, $i = 1, 2, \ldots, k-1$, terminated by a *positive semidefinite* matrix M_k. At each stage in the construction of such laminates, component 1 (or component 2) is layered with the composite in two orthogonal directions before the other component is layered with the resulting composite. Using this laminating procedure we obtain the function $A_m(a_1, a_2)$ of degree m using a laminate of rank $r = m$.

Other decompositions of the function $A_m(a_1, a_2)$ are clearly possible. For example by taking

(3.39) $$M_m = \tfrac{1}{2} I$$

and decreasing θ until one of the eigenvalues of $S_p^{(2)}(0)$ vanishes and then defining $S_q^{(1)}(t)$ via (3.35)-(3.37) and repeating this procedure (at the next stage with $M_q = \tfrac{1}{2} I$) we obtain a different prescription for constructing a sequentially layered laminate material with conductivity function $A_m(a_1, a_2)$. These other decompositions can be excluded by requiring all the matrices M_i, $i = 1, 2, \ldots, k-1$, (but not the final matrix M_k) to be non-singular.

4. Generalizations. The form of the relation (2.1) between the effective conductivity A' and the component conductivities a and A applies not only to laminated materials of the type described in Section 2, but also to assemblages of aligned coated ellipsoids. In such assemblages, which are generalizations of the Hashin-Shtrikman coated sphere assemblages [13], each coated ellipsoid is comprised of a core of material A surrounded by a confocal coating of material a, in volume fractions θ and $1-\theta$ respectively, and is similar in all aspects except size to any other coated ellipsoid in the assemblage: the coated ellipsoids must of course have a variety of sizes ranging to the infinitesimal so they can be packed to fill all space.

Milton [9], Bergman [14] and Tartar and Murat [11] independently calculated the effective conductivity of such assemblages in 2-dimensions, 3-dimensions and arbitrary dimension respectively, basing their calculations on arguments given by Hashin and Shtrikman [13]. While they restricted their analysis to conductivity tensors A that are isotropic or that have eigenvectors aligned with the ellipsoidal axes it can be established that the formula (2.1) is valid even when A has an arbitrary orientation. (The lower and upper Beran bounds [15] when extended to anisotropic composites with one anisotropic phase coincide and imply the formula (2.1) because the pair of weight matrices $W_a^{(1)}$ and $W_A^{(1)}$ associated with the geometry of the regions occupied by materials A and a in the assemblage equal zero and the identity matrix. These weight matrices are defined in [1]).

The matrix M has eigenvectors in the directions of the principal axes of the coated ellipsoids and positive eigenvalues

$$m_i = (n_{A,i} - \theta n_{a,i})/(1-\theta), \quad i = 1,2,\ldots,d, \tag{4.1}$$

where $n_{A,i}$ and $n_{a,i}$ are the depolarization factors of the core surface and the outer shell surface in the direction of the ith eigenvector [14]. Conversely Tartar and Murat [11] establish that for every symmetric positive semidefinite matrix M of unit trace and for every parameter θ in the range (0,1) there exists a coated ellipsoid assemblage with effective conductivity A' given by (2.1).

Hierarchical assemblages of coated ellipsoids are constructed as follows. We start with component 1 and use it as the core material and component 2 as the shell material to form a coated ellipsoid assemblage of effective conductivity A_1. Next, on a much larger scale, we take this composite as the core material and component 1 as the shell material in another coated ellipsoid assemblage, in which the ellipsoids are not necessarily oriented in the same way as the ellipsoids in the core material. In turn this composite of effective conductivity A_2 is taken as the core material and component 2 as the shell material in yet another coated ellipsoid assemblage, and this procedure is repeated until we obtain a hierarchical coated ellipsoid assemblage of effective conductivity A_k. In two dimensions we obtain a hierarchical coated ellipse assemblage.

From the above analysis, any rational function that can be expressed as a continued fraction of the form implied by (2.8) and (2.9), where the parameters θ_i and matrices M_i satisfy the usual constraints, is the effective conductivity function of a hierarchical coated ellipse assemblage. Therefore in two dimensions the conductivity functions of such assemblages include all possible conductivity functions of composites of two isotropic components. In particular the conductivity functions of multicoated cylinder assemblages include all possible isotropic conductivity functions of two-phase, two-dimensional composites [9].

In dimensions d greater than 2 the conductivity functions of sequentially layered laminates, or equivalently the conductivity functions of hierarchical assemblages of coated ellipsoids, do not include all possible rational conductivity functions of two phase composites. For example in three dimensions Schulgasser [16] constructs a composite with isotropic effective conductivity

$$a_*(a_1,a_2) = \tfrac{2}{3}[\theta a_1 + (1-\theta)a_2] + \tfrac{1}{3}[\theta/a_1 + (1-\theta)/a_2]^{-1} \tag{4.2}$$

by layering component 1 with component 2 in proportions θ and $1-\theta$ and laminating the resulting composite with *itself*. Note that $a_*(a_1,a_2)$ remains non-zero both when $(a_1,a_2)=(0,1)$ and when $(a_1,a_2)=(1,0)$, which reflects the fact that the geometry is biconnected like a porous rock. By contrast no sequentially layered laminate (of finite rank) can have an isotropic conductivity function $a_*(a_1,a_2)$ that remains non-zero in both these limits. If the last layering is done with component 1 the current will be blocked in the direction perpendicular to the layers when $a_1=0$. Alternatively if the last layering is done with component 2 the current will be blocked when $a_2=0$. Thus we need to consider other types of laminates and perhaps completely different geometries to produce all possible rational effective conductivity functions.

As first step towards characterizing the set of all possible conductivity functions of two phase laminates let us consider the class of sequentially layered laminates that have isotropic conductivity functions $a_*(a_1,a_2)$. Such laminates are characterized by matrices

$$\boldsymbol{M}_i = \boldsymbol{I}/d\,, \quad i=1,2,\ldots,k\,, \tag{4.3}$$

and have conductivity functions $a_*(a_1,a_2)$ satisfying the generalized Keller duality relationship

$$\frac{a_*(a_1,a_2)\,a_*(a_2,a_1)}{a_1a_2} + (d-2)\left[\frac{a_*(a_1,a_2)+a_*(a_2,a_1)}{a_1+a_2}\right] = d-1\,. \tag{4.4}$$

This last relation takes the equivalent forms,

$$(dt-1)\,s_*^{(2)}(s) - (ds-1)\,s_*^{(2)}(t) = t-s\,, \tag{4.5}$$

$$(ds-1)\,s_*^{(1)}(t) - (dt-1)\,s_*^{(1)}(s) = s-t\,, \tag{4.6}$$

in terms of the functions

$$s_*^{(2)}(s) = (1-a_*(a_1,a_2)/a_2)^{-1}, \tag{4.7}$$

$$s_*^{(1)}(t) = (1-a_*(a_1,a_2)/a_1)^{-1}, \tag{4.8}$$

and is easily proved by induction from the recursion relations (2.4) and (2.5) once it is noted the identity

$$(dt-1)\,\boldsymbol{M}_i - (ds-1)\,\boldsymbol{M}_i = (t-s)\,\boldsymbol{I} \tag{4.9}$$

is satisfied for all matrices $\boldsymbol{M}_i$ of the form (4.3). Conversely the analysis of Section 3 implies that any *isotropic* rational function $\boldsymbol{A}_*(a_1,a_2)=a_*(a_1,a_2)\,\boldsymbol{I}$ satisfying (1.1)-(1.3) and (4.4), can be identified with the conductivity function of a sequentially layered laminate. An analogous result was stated elsewhere [9] for assemblages of multicoated spheres, based on a similar but unpublished argument. For most other types of composite geometry the generalized Keller duality relationship (4.4) is not satisfied. Many years ago I conjectured [9] that the left hand side of equation (4.4) might always be greater than or equal to $d-1$. This has never been proved: only Schulgasser's inequality [17] has been established.

In summary (1.1)-(1.3) and (4.3) provide a simple characterization of all possible isotropic conductivity functions of sequentially layered laminates. I doubt the result can be extended to anisotropic conductivity functions: the simplest characterization of these functions in dimensions $d \geq 3$ might be in terms of the continued fraction formulation implied by the recursion relations (2.8) and (2.9).

Acknowledgements The author gratefully acknowledges the support of the California Institute of Technology through the provision of facilities and the award of a Weingart Fellowship.

REFERENCES

[1] G. W. MILTON, *Multicomponent composites, electrical networks and new types of continued fractions-I and II*, submitted to Comm. Math. Phys.

[2] D. J. BERGMAN, *The dielectric constant of a composite material - a problem in classical physics*, Phys. Rep. C, 43 (1978), pp. 377-407.

[3] K. GOLDEN AND G. PAPANICOLAOU, *Bounds for effective parameters of homogeneous media by analytic continuation*, Comm. Math. Phys., 90 (1983), pp. 473-491.

[4] J.B. KELLER, *A theorem on the conductivity of a composite medium*, J. Math. Phys., 5 (1964), pp. 548-549.

[5] A.M. DYKHNE, *Conductivity of a two-dimensional two-phase system*, Zh. Eksp. Teor. Fiz., 59 (1970), pp. 110-115 [Soviet Phys. JETP, 32 (1971), pp. 63-65.]

[6] K.S. MENDELSOHN, *A theorem on the effective conductivity of a two-dimensional heterogeneous medium*, J. Appl. Phys., 46 (1975), pp. 4740-4741.

[7] G.W. MILTON, *Modelling the properties of composites by laminates*, in *Homogenization and Effective Moduli of Materials and Media*, J.L. ERICKSEN, D. KINDERLEHRER, R. KOHN, AND J.-L. LIONS, eds., Springer-Verlag, New York, 1986, pp. 150-174.

[8] D.A.G. BRUGGEMAN, *Elastizität Konstanten von Kristallaggregaten*, Ph.D. Thesis, Utrect, 1930, pp. 19-38.

[9] G.W. MILTON, *Bounds on the complex permittivity of a two-component composite material*, J. Appl. Phys., (1981), pp. 5286-5293.

[10] ———, *Bounds on the transport and optical properties of a two-component composite material*, J. Appl. Phys., (1981), pp. 5294-5304.

[11] L. TARTAR, *Estimations fines des coefficients homogénéises*, in *Ennio De Giorgi Colloquium*, P. Krée, ed., Pitman Press, 1985, pp. 168-187.

[12] K.A. LURIE AND A.V. CHERKAEV, *Exact estimates of conductivity of composites formed by two isotropically conducting media taken in prescribed proportion*, Proc. Royal Soc. Edinburgh, 99A (1984), pp. 71-87.

[13] Z. HASHIN AND S. SHTRIKMAN, *A variational approach to the theory of the effective magnetic permeability of multiphase materials*, J. Appl. Phys., 33 (1962), pp. 3125-3131.

[14] D.J. BERGMAN, *Rigorous bounds for the complex dielectric constant of a two-component composite*, Annals of Physics, 138 (1982), pp. 78-114.

[15] M.J. BERAN, *Use of the variational approach to determine bounds for the effective permittivity in random media*, Nuovo Cimento, 38 (1965), pp. 771-782.

[16] K.S. SCHULGASSER, *Bounds on the conductivity of statistically isotropic polycrystals*, J. Phys. C, 10 (1977), pp. 407-417.

[17] ———, *On a phase interchange relationship for composite materials*, J. Math. Phys., 17 (1976), pp. 378-381.

Percolation Theory: A Selective Survey of Rigorous Results

CHARLES M. NEWMAN*

Abstract. We survey some of what has (and has not yet) been rigorously proved about the percolation phase transition. Our discussion involves comparisons between: a) short and long range percolation, b) static and dynamic percolation, c) percolation and other phase transitions and d) phase transitions and shock waves. Related to item d) and included in an appendix is a previously unpublished preprint about the connection between Burgers' equation and phase transitions.

1. Introduction. Percolation is the phenomenon in which randomly placed "microscopic" geometric objects (occupied sites, passable bonds, overlapping spheres, ...) connect together to form an infinite "macroscopic" network. In percolation models, there is usually a free parameter, say β, which determines the average density of the microscopic objects. A percolation phase transition occurs if there is a threshold or critical value β_c (not at an endpoint of the allowed range of β) such that percolation does not occur for $\beta < \beta_c$ but does for $\beta > \beta_c$.

Questions and answers about the nature of percolation models can be loosely divided into three categories according to which of the following three regimes they are concerned with: i) the <u>disordered regime</u> $(\beta < \beta_c)$, ii) the <u>ordered regime</u> $(\beta > \beta_c)$ or iii) the <u>critical regime</u> $(\beta = \beta_c$ or β approaching $\beta_c)$. In this survey we will discuss some of what has (and has not yet) been rigorously proved about these three regimes.

---

*Dept. of Mathematics, Univ. of Arizona, Tucson, AZ 85721. Research supported in part by NSF Grant DMS-8514834.

Among percolation models, the classic site and bond percolation models on regular d-dimensional lattices are the ones for which the most extensive rigorous results have been obtained. For concreteness, we will focus our attention mainly on bond percolation on the hypercubic lattice Z^d with nearest neighbor bonds. (Essentially all results for these models extend to other lattices and to site percolation.) However, we will begin by describing a more general class of bond models which includes both short range (such as nearest neighbor) and long range models as special cases. We do this because some long range models are of intrinsic interest and in order to emphasize that there are some open problems (see Questions 3 and 4 below) for short range models whose solution will require techniques that distinguish these models from certain long range models.

The bond models we consider have (undirected) bonds between every pair $\{x,y\}$ of sites in Z^d, which are independently chosen to be either passable or blocked with

$$\text{(1)} \qquad \text{Prob(the } \{x,y\} \text{ bond is passable)} = 1 - \exp(-\beta J_{y-x}).$$

Here $0 < \beta < \infty$ and $J_{y-x} \geq 0$. There are two examples of special interest which we will often refer to:

The nearest neighbor models. Here $J_{y-x} = 1$ if x and y are nearest neighbors (i.e., are separated by unit Euclidean distance, $||x - y|| = 1$) and otherwise $J_{y-x} = 0$. For this classic model, the parameter β is often replaced by the bond density, $p = 1 - \exp(-\beta)$. Note that p increases with β and ranges over $(0,1)$ as β ranges over $(0,\infty)$.

The $1/r^s$ models. Here $d = 1$, and the J_z's are such that for some real s (usually $s > 1$),

$$\text{(2)} \qquad \lim_{|y-x|\to\infty} |y - x|^s J_{y-x} = 1.$$

As we shall see below, the case $s = 2$ is quite interesting.

To avoid future caveats, we will henceforth assume that the model is irreducible in the sense that there is no partition of Z^d into two (or more) subsets with $J_{y-x} = 0$ whenever x and y are in different subsets. This assumption implies, for example, that a model on Z^d is really d-dimensional.

Before beginning our survey, we need to introduce a small amount of notation. Two sites are said to be connected if there is a path of passable bonds between them (a site is always connected to itself). A cluster is a maximal set of mutually connected sites. The cluster containing a site x is denoted $C(x)$ and the number of sites in it is denoted $|C(x)|$. The passable bonds give a (random) partition of

Z^d into clusters. Percolation occurs when (at least) one of these clusters is infinite; by ergodicity, the probability of percolation is either zero or one and is one if and only if the percolation density M (sometimes denoted P_∞), defined as

$$M = M(\beta) = \text{Prob}(C(0) \text{ is infinite}) \tag{3}$$

is strictly positive. M also equals the spatial density of the percolating network (i.e., the fraction of sites whose cluster is infinite). The critical point is formally defined as

$$\beta_c = \inf\{\beta : M(\beta) > 0\} \tag{4}$$

($\beta_c = \infty$ if $M(\beta) = 0$ for all $\beta < \infty$).

The first issue to be settled about any percolation model is whether there is a phase transition, i.e., whether $0 < \beta_c < \infty$. If $\sum_z J_z = \infty$, then it is easily seen that $\beta_c = 0$ since for any $\beta > 0$, there will be (with probability one) infinitely many passable bonds touching the origin (i.e., of the form $\{0,x\}$). Henceforth we assume that $\sum_z J_z < \infty$.

Our survey of rigorous results and open problems will be presented as a series of questions and answers. At the end of most of the questions we include in parentheses a phrase, which describes the question in percolation shoptalk. We conclude this section with our first question, numbered zero, and its answer.

Question 0. Is $0 < \beta_c < \infty$ (existence of a phase transition)?

Under our assumption that $\sum_z J_z < \infty$, it can be shown that $\beta_c > 0$ always, both for short range [1] and long range [2, 3] models. Indeed, a simple lower bound on β_c follows from the fact (based on a comparison to generalized Bethe lattice models) that

$$\sum_z [1 - \exp(-\beta J_z)] < 1 \text{ implies } M = 0. \tag{5}$$

It can also be demonstrated that $\beta_c < \infty$, at least for $d > 1$. This was originally done by Broadbent and Hammersley for nearest neighbor models [1] and their result immediately implies the same for our general models. When $d = 1$, the situation is more interesting. For example, in $1/r^s$ models, $\beta_c = \infty$ for $s > 2$ [2] while $\beta_c < \infty$ for $s \leq 2$ [4]. An improved version [5] of the $s > 2$ result shows that $\beta_c = \infty$ whenever $\lim z^2 J_z = 0$, e.g., if J_z is asymptotic to $(z^2 \log(\log |z|))^{-1}$. We remark that the proof in [4], which is of a renormalization group character, is unfortunately somewhat complex, especially for the case $s = 2$; a simpler proof ought to be possible.

2. Questions and Answers. Our next question concerns the disordered regime. The expected cluster size χ, is defined as

$$\chi = \chi(\beta) = E(|C(0)|) = \sum_{n \leq \infty} nP_n \tag{6}$$

where P_n is defined as

$$P_n = P_n(\beta) = \text{Prob}(|C(0)| = n). \tag{7}$$

Clearly $\chi = \infty$ if $M(= P_\infty)$ is strictly positive, but it is a priori possible that χ might diverge at some other critical value strictly below β_c. In that case, there would be an "intermediate disordered phase" regime - i.e., an interval of values of β in which χ was infinite while M was zero.

Question 1. <u>Can $\chi = \infty$ for some $\beta < \beta_c$ (existence of an intermediate (disordered) phase)</u>?

It was very recently proved by Aizenman and Barsky [6] that the answer to this question is always no. This was previously proved for the $d = 2$ nearest neighbor model [7]. We will have a little to say about the proof of the Aizenman-Barsky result in Section 3 below (in the paragraph following equation (26)). Meanwhile we note that as long as $\chi(\beta) < \infty$, it follows that for some constant $c = c(\beta)$, $P_n(\beta) \leq \exp(-cn)$ both for short range [8] and long range [3] models. These various results about the disordered regime show that it is reasonably well understood.

Our next three questions concern mostly the ordered regime and slightly the critical regime; the last four questions concern the critical regime. As we shall see, the ordered regime is partly understood and the critical regime is only slightly understood.

Question 2. <u>Can there be coexisting distinct infinite clusters (nonuniqueness)</u>?

It was recently proved [9] that the answer to this question is always no; i.e., if $M > 0$, then (with probability one) there is exactly one infinite cluster. This was previously proved for the $d = 2$ nearest neighbor model [10; 11]. The proof of [9] involves the differentiability (in β) of the "free energy",

$$f = f(\beta) = E(|C(0)|^{-1}) = \sum_{n<\infty} n^{-1}P_n; \tag{8}$$

this differentiability turns out to be related to the surface-to-volume ratio of large finite clusters.

Remarks. i) The uniqueness proved in [9] is applicable even to

$1/r^2$ models at their critical point (see the remark following Question 3 below). The $1/r^2$ models had been the best candidates for nonuniqueness among independent models.

ii) It is an open problem whether nonuniqueness occurs for any natural dependent percolation models. It has been shown in [12] under mild assumptions that if nonuniqueness does occur, then there are infinitely many distinct infinite clusters (with probability one).

iii) The results of [9] do not apply to so-called "incipient infinite clusters" (see [13; 14]). Indeed, these are expected to be nonunique, in some appropriate sense (see e.g., Section 3.2 of [3]), above the "upper critical dimension" (see Question 6 below).

Question 3. Is $M(\beta)$ continuous for $\beta > \beta_c$? Is $M(\beta)$ continuous at $\beta = \beta_c$ (continuity of the phase transition)?

The answer to the first part of the question is always yes; this is a consequence of the answer to Question 2 [9] and the earlier result [15] that uniqueness for $\beta > \beta_c$ implies continuity of M. The status of the second part of the question is more interesting. The answer is yes for the $d = 2$ nearest neighbor model [16] but is no for $1/r^2$ models [5]. It is an open problem to prove continuity at β_c for $d > 2$ nearest neighbor models or for $1/r^s$ models with $1 < s < 2$.

Remarks. i) Since $M(\beta)$ is always continuous from the right, the discontinuity of M in $1/r^2$ models implies the existence of an infinite cluster (i.e., the occurrence of percolation) exactly at the critical point. Although this infinite cluster is unique and hence has positive density, its precarious existence suggests it should have some interesting qualitative properties.

ii) The discontinuity in M for $1/r^2$ percolation models is an analogue of a phenomenon originally proposed for Ising models [17; 18]. The Ising model result has recently been rigorously proved by percolation methods [19].

The next question is the disordered regime analogue of Question 1. We define χ', the expected size of finite clusters as

$$\chi' = \chi'(\beta) = \sum_{n<\infty} nP_n(\beta). \tag{9}$$

Question 4. Can $\chi' = \infty$ for some $\beta > \beta_c$ (existence of an intermediate (ordered) phase)?

The answer is no for the $d = 2$ nearest neighbor model [20] but is yes for (at least some) $1/r^2$ models [21]. It is an open problem to prove the absence of such an intermediate phase for $d > 2$ nearest neighbor models, but there are some interesting partial results (with a connection to dynamic percolation and multiphase flow) which we will now describe.

For $d > 2$ nearest neighbor models, let us define $\beta_c(L)$ to be the critical value for the model in which Z^d is replaced by the slab,

(10) $$Z^{d-1} \times \{-L, -L+1, \ldots, L\},$$

and then define

(11) $$\beta_c' = \lim_{L\to\infty} \beta_c(L).$$

β_c' was introduced in [22] and conjectured to coincide with β_c (clearly $\beta_c' \geq \beta_c$). It was shown in [23] that $\chi' < \infty$ for $\beta > \beta_c'$; thus the open problem from Question 4 for nearest neighbor models is to prove that indeed $\beta_c' = \beta_c$ (for $d > 2$).

The finiteness of χ' is not the only result that has been proved for the disordered regime above β_c'. For example, it is known [23] that for $\beta > \beta_c'$, the quantity $\tau'(x,y)$, defined as

(12) $\tau'(x,y) = \text{Prob}(x$ and y belong to the same <u>finite</u> cluster$)$,

decays exponentially as $||x - y|| \to \infty$. The techniques used in [23] to study the regime above β_c' are themselves of interest. They are based on a dynamic percolation model, called Invasion Percolation, which was originally invented as a model for the microscopic growth of two-fluid interfaces in a porous medium [24; 25; 26].

In the present context, Invasion Percolation may be defined as follows. Assign independent random variables $W_{\{x,y\}}$ to each nearest neighbor bond $\{x,y\}$ with each random variable having an exponential distribution of mean one (so $\text{Prob}(W_{\{x,y\}} \leq w) = 1 - \exp(-w)$). Imagine that fluid no. 1 is injected at the origin and it then invades neighboring bonds (channels) by pushing out the previously present fluid no. 2; $W_{\{x,y\}}$ represents the difficulty of invading the bond $\{x,y\}$. The first bond invaded is the one touching the origin which has the smallest W-value, ..., the kth bond invaded is the one with the smallest W-value, which has not been invaded yet but which touches a previously invaded bond.

A key to the results of [23] is that there is a close relationship between Invasion Percolation and standard percolation (see also [27]). For example, with a given β,

(13) $\tau'(0,y) \leq \text{Prob}($at some time after y is reached by the invading region, a bond with W-value greater than β will be invaded$)$.

Inequality (13) is easily derived by imagining that all (nearest neighbor) bonds $\{x,y\}$ with $W_{\{x,y\}} \leq \beta$ are colored red and realizing that i) the red bonds determine a standard percolation model with parameter β and ii) the (finite) red cluster of the origin must be entirely invaded before any non-red bond is invaded.

The remaining questions we will discuss all involve the critical regime - in particular the critical exponents γ, b (usually denoted β) and δ. There are other critical exponents which we do not discuss for the sake of brevity. Henceforth, we restrict attention to nearest neighbor models, although there are of course interesting questions to be asked concerning the critical regime of $1/r^s$ and other models.

Question 5. Is $\chi(\beta)$ asymptotic, in some appropriate sense, to $(\beta_c - \beta)^{-\gamma}$ for some γ as $\beta \uparrow \beta_c$ (existence of the critical exponent γ)? Is $M(\beta)$ asymptotic to $(\beta - \beta_c)^b$ for some b as $\beta \downarrow \beta_c$? Is $P_n(\beta_c)$ asymptotic to $n^{-(1+1/\delta)}$ for some δ as $n \to \infty$?

Unfortunately, it is still an open problem to prove power law behavior of any of these quantities in any nontrivial models. There are certain simple exactly soluble models, such as percolation on a Bethe lattice (or Cayley tree), where the exact exponents can be computed; the resulting "mean field" values are $\gamma = 1$, $b = 1$ and $\delta = 2$. The next question concerns the belief that above an "upper critical dimension" (whose value is computed via nonrigorous comparisons between percolation and ϕ^3 field theories), short range model exponents should equal these mean field values. The upper critical dimension is usually denoted by d_c.

Question 6. Are $\gamma = 1$, $b = 1$ and $\delta = 2$ for $d \geq 6$ nearest neighbor models (is the upper critical dimension 6)?

This is largely an open problem although there are some potentially useful partial results, as follows. There is a sufficient condition, introduced in [3] (related to a different condition used for Ising models in [28]), which is known to imply that $\gamma = 1$ [3] and also that $b = 1$ and $\delta = 2$ [29]. The condition is that

$$\lim_{\beta \uparrow \beta_c} \sum_x \sum_y \tau'(0,x)\tau'(x,y)\tau'(y,0) < \infty. \tag{14}$$

If it were true that

$$\lim_{\beta \uparrow \beta_c} \tau'(0,x) = O(||x||^{-(d-2)}) \quad \text{as} \quad ||x|| \to \infty \tag{15}$$

(and there are reasons for believing this to be the case for sufficiently large d), then it would follow that condition (14) would be satisfied for $d > 6$. Thus the open problem of Question 6 is (for $d \neq 6$) "reduced" to showing that (15) is valid above dimension six.

There are a number of results which are special to $d = 2$. For example, the critical value is exactly given by $\exp(-\beta_c) = 1/2$ [7]. Although the $d = 2$ nearest neighbor model, unlike the corresponding Ising model, has not been exactly solved (e.g., in the sense of obtaining an exact expression for $M(\beta)$), there does exist a nonrigorous deviation of the presumably exact critical exponents - confirmed numerically (see [30] and the references there) - which is the basis for the following question. The notion of "universality" suggests that these exponent values should also be valid for other $d = 2$ models (of not too long range).

Question 7. <u>For the $d = 2$ nearest neighbor model, is it true that $\gamma = 43/18$, $b = 5/36$ and $\delta = 91/5$?</u>

This is an open problem; no single exponent has been shown to equal its predicted value (or even to exist - see Question 5) for $d = 2$. However, there are a number of identities, called hyperscaling laws or relations, among the exponents which are expected to be valid for $d \leq 6$ and which are consistent with the predicted $d = 2$ values. (Related scaling laws, such as $\gamma = \beta(\delta - 1)$ should be valid for all d.) Most of these relations have been proved for $d = 2$ (assuming the exponents exist) [31; 32]; proving them for $2 < d \leq 6$ is an open problem.

Since the answers to the previous three questions about critical exponents have all been essentially that they are open problems, we ask the following.

Question 8. <u>What then has been proven about critical exponents?</u>

The answer to this question is - mostly, inequalities. Four such inequalities are $\gamma \geq 1$ [3], $b \leq 1$ [33], $\delta \geq 2$ and $b \geq 1/(\delta - 1)$ [6]. These inequalities are valid for any d and for both short and long range models. The first three inequalities, which bound the exponents by their mean field values (see Question 6), are naturally called mean field inequalities. The fourth inequality is also a mean field inequality in the sense that it saturates for the mean field values of b and δ. We discuss this last inequality and its proof in the next section together with some interesting related matters.

3. Phase Transitions as Shock Waves. The proof of the inequality $b \geq 1/(\delta - 1)$ for percolation [6] is closely related to the proof of the analogous inequality for Ising models, which appears in a previously unpublished "announcement of results". A copy of that announcement is included as an appendix to this paper. The proof of the inequality is based on a connection between phase transitions and Burgers' equation which is probably of more interest than the exponent inequality itself. The connection is explored later on in this section.

We begin by briefly describing the relation between the percolation and Ising model phase transitions. Let us first define, in percolation, the quantity

(16) $$M(h,\beta) = 1 - \sum_{n\leq\infty} P_n(\beta)e^{-nh}.$$

Note that $M(0,\beta) = 0$, while

$$M(0+,\beta) = \lim_{h\downarrow 0} M(h,\beta) = P_\infty(\beta) = M(\beta).$$

Thus the phase transition is manifested by the appearance of a jump discontinuity at $h = 0$ in $M(h,\beta)$ (regarded as a function of h in $[0,\infty)$) when β exceeds β_c. The Ising model phase transition is manifested in exactly the same way for a (different) function $M(h,\beta)$ with the following interpretation of the various quantities: M is the magnetization (per site), β is the inverse temperature and h is the external magnetic field.

For both percolation and Ising models, the critical exponents b and δ may be "defined" by

(17) $$M(0+,\beta) \approx (\beta_c - \beta)^b \quad \text{as} \quad \beta_c \downarrow \beta,$$

(18) $$M(h,\beta_c) \approx h^{1/\delta} \quad \text{as} \quad h \downarrow 0.$$

The mean field values of these two exponents for Ising models are $b = 1/2$ and $\delta = 3$ (and the upper critical dimension is 4 [28; 34] rather than 6). Note that in spite of the difference in mean field values between percolation and Ising models, the same inequality, $b \geq 1/(\delta - 1)$, is valid for both and it saturates for both sets of mean field values! The key to understanding the universality of this mean field inequality lies in the relation between phase transitions and Burgers' equation (see Sections I and II of the appendix) as follows.

The mean field Ising model magnetization $M_{MF}(h,\beta)$ may be thought of as an approximation to the actual magnetization of some original (e.g., nearest neighbor) model. M_{MF} can be obtained as the infinite "volume" limit of the magnetization, $M_{CW}(h,\beta)$, of finite volume Curie-Weiss models. As pointed out in the appendix, M_{CW} satisfies exactly Burgers' equation,

(19) $$\partial u/\partial t = -u\, \partial u/\partial x + (\nu/2)\partial^2 u/\partial x^2,$$

providing u is interpreted as the magnetization M, x as (the negative of) the external field h, t as (proportional to) the inverse temperature β, and ν as the inverse volume (number of sites) of the Curie-Weiss model. Burgers' equation for the Curie-Weiss model is an immediate consequence of the fact that

(20) $$M_{CW}(h,\beta) = n^{-1}(\partial/\partial h)\log Z(h,\beta)$$

where (see the appendix)

(21) $$Z(h,\beta) = \int \exp[h(\sum_i s_i) + (\frac{\beta J}{2n})(\sum_i s_i)^2]\Pi\rho(ds_i)$$

satisfies the heat equation,

(22) $$\partial Z/\partial\beta = (\frac{J}{2n}) \cdot \partial^2 Z/\partial h^2.$$

The initial data of (19) or (22) is determined by ρ, which in usual (i.e., "spin 1/2") Ising models, is

(23) $$\rho(ds) = [\delta(s - 1) + \delta(s + 1)]ds/2,$$

leading to $u(x,0) = -\tanh(x)$.

In the limit $n \to \infty$ $(\nu \to 0)$, the mean field magnetization satisfies the conservation law

(24) $$\partial u/\partial t = -u\ \partial u/\partial x.$$

Furthermore, the original Ising model satisfies the sub-conservation law,

(25) $$\partial u/\partial t \leq -u\ \partial u/\partial x.$$

Interpreting the exponents b and δ given by (17)-(18) as describing the growth of a shock just after formation and the "pre-shock" profile just at formation, it is not hard to see that the conservation law (24) leads to the equality, $b = 1/(\delta - 1)$, while the sub-conservation law (25) leads to the inequality, $b \geq 1/(\delta - 1)$. This completes the Ising model derivation, as given in the appendix.

For the case of percolation, it was shown in [6] that <u>again the sub-conservation law is satisfied!</u> Hence the same inequality, $b \geq 1/(\delta - 1)$, results. To see why the mean field percolation exponents saturate the inequality, we explain which (mean field) percolation model obeys the conservation law (24). The percolation analogue of the (Ising) Curie-Weiss model of "volume" n, has n sites with the probability of a passable bond given by our usual formula (1), $1 - \exp(-\beta J_{y-x})$, except that J_{y-x} is taken to be $1/n$ for all x and y. Although we do not know whether the percolation density of the model (or some modification of it) satisfies Burgers' equation with viscosity, it is true that the $n \to \infty$ limit, which we denote M_{MF}, does satisfy the zero-viscosity conservation law (24). M_{MF} may be interpreted as the percolation density of a generalized Bethe lattice model in which the number of passable bonds leading forward from a given site has a Poisson distribution of mean β. It follows that $M = M_{MF}(h,\beta)$ satisfies the self-consistency equation,

(26) $$1 - M = \exp(-h - \beta M),$$

which can be shown to imply the conservation law.

The first two exponent inequalities mentioned after Question 8 above, namely $\gamma \geq 1$ [3] and $b \leq 1$ [33], are also consequences of (other) differential inequalities (but not involving any h dependence). The third inequality, $\delta \geq 2$, can be obtained by combining $b \leq 1$ and $b \geq 1/(\delta - 1)$. It is quite interesting that the Aizenman-Barsky proof of the absence of a disordered intermediate phase [6] is also based on a differential inequality (involving both h and β dependence) - one which simultaneously generalizes the sub-conservation inequality (25) and the differential inequality used in [33].

We remark that there are other critical exponent inequalities for percolation whose proofs are not based on differential inequalities. One such is $\gamma \geq 2(1 - 1/\delta)$ [35]. This inequality, which is saturated by the percolation mean field values ($\gamma = 1$, $\delta = 2$), is not valid for the Ising mean field values ($\gamma = 1, \delta = 3$). Exponent inequalities for percolation, in addition to the ones mentioned in this survey, may be found in [36; 9; 35].

Before concluding this section, we pursue further the relation between the conservation law (24) and mean field phase transitions by adding a few explanatory details to the comment in Section I of the appendix concerning the van der Waals liquid-gas phase transition.

The van der Waals equation of state (with Maxwell's equal area rule automatically taken into account) may be obtained as an infinite volume limit of a Curie-Weiss type model of the following form. Take the partition function $Z(h,\beta)$, where β is the inverse temperature and h represents the <u>chemical potential</u> (times β), in a finite region Λ in $\mathbf{R}^d$, of volume $|\Lambda|$, to be

$$(27)\qquad Z = \sum_{n=0}^{\infty} (e^{hn}/n!) \int_{\Lambda^n} \exp[-\beta \sum_{i,j=1}^{n} V(x_i - x_j)] dx_1 \ldots dx_n .$$

Here $V(x - y)$ is the pair potential between particles at x and y in Λ. The (finite volume) <u>pressure</u> P and <u>density</u> M are given by

$$(28)\qquad P = (\beta|\Lambda|)^{-1} \log Z; \quad M = \beta \cdot \partial P/\partial h.$$

For a fluid consisting of hard spheres (of diameter D) with a Curie-Weiss type attractive interaction,

$$(29)\qquad V(x) = \begin{cases} +\infty, & \text{for } ||x|| < D \\ -a/|\Lambda|, & \text{for } ||x|| \geq D \end{cases},$$

and the partition function may be expressed as

$$(30)\qquad Z(h,\beta) = \int \exp[hN + (\beta a/|\Lambda|)N^2] \rho(dN)$$

where N represents the total number of hard spheres with centers in Λ and ρ is the (unnormalized) measure which assigns to N=n a weight equal to

$$(31)\qquad (n!)^{-1}\int_{\substack{\text{each } x_i\in\Lambda \\ \text{each } ||x_i-x_j||\geq D}} dx_1 \ldots dx_n .$$

As in the Ising model case (compare (21)-(22)), this representation for the partition function leads to a heat equation for Z, a Burgers' equation with viscosity for the finite volume density $M(h,\beta)$,

$$(32)\qquad \partial M/\partial\beta = 2aM \cdot \partial M/\partial h + (a/|\Lambda|) \cdot \partial^2 M/\partial h^2,$$

and a conservation law in the infinite volume limit,

$$(33)\qquad \partial M/\partial\beta = 2aM \cdot \partial M/\partial h.$$

In order to derive an equation which can be directly compared with the usual van der Waals equation of state, we must consider P as a function of M and β rather than M as a function of h and β. To do this let us define $\tilde{P}$ and F by

$$(34)\qquad \tilde{P}(h,\beta) = \beta\, P(h,\beta) = F(M(h,\beta),\beta).$$

To obtain a differential equation for $F(M,h)$, we note first that (from (32) or directly from the heat equation for Z, in the limit $|\Lambda| \to \infty$)

$$(35)\qquad \partial\tilde{P}/\partial\beta = a[\partial\tilde{P}/\partial h]^2 = a\, M^2$$

and second that the (M = constant) characteristics of the conservation law (33) are given by

$$(36)\qquad (dh/d\beta)|_M = -2aM.$$

Combining the last three equations yields

$$(37)\qquad \partial F/\partial\beta = \partial\tilde{P}/\partial\beta + (\partial\tilde{P}/\partial h) \cdot (dh/d\beta)|_M = -a\, M^2$$

and the equation of state,

$$(38)\qquad \tilde{P} = F(M,0) - a\, M^2\beta.$$

Changing into the conventional variables of pressure, $P = \tilde{P}/\beta$, temperature, $T = 1/\beta$, and specific volume, $v = 1/M$, we have

$$(39)\qquad P = T\, f(v) - a/v^2,$$

where $f(v) = F(1/v,0)$ is the initial data for (37), which is determined by the ($|\Lambda| \to \infty$ limit of the) hard sphere measure ρ as defined in (30)-(31). f is not calculable in closed form except when the spatial dimension $d = 1$, in which case

$$f(v) = 1/(v - D). \tag{40}$$

Inserting (40) into (39) gives the standard (modulo choice of units) van der Waals equation of state (see e.g., [37]).

Acknowledgments. Much of the material in this paper was presented on June 3, 1986 at the SIAM workshop on multiphase flow held at Leesburg, VA. Much of the material in the appendix was presented on December 17, 1981 at the 46th statistical mechanics meeting held at Rutgers Univ. [49] and on August 9 and 10, 1982 at the functional integral methods workshop held at the Institute for Theoretical Physics, Santa Barbara, CA. The author thanks the organizers of these workshops and meetings (in particular, George Papanicolaou, Joel Lebowitz and John Klauder) for the opportunity to present this work. He also thanks Mark Kac and Jim Glimm for their comments and suggestions, in 1981, about the results in the appendix. Finally, he thanks Hermann Flaschka and Alan Newell for explaining to him, at an opportune moment, all about the Cole-Hopf transformation. Proposition 1 of the appendix was a consequence of that explanation.

REFERENCES

[1] S. R. Broadbent and J. M. Hammersley, Percolation processes. I.Crystals and mazes, Proc. Camb. Phil. Soc, 53 (1957), pp. 629-641.

[2] L. S. Schulman, Long range percolation in one dimension, J. Phys. A Lett., 16 (1983), pp. L634-L641.

[3] M. Aizenman and C. M. Newman, Tree graph inequalities and critical behavior in percolation models, J. Stat. Phys., 36 (1984), pp. 107-143.

[4] C. M. Newman and L. S. Schulman, One-dimensional $1/|j-i|^s$ percolation models: the existence of a transition for $s \leq 2$, Commun. Math. Phys., 104 (1986), pp. 547-571. See also C. M. Newman and L. S. Schulman, Infinite chains and clusters in one dimensional directed and undirected percolation, in Particle Systems, Random Media and Large Deviations (R. Durrett, Ed.), Contemporary Mathematics, 41 (1985), pp. 297-313.

[5] M. Aizenman and C. M. Newman, Discontinuity of the percolation density in one-dimensional $1/|x-y|^2$ percolation models, Rutgers Univ./Univ. of Arizona preprint (1986), Commun. Math. Phys., submitted.

[6] M. Aizenman and D. J. Barsky, Proof of the sharpness of the phase transition in translation invariant percolation models, Rutgers Univ. preprint, in preparation.

[7] H. Kesten, The critical probability of bond percolation on the square lattice equals 1/2, Commun. Math. Phys., 74 (1980), pp. 41-59.

[8] J. M. Hammersley, Percolation processes. Lower bounds for the critical probability, Ann. Math. Statist., 28 (1957), pp. 790-795.

[9] M. Aizenman, H. Kesten and C. M. Newman, Uniqueness of the infinite cluster and continuity of connectivity functions for short and long range percolation, Rutgers Univ./Cornell Univ./Univ. of Arizona preprint, in preparation.

[10] T. E. Harris, A lower bound for the critical probability in a certain percolation process, Proc. Camb. Phil. Soc., 56 (1960), pp. 13-20.

[11] M. E. Fisher, Critical probabilities for cluster size and percolation problems, J. Math. Phys., 2 (1961), pp. 620-627.

[12] C. M. Newman and L. S. Schulman, Infinite clusters in percolation models, J. Stat. Phys., 26 (1981), pp. 613-628. See also C. M. Newman and L. S. Schulman, Number and density of percolating clusters, J. Phys. A, 14 (1981), pp. 1735-1743.

[13] A. Coniglio, Shapes, surfaces, and interfaces in percolation clusters, in Physics of Finely Divided Matter (M. Daoud, Ed.), Proc. Les Houches Conf. of March, 1985, to appear.

[14] H. Kesten, The incipient infinite cluster in two-dimensional percolation, Theor. Probab. Rel. Fields, to appear.

[15] J. van den Berg and M. Keane, On the continuity of the percolation probability function, Contemporary Mathematics, 26 (1984), pp. 61-65.

[16] L. Russo, On the critical percolation probabilities, Z. Wahrsch. verw. Geb., 56 (1981), pp. 229-237.

[17] D. J. Thouless, Long range order in one-dimensional Ising systems, Phys. Rev., 187 (1969), pp. 732-733.

[18] P. W. Anderson, G. Yuval and D. R. Hamann, Exact results in the Kondo problem. II. Scaling theory, qualitatively correct solution and some new results on one-dimensional classical statistical models, Phys. Rev. B, 1 (1970), pp. 4464-4473; P. W. Anderson and G. Yuval, Some numerical results on the Kondo problem and the inverse square one-dimensional Ising model, J. Phys. C, 4 (1971), pp. 607-620.

[19] M. Aizenman, J. T. Chayes, L. Chayes and C. M. Newman, Discontinuity of the order parameter in one dimensional $1/|x-y|^2$ Ising and Potts models, Rutgers Univ./Cornell Univ./Univ. of Arizona preprint, in preparation.

[20] H. Kesten, Percolation Theory for Mathematicians, Birkhäuser, Boston, 1982.

[21] M. Aizenman, J. T. Chayes, L. Chayes, J. Imbrie and C. M. Newman, An intermediate phase with slow decay of correlations in one-dimensional $1/|x-y|^2$ percolation, Ising and Potts models, Rutgers Univ./Cornell Univ./Harvard Univ./Univ. of Arizona preprint, in preparation.

[22] M. Aizenman, J. T. Chayes, L. Chayes, J. Fröhlich and L. Russo, On a sharp transition from area law to perimeter law in a system of random surfaces, Commun. Math. Phys., 92 (1983), pp. 19-69.

[23] J. T. Chayes, L. Chayes and C. M. Newman, Bernoulli percolation above threshold: An invasion percolation analysis, Ann. Prob., to appear.

[24] R. Lenormand and S. Bories, Description d'un mécanisme de connexion de liaison destiné à l'étude du drainage avec piégeage en milieu poreux, C. R. Acad. Sc. Paris, 291 Série B (1980), pp. 279-282.

[25] R. Chandler, J. Koplick, K. Lerman and J. F. Willemsen, Capillary displacement and percolation in porous media, J. Fluid Mech., 119 (1982), pp. 249-267.

[26] D. Wilkinson and J. F. Willemsen, Invasion percolation: A new form of percolation theory, J. Phys. A, 16 (1983), pp. 3365-3376.

[27] J. T. Chayes, L. Chayes and C. M. Newman, The stochastic geometry of invasion percolation, Commun. Math. Phys., 101 (1985), pp. 383-407.

[28] M. Aizenman, Geometric analysis of ϕ^4 fields and Ising models. Parts I and II, Commun. Math. Phys., 86 (1982), pp. 1-48.

[29] M. Aizenman and D. J. Barsky, in preparation.

[30] D. Stauffer, Scaling properties of percolation clusters, in Disordered Systems and Localization (C. Castellani, C. DiCastro and L. Peliti, Eds.), Springer-Verlag, New York, 1981, pp. 9-25.

[31] H. Kesten, A scaling relation at criticality for 2D-percolation, Cornell Univ. preprint (1985).

[32] H. Kesten, Scaling relations for 2D-percolation, I.M.A. (Minneapolis) preprint (1986).

[33] J. T. Chayes and L. Chayes, An inequality for the infinite cluster density in Bernoulli percolation, Phys. Rev. Lett., 56 (1986), pp. 1619-1622.

[34] J. Fröhlich, On the triviality of $\lambda\phi_d^4$ theories and the approach to the critical point in $d \underset{(=)}{>} 4$ dimensions, Nucl. Phys. B, 200 [FS4] (1982), pp. 281-296.

[35] C. M. Newman, Some exponent inequalities for percolation, J. Stat. Phys., to appear. See also C. M. Newman, Inequalities for γ and related critical exponents in short and long range percolation, in Proc. I.M.A. Workshop on Percolation Theory and Ergodic Theory of Infinite Particle Systems of Feb., 1986 (H. Kesten, Ed.), IMA Volumes in Mathematics and its Applications, Springer-Verlag, to appear.

[36] R. Durett and B. Nguyen, Thermodynamic inequalities for percolation, Commun. Math. Phys., 99 (1985), pp. 253-269.

[37] M. Slemrod, in Proc. SIAM Workshop on Multiphase Flow of June, 1986 (G. Papanicolaou, Ed.), to appear.

[38] C. J. Thompson, Upper bounds for Ising model correlation functions, Commun. Math. Phys., 24 (1971), pp. 61-66.

[39] P. A. Pearce, Mean field bounds on the magnetization for ferromagnetic spin models, J. Stat. Phys., 25 (1981), pp. 309-320.

[40] J. Slawny, On the mean field theory bound on the magnetization, J. Stat. Phys., 32 (1983), pp. 375-388.

[41] H. Tasaki and T. Hara, Mean field bound and GHS inequality, J. Stat. Phys., 35 (1984), pp. 99-107.

[42] A. D. Sokal, More inequalities for critical exponents, J. Stat. Phys., 25 (1981), pp. 25-50.

[43] M. Aizenman and R. Fernández, On the critical behavior of the magnetization in high-dimensional Ising models, Rutgers Univ. preprint (1986).

[44] M. Aizenman, D. J. Barsky and R. Fernández, in preparation.

[45] D. C. Brydges and T. Kennedy, Mayer expansions and Burgers' equation, Univ. of Virginia/I.H.E.S preprint (1986).

[46] R. S. Ellis, C. M. Newman and M. R. O'Connell, The GHS inequality for large external field, J. Stat. Phys., 26 (1981), pp. 37-51.

[47] D. Ruelle, Some remarks on the location of zeroes of the partition function for lattice systems, Commun. Math. Phys., 31 (1973), pp. 265-277.

[48] E. H. Lieb and A. D. Sokal, A general Lee-Yang theorem for one-component and multi-component ferromagnets, Commun. Math. Phys., 80 (1981), pp. 153-179.

[49] C. M. Newman, Phase transitions, shock waves and mean field bounds, J. Stat. Phys., 27 (1982), p. 836.

Appendix. The next four pages are a copy of an announcement which was first distributed at the 46th statistical mechanics meeting held at Rutgers University on December 17-18, 1981. It has not been previously published.

To enhance the usefulness of the announcement, we give some related references: For the results of Thompson and Pearce mentioned before Theorem 2, see [38; 39]. For related results on mean field magnetization bounds, see [40; 41]. For an earlier derivation of the sub-conservation law for the magnetization (but without any tie-in to shock waves) and use of it to obtain critical exponent inequalities, see Section 3 of [42]. For recent applications and extensions of the approach of Section II of the announcement, see [6; 43; 44]. For other uses of Burgers' equation techniques in statistical mechanics, see [45]. For the results of Ellis-Newman-O'Connell mentioned in Section III, see [46]. The result of Ruelle mentioned in Section IV is unpublished work related to [47]; for the results of Lieb-Sokal, see [48].

To facilitate the reading of the announcement, we remark that the fonts used for "greater than or equal" and "less than or equal" signs, as in Theorem 2, appear at first glance to be strict inequality signs, but they are not. The fonts used for strict inequality signs, as in Proposition 1, are different. To check your ability to distinguish between the different fonts, see the first lines of Theorems 5 and 6 (hint: only the inequality signs within the parentheses are strict inequalities).

Shock Waves and Mean Field Bounds: Concavity and Analyticity of the Magnetization at Low Temperature

C. M. Newman

I. The Curie-Weiss Model and Burger's Equation.

Prop. 1 Let $m = m_n(h,\beta) = E(S_1)$ where $\{S_i: i=1,\dots,n\}$ has joint distribution $Z^{-1}\exp[(\beta/2)\sum J_{ij}s_is_j + \sum h_is_i]\,\Pi\rho(ds_i)$ with $J_{ij} = (J/n)$, $h_i = h$ for all i,j and $\int\exp(Ks^2)\,\rho(ds) < \infty$ for all $K > 0$; then

$$(1)\qquad \partial m/\partial\beta = Jm(\partial m/\partial h) + (J/2n)(\partial^2 m/\partial h^2);\ m(h,o) = (d/dh)\int e^{hs}\rho(ds).$$

Remark. Equation (1), whose proof is trivial, is identical to Burger's equation describing the velocity $u(x,t)$ at position x and time t in a one-dimensional compressible gas of viscosity ν if we make the identifications, $m = u$, $h = -x$, $\beta = t/J$, $n = (1/\nu)$. In the limit $n \to \infty$, the mean field magnetization satisfies the conservation law,

$\partial m/\partial\beta = m(\partial m/\partial h)$, so that infinite volume ~ zero viscosity, phase transition ~ shock wave, critical point ~ shock formation, Maxwell's rule ~ jump condition, triple point ~ confluence of two shocks, etc. Similar remarks apply to a van der Waals gas with m replaced by density and h by the chemical potential.

II. Mean Field Bounds

Suppose $h_i = h$, $J_{ij} = J(i-j)$ and the i's denote sites in a rectangle in Z^d with periodic boundary conditions; then $m(h,\beta)$ satisfies

$$\partial m/\partial\beta = Jm(\partial m/\partial h) + \frac{1}{2}\sum_{i,j} J(i-j)\, u_3\,(k,i,j) \tag{2}$$

where $J = \sum_i J(i)$, $u_3(k,i,j) = (\partial^2/\partial h_i \partial h_j)\, E\,(S_k)$ is the third Ursell function, and the last term in (2) is independent of k. Equation (2) can be easily extended to a kind of generalized Burger's equation for the system of $E(S_i)$'s as functions of the h_i's; we do not pursue that point of view here but rather note that the GHS inequality, which is valid when $J(i) > 0$ for certain ρ's, states that $u_3 < 0$ when $h > 0$, so that

$$\partial m/\partial\beta < Jm(\partial m/\partial h) \quad \text{for} \quad h > 0. \tag{3}$$

Inequality (3) and the corresponding equality for mean field models lead easily to a number of mean field bounds. The first extends results of Thompson and of Pearce to a new class of ρ's .

Theorem 2. Suppose $J(i) > 0$ and ρ is such that the GHS inequality is valid, then the infinite volume magnetization $m(h,\beta)$ satisfies the inequality,

$$m(h,\beta) < m_{MF}(h,\beta), \quad \text{for} \quad \beta > 0,\ h > 0+, \tag{4}$$

where m_{MF} is the mean field magnetization with $J = \sum J(i)$ and the same ρ.

The next theorem is of interest, for example, applied to the two-dimensional Ising Model where the spontaneous magnetization is known explicitly.

Theorem 3. Suppose, in addition to the hypotheses of Theorem 2, that $m^+(\beta) \equiv m(0+,\beta) > 0$ for $\beta > \beta_c$; then

(5) $$m(h,\beta_c) \geq m_o(h), \quad \text{for} \quad h > 0+,$$

where $m_o(h)$ is chosen so that the solution of

(6) $$\partial\bar{m}/\partial\beta = J\bar{m}(\partial\bar{m}/\partial h) \quad \text{for} \quad \beta > \beta_c,\ h > 0+;\ \bar{m}(h,\beta_c) = m_o(h)$$

has $\bar{m}(0+,\beta) = m^+(\beta)$. m_o can be determined by the mean field equation,

(7) $$m_o(J(\beta-\beta_c)\ m^+(\beta)) = m^+(\beta).$$

Corollary 4. Define the critical exponents, δ and b (usually denoted β) by

(8) $$m^+(\beta) \sim (\beta-\beta_c)^b \quad \text{as} \quad \beta \to \beta_c+\ ;\ m(h,\beta_c) \sim h^{1/\delta} \quad \text{as} \quad h \to 0+\ ;$$

then

(9) $$b \geq 1/(\delta-1).$$

Remark. For a mean field critical point, $\delta = 3$ and $b = 1/2$; for a mean field tricritical point, $\delta = 5$ and $b = 1/4$; etc. In the two dimensional Ising model, $\delta = 15$ and $b = 1/8$.

III. The GHS Inequality at Low Temperature.

It is known (Ellis-Newman-O'Connell) that for a large class of ρ's, the GHS inequality is valid for $h \geq \bar{h}$ where $\bar{h}$ depends only on ρ. An analysis of mean field models in terms of characteristics (i.e., isomagnetization lines) motivates the following theorem whose proof is based on the techniques of Ellis-Newman-O'Connell.

Theorem 5. Suppose $J(i) \geq 0$ (> 0 for some i) and ρ satisfies the hypotheses of Lemma 1 of Ellis-Newman-O'Connell; then the GHS inequality is valid outside of a certain bounded region in the $h \geq 0$, $\beta \geq 0$ quadrant so that in particular $m(\beta,h)$ is a concave function of $h \geq 0+$ for $\beta \geq \beta_0$ where β_0 depends on ρ and $\{J(i)\}$.

Remark. ρ's which satisfy the required hypotheses include: $\rho(ds) = \exp(-V(s))ds$ with V smooth, even, dV/ds concave for large s and $d^2V/ds^2 \to \infty$ as $s \to \infty$; and

$$\rho(ds) = \sum_{i=-n/2}^{n/2} W_i \delta(s-i), \quad W_i = W_{-i} , \tag{10}$$

with $W_i > 0$ for all i. We also note that for the ρ's of Theorem 5, the inequality of (4) will be valid outside of a (possibly different) bounded region in the quadrant.

IV. The Lee-Yang Property at Low Temperature.

It is known (Ruelle, Lieb-Sokal) that if

$$\text{the zeros of } \int e^{hs}\rho(ds) \text{ lie in } \operatorname{Re} h \leq \bar{h} , \tag{11}$$

then so do the zeros of the partition function. Theorem 5 motivates the following extension whose proof is based on the techniques of Lieb-Sokal.

Theorem 6. Suppose $J(i) \geq 0$ (> 0 for some i) and ρ satisfies both (11) and

(12) for some positive K, $\iint \exp(hs+h's' + Kss')\rho(ds)\rho(ds') \neq 0$ when $\operatorname{Re} h$, $\operatorname{Re} h' > 0$;

then the finite volume (periodic) partition function $Z(\beta,h)$ has no zeros when β and $\operatorname{Re} h$ are outside of a certain bounded region in the $\operatorname{Re} h > 0$,

$\beta > 0$ quadrant so that in particular $m(\beta,h)$ is an analytic function of h in the right half plane, $\text{Re}h > 0$, for $\beta > \beta_o$ where β_o depends on ρ and $\{J(i)\}$.

Remark. One class of ρ's which satisfy the required hypotheses are those given by (10). We conjecture that (11) implies (12) for even ρ's with faster than Gaussian fall-off.

Properties of Time-Dependent PDE Describing Granular Flow

E. BRUCE PITMAN* AND DAVID G. SCHAEFFER**

Abstract. The time dependent partial differential equations of Critical State Theory in soil mechanics are subjected to a linear stability analysis. It is found that, in two space dimensions, these equations are well posed or ill posed according as the stresses lie on the consolidation or expansion side of the yield locus, respectively.

Since this work is being published elsewhere, we only present a summary of the talk. We study the PDE of Critical State Theory [1] in soil mechanics. This theory provides a continuum description of flowing granular materials such as sand, grain, coal, or other raw materials used by manufacturing industries. The partial differential equations in this theory come from conservation of mass and momentum and from constitutive laws that represent rigid-perfectly plastic behavior. We restrict our attention to two space dimensions.

The specific focus of this work is on *time dependent* equations; most earlier work studied the steady state equations. In [4] it was shown that for an *incompressible* granular material, the linearized evolution equations are badly ill posed, in a manner analogous to the mixed forward/back-

---

*Courant Institute of Mathematical Sciences, 251 Mercer Street, New York, NY 10012. Supported by NSF grant DMS-8312229.

**Department of Mathematics, Duke University, Durham, NC 27706.
Supported in part by NSF grant MCS-8401591.

wards heat equations $u_t = u_{xx} - u_{yy}$. Critical State Theory allows for density changes and the resulting changes in yield strength. Although we find that the equations of Critical State Theory still suffer from an instability, the instability is greatly mollified and moreover corresponds to an observed physical phenomenon. In particular, it seems that, in spite of the instability, numerical solution of the equations may reasonably be attempted.

The following 2×2 system (in two space dimensions) provides a model of the kind of instability encountered in Critical State Theory:

$$(1) \qquad \begin{aligned} &(a) \quad u_t = u_{xx} + v_y \\ &(b) \quad v_t = b\, u_y \end{aligned}$$

where b is a constant. Note that (1) is a degenerate parabolic system -- degenerate both because (1b) contains no second order terms and because in (1a) the second order operator u_{xx} is degenerate elliptic. The general solution of (1) is a linear superposition of exponentials

$$e^{i(\xi x + \eta y) + \lambda t}$$

where λ is an eigenvalue of the symbol matrix

$$(2) \qquad S(\xi,\eta) = \begin{bmatrix} -\xi^2 & i\eta \\ ib\eta & 0 \end{bmatrix} .$$

Equation (1) is well posed if and only if

$$(3) \qquad \sup_{\xi,\eta} \max_{i=1,2} \operatorname{Re} \lambda_i(\xi,\eta)$$

is finite. The following proposition shows that well posedness for (1) is determined by the sign of b.

Proposition. If $b < 0$, the supremum in (3) is infinite; while if $b > 0$, it is zero (i.e., $\operatorname{Re} \lambda_i \le 0$).

On an intuitive level, the importance of the parameter b can be seen by considering solutions of (1) which are independent of x. (These are the most problematic since for them the second order term u_{xx} provides no smoothing.) If $b < 0$, then (1) is equivalent to the Cauchy-Riemann equations, while if $b > 0$, then (1) is equivalent to the wave equation written as a first order system.

We shall discuss the proof of the proposition only so far as it clarifies the phenomena. Since the line $\{\xi = 0\}$ is degenerate, we shall for the moment restrict ξ, η to lie in a wedge

$$|\xi| \geq \epsilon|\eta| , \tag{4}$$

where $\epsilon > 0$. Now the finiteness of (3) is determined by the behavior of $\lambda_i(\xi,\eta)$ at infinity. As $(\xi,\eta) \to \infty$ while satisfying (4), the quadratic term on the diagonal of (2) dominates the linear off diagonal terms. Thus we may compute from perturbation theory that

$$\lambda_1(\xi,\eta) = -\xi^2 + O(|\xi| + |\eta|) , \tag{5}$$

so that, regardless of the sign of b, this eigenvalue is bounded from above. Next we substitute (5) into the relation $\lambda_2 = (\det S)/\lambda_1$ to conclude that

$$\lambda_2(\xi,\eta) = \frac{b\eta^2}{\lambda_1} = -\frac{b\eta^2}{\xi^2} + O\left(\frac{1}{|\xi|+|\eta|}\right) . \tag{6}$$

(Note that the leading term in (6) is homogeneous of degree zero.) We see from (6) that if $b < 0$, then

$$\limsup_{\substack{(\xi,\eta) \to \infty \\ \text{satisfying (4)}}} \operatorname{Re} \lambda_2(\xi,\eta) = \frac{|b|}{\epsilon^2} .$$

Letting $\epsilon \to 0$ we reach the stated conclusion concerning the case $b < 0$. If $b > 0$, the RHS of (6) is negative over the set (4). Further analysis is required to show that when $b < 0$, $\operatorname{Re} \lambda_2$ is bounded even without the restriction (4), but we do not give this analysis here.

Note that in the ill posed case $b < 0$, the instability is concentrated along the line $\{\xi = 0\}$; i.e., only exponentials with wave number lying close to this axis are amplified uncontrolably.

After appropriate manipulations, the two dimensional equations of Critical State Theory can be written as a 3 × 3 nonlinear degenerate parabolic system with the density and the two components of velocity as unknowns. This system exhibits behavior similar to (1) in that it is linearly ill posed or well posed according as the stresses lie on the expansion or consolidation side of the yield locus, respectively (terminology of [1]). In the ill posed case, the instability is concentrated along two rays in Fourier transform space which are double characteristics of the steady state equations. This instability presumably corresponds to the observed tendency for overconsolidated granular material to shear discontinuously along these characteristics.

(**Remark**. To resolve discontinuities arising from the instability, one needs to consider a more general theory which accounts for collisions in rapid shearing; Critical State Theory considers only friction--cf. [2].)

Preliminary investigation of the equations in three dimensions suggests that for fully three dimensional flow (i.e., all three eigenvalues of the strain rate tensor nonzero), the instability is further mollified and the equations become linearly well posed.

REFERENCES

[1] R. JACKSON, Some mathematical and physical aspects of continuum models for the motion of granular materials, in Theory of Dispersed Multiphase Flow, R. Meyer, ed. (MRC Seminar, May 1982), Academic Press.

[2] R. JACKSON, Flow of granular materials and aerated granular materials, these Proceedings.

[3] A. JENIKE, Gravity flow of bulk solids, Bull. No. 108 of Utah Eng. Expt. Station, University of Utah, 1961.

[4] D. SCHAEFFER, Instability in the evolution equations describing granular flow, J. Diff. Eq., to appear.

Instabilities in Steady Shearing Flows of Two Fluids

YURIKO RENARDY*

Abstract. We consider two incompressible Newtonian fluids with different viscosities and densities in steady shearing motion. Many interface positions are theoretically allowed but only a few are observed experimentally, thus motivating a study of their stability. Numerical computations, asymptotic analyses and open problems for the stability of various arrangements are discussed.

1. Formulation of equations. The formulation of the equations governing the steady flow of two incompressible Newtonian fluids with different viscosities and densities is presented in detail by Joseph, Nguyen and Beavers [1]. The volume ratio of the fluids is given but the position of the interface is unknown. At the unknown interface, we have the kinematic free surface condition, the continuity of velocity and shear stress, and the jump in the normal stress is balanced by surface tension. Appropriate boundary conditions are imposed to complete this formulation, e.g., the no-slip condition at solid walls, periodicity in the unbounded direction. There are at least five dimensionless parameters: a Reynolds number, a surface tension parameter, volume ratio, viscosity ratio and density ratio. The problem is to find out where the fluids end up, and what physical process governs the arrangement of the fluids and selection of interface shapes.

There has been considerable interest in concentric pipe flow of two fluids as a model for the transport of oil with the addition of water [2,3], as well as the flow

---

* Department of Mathematics, 460 McBryde Hall, Virginia Polytechnic Institute and State University, Blacksburg, VA 24061 - 4097. This research was supported at the Mathematics Research Center, University of Wisconsin - Madison, under the U. S. Army Research Office Contract No. DAAG29-80-C-0041.

of bicomponent polymers [4,5,6,7,8]. Experiments have indicated that the less viscous fluid tends to migrate to the walls and eventually encapsulates the more viscous fluid. On the other hand, there are many theoretically possible flows for this problem, so that we need to explain why certain flows are preferred. A second example of two-fluid flow occurs in ink-jet printing [9] where an electrically charged ink is forced out of a nozzle and the direction of the jet is guided by an electric field. In order to prevent pigments in the ink from clogging up the nozzle, it is bathed in a second fluid, so that the resulting jet consists of two fluids. Experiments indicate that if the outer fluid is the less viscous, the jet tends to be more stable than otherwise. A third example occurs in rheological experiments in extensional flow [10] where hyperbolic streamlines are to be set up inside a rheometer. The boundaries of the apparatus reflect the shape of the desired streamlines but the no-slip boundary condition prevents the streamlines from being hyperbolic at the walls. A small amount of a much less viscous fluid is injected at the walls, migrates along the walls taking up the shear, so that the more viscous test fluid produces the desired flowfield. The question of why the less viscous fluid tends to stay at the walls is not completely understood.

In order to find out why only certain arrangements are observed, there are a couple of routes we can take. The first route is to pick a particular steady flow and study its stability and bifurcation. A second route is to solve an initial value problem for the full set of equations with the time-dependent Navier-Stokes equations, or solve some simplified set of equations. A third route is to use intuitive reasoning based on experience. This paper concerns mainly the first route. In addition, by studying the stability of several different arrangements and flows, we felt we might have enough experience in order to make general conclusions about where the fluids would end up, given a totally new situation. We will see in the following the outcome of such attempts.

2. Two-layer Couette flow. The main part of this paper concerns two-layer Couette flow because we have a more complete picture about this flow than we have of other flows, and it displays some phenomena which occur in more complicated flows. A steady shearing flow solution to the two-layer problem is given by a velocity profile which is linear in each fluid and an interface which is flat. We first examine the linear stability of this arrangement by adding a small perturbation to the velocity field and interface position; the perturbation is to have wavenumber α in the streamwise direction and depend on time through the factor $\exp(\sigma t)$. The linear stability analysis is set up in the usual way as an eigenvalue problem, with σ as the complex-valued eigenvalues. What do we know about σ? The one-fluid Couette flow with no interface is known to be linearly stable at all Reynolds numbers [11]. When there are two fluids with identical properties separated by an interface, the eigenvalues consist of the one-fluid eigenvalues plus what was referred to as the *interfacial mode* by Yih [12]. Here, the interfacial eigenvalue corresponds to the eigenfunction for which the perturbation to the velocity is zero, and the perturbation to the flat interface is wavy. This eigenvalue is neutrally stable when the two fluids have identical properties and there is no surface tension; when the fluids have different properties, the interfacial eigenvalue can become unstable. This can occur at any Reynolds number. In tracking the interfacial eigenvalue numerically, there may be problems at higher Reynolds numbers, such as 1000, because the eigenvalue can cross other eigenvalues in such a way that it becomes impossible to distinguish between it and the others. Therefore, the discussion below will be confined to low Reynolds numbers.

The two-layer Couette flow has three length scales: the plate separation, the thickness of one of the fluids, and a viscous diffusion length scale which is defined using the kinematic viscosity and shear rate in one of the fluids. The long-wave and short-wave asymptotics below are to be interpreted with respect to all three length scales. Yih analyzed the long wave asymptotics of the interfacial eigenvalue [12] by expanding the eigenfunction and eigenvalue in the parameter: α times the Reynolds number for when that and α are small. He had developed this method to look at the stability of film flow down an inclined plane for which the first instability is a long-wave one. For the two-layer problems, however, the first instability is not necessarily a long-wave one. Yih's method has been applied to many problems since, including 2-fluid flows [13, 14, 15, 16, 17, 18, 19]. The effect of the boundaries is important in the long-wave analysis and although the calculations yield closed form expressions for the eigenfunctions and eigenvalues at leading order, they are too complicated to deduce any qualitative behavior immediately, so that computing is necessary. Yih presented computed results when the densities are the same, and they indicate that the arrangement with the less viscous fluid in a thin layer is stable to long waves, although he did not emphasize this. Recently, Hooper [20] analyzed the long wave asymptotics of the semi-infinite Couette flow, in which the lower fluid lay in a horizontal strip and the upper fluid occupied a half-plane. She showed that when the lower fluid is the less viscous, the interfacial eigenvalue is stable to long waves and is unstable otherwise. She allowed the density stratification to be the stable one.

Hooper and Boyd [21] have given a formal method for finding the shortwave asymptotics of the interfacial mode in closed form for shearing flows in general. In this limit, the eigenfunction decays exponentially away from the interface, so that the other boundaries can be neglected in the calculation of the interfacial eigenvalue. Thus, the expression for the dominant terms in σ turn out to be simpler than in the long-wave case. The growth/decay rate (the real part of σ)at leading order consists of essentially three terms: a 'viscous' term of $O(1/\alpha^2)$ which involves the Reynolds number, viscosities, densities and volume ratio of the fluids, a term of $O(1/\alpha)$ involving gravity, and a term of $O(\alpha)$ involving surface tension. The latter term is always negative and thus stabilizing. From what we know about waves in one fluid, we might have expected short waves to be damped by a viscous term, but here they are damped by surface tension. If there is no surface tension, a density difference can influence stability to short waves through the gravity term. In the absence of surface tension or a density difference, the term of $O(1/\alpha^2)$ is positive. Hence, if we consider viscosity stratification alone, the interfacial eigenvalue is unstable to sufficiently short waves at all Reynolds numbers. The physical meaning of this instability is not clear.

In summary, there is a rule of thumb for finding a linearly stable arrangement, at least for the case of equal densities: stabilize the long waves by putting the less viscous fluid in a thin layer, and stabilize the short waves by including enough surface tension. In the case of a slight density difference, we find that the rule of thumb can still give a stable arrangement even when the upper fluid is the heavier. We call this the *thin-layer* effect. An example is presented in Figure 2 in [22] where the eigenvalues were computed using a spectral method based on Chebyschev polynomials [23, 24] to discretize the dependent variables. In the example, the upper more viscous fluid is the heavier. The Froude number is chosen to be not too small because gravity must not dominate the problem; otherwise, the adverse density stratification would destabilize the long waves. When the depth of the lower fluid is small, there is linear stability. As that layer becomes thicker, the flow becomes unstable resulting from gravity. We point out that the above *rule of thumb* does not always work because there can be instabilities that have nothing to do with either of the asymptotic limits. Numerical calculations have revealed such instabilities (see, for example, Figure 1 in [22], and for the semi-infinite Couette flow, [25]).

The above example of thin-layer effect was found numerically. In addition, this effect has been examined analytically for the case of two-layer Couette flow when the two fluids are only slightly different [26]. This is a perturbation problem, where the basic unperturbed problem is the linear stability of two-layer Couette flow where the two fluids have identical properties, for which we know the value of the interfacial eigenvalue. In the perturbed problem, the viscosities of the fluids differ by a small amount, say of $O(\epsilon)$, the densities differ by an $O(\epsilon)$, and surface tension is $O(\epsilon)$. We then expand the interfacial mode for small ϵ. This is a regular perturbation of a simple eigenvalue, and the leading terms can be written down in closed form. The thin-layer limit of this was examined, i.e. a parameter l_1 measuring the thickness of the lower fluid is assumed small compared with other lengths. The result shows that the dominant term, at $O(l_1^2)$, involves the viscosity difference: it is negative and therefore stabilizing if the thin layer is the less viscous and destabilizing otherwise. Terms involving surface tension and density difference are an $O(l_1)$ less. This indicates the importance of viscosity stratification in the stability of thin layers to disturbances that are not too short compared with the l_1.

3. Some other two-fluid flows. We mention briefly further two-fluid flows to illustrate the pronounced effect of viscosity stratification when one of the fluids is in a thin layer. An example is density-matched concentric pipe flow [16, 27]: if the less viscous fluid is in a thin layer next to the wall and there is some surface tension, the flow is linearly stable, whereas if there is too much of the less viscous fluid, then it is unstable, and if the less viscous fluid is centrally located, the flow is unstable. This tendency is reflected in experiments. However, there has been no direct comparison made between the stability analysis and experimental data. One reason may be the difficulty in deducing from the literature on experiments what periodicity if any is in the axial direction, and how the interface is shaped azimuthally. An open problem for vertical pipe flow is to look at the effect of density difference as well as viscosity difference. We expect that the heavier fluid should want to fall faster so that it should tend to move to the center of the pipe, but this may be counteracted by the viscosity stratification.

In two-layer Taylor flow, the fluids are assumed to lie concentrically between concentric cylinders [28]. The outer cylinder is fixed and the inner one rotates. We examined the linear stability of the cylindrical interface. At low speeds, a thin layer effect was found. Close to criticality, what happens can be counter to intuition. i.e. one may expect the flow to become more stable as the amount of the more viscous or stabilizing fluid is increased, and to become more unstable as the amount of the less viscous fluid is increased. However, the onset of Taylor instability in one-fluid flow is promoted by the addition of a thin layer of a more viscous fluid next to the inner cylinder and delayed by the addition of a thin layer of a less viscous fluid there.

The concentric rigid-body rotation is the above Taylor problem but with both cylinders rotating at the same rate [29]. This problem is in a way a rare problem because the stability of the interface is governed by a variational problem which derives from the energy equation. If one uses the energy equation to look at stability, there are integrals along the interface that cannot usually be integrated exactly so that the signs of those integrals are not known, but here, those terms could be integrated exactly. We have derived criteria for when the concentric interface is a local or a global minimizer of the variational problem. It turns out that criticality depends only on the centrifugal term and surface tension, and is independent of the viscosity difference, which makes sense from an analogy with the centrifuge.

Vertical plane Poiseuille flow in three layers [19, 30] may be expected to model vertical concentric pipe flow. Intuitively, we draw an analogy between the plane flow and a cross section of the pipe. However, results concerning linear stability are quite different for the two flows. For the plane flow, consider the case of equal densities. The arrangement with the less viscous fluid at the walls is stable to long waves no matter what the volume ratio is, and otherwise it is unstable. Hence, the instability of a thick layer of the less viscous fluid is absent. Next, consider pure density stratification [30]. Intuition would suggest that for flow downward, the heavier fluid should prefer to be at the center and for flow upward, it should prefer to be at the walls in order to flow in the easiest way, i.e., to maximize the flow rate for a given pressure gradient. However, there are situations where either arrangement is unstable, and there are situations that are counter to expectation which turn out to be linearly stable. This shows how helpful heuristic reasoning can be.

4. Nonlinear effects. Surface tension is an important factor in determining what results from an interfacial instability. If there is a sufficient amount of surface tension, then an instability, if it occurs, involves the longer waves. If periodic boundary conditions are posed, for example, in the streamwise direction for the two-layer Couette flow, then the interaction of a finite number of modes determines what solutions bifurcate from them. On the other hand, if there is no surface tension, then waves of arbitrarily short wavelengths are unstable. In this case, standard methods of bifurcation theory are not applicable, and it seems possible that no smooth interface, steady or unsteady, would be stable. D. D. Joseph has suggested that this might be a mechanism for the formation or sustenance of emulsions in shearing flows.

We will assume in what follows that there is some surface tension. Yih [12] had conjectured that finite-amplitude traveling waves may occur at the interface. For two-layer Couette flow with surface tension, Hooper and Grimshaw [31] have done a formal analysis for weakly nonlinear interaction of long waves. They use multiple scaling to find that the amplitude is governed by the Kuramoto-Sivashinsky equation which also occurs in theories for flame instability. Surface tension must be included in order that unstable waves have a high wavenumber cutoff. As the initial condition, they take the case of linear instability and find that only a finite number of Fourier modes are unstable. These unstable modes are stabilized by energy transfer to higher (stable) harmonics. It is deduced that the final state is either a *two-mode equilibrium* where the mode with the highest growth rate is stabilized by its second harmonic, or a *bouncy* state in which energy is continually exchanged between several two-mode states.

Joseph and M. Renardy [32] have proved a Hopf bifurcation theorem, showing that when one mode becomes unstable, there exist new bifurcating solutions which are traveling interfacial waves. The main part of their work was to derive coercive estimates for the underlying partial differential equations. They give an algorithm for computing the bifurcating solutions but these have not yet been computed and the question of whether they are stable is open.

Bifurcation theory is useful for studying the solutions close to the onset of instability. For a more global understanding of the behavior of solutions, we would like to compute the fully nonlinear evolution of the interface as an initial value problem. One could begin by taking an existing code for free surface flows and modify it to handle a two-layer problem. A possibility is to use parts of a package such as FIDAP [33] which uses some of the finite element techniques for time-dependent free surface flows developed by Scriven et al [34, 35, 36].

Some time-dependent computational works have been done [37, 38, 39]. Dervieux and Thomasset [37] showed the results of a code written for the package MODULEF depicting a two-layer flow undergoing a Rayleigh-Taylor type instability. They presented one computed situation, but due to misprints in their paper, it is not possible to make comments on the significance of their results. Ooms, Segal, Cheung and Oliemans [38] used an order of magnitude argument to derive an equation for the evolution of finite amplitude long waves. They sought periodic solutions superposed on plane Couette-Poiseuille flow with sinusoidal waves as an initial value. Their computations yielded periodic sawtooth-shaped waves that travelled with a constant velocity in the streamwise direction and appeared not to change with time. This solution may be related to sawtooth shaped waves that Hooper and Grimshaw computed. Ooms, Segal, van der Wees, Meerhoff and Oliemans [39] have looked for a new solution to the two-fluid pipe flow problem by applying a lubrication approximation to the steady Navier-Stokes equations with periodicity in the axial direction. Their practical interest concerned oil transport through horizontal pipes with the addition of water. They observed an eccentric arrangement in experiments: in a cross section, the central core was not circular. There was more water below the core than above. In the axial direction, the core was rippled with sawtooth shaped ripples. If the streamlines were all parallel to the axis of the pipe, it would not possible for a heavier fluid to be supported above a lighter fluid. Hence, the irregular shape of the core was required in order to support the layer of water above the core. The core was modelled as a solid, so that the shape of the interface could be chosen, and they chose to use the experimentally observed shape. The velocity field in the water was computed. It was found that when the water layer above was thin, the motion of the rippled core with respect to the pipe wall induced an uneven pressure distribution around the core so that an azimuthal flow was superposed on axial flow. This type of flow may be a steady solution to the original problem.

REFERENCES

[1] D. D. JOSEPH, K. NGUYEN AND B. S. BEAVERS, *Nonuniqueness and stability of the configuration of flow of immiscible fluids with different viscosities*, J. Fluid Mech., 141 (1984), pp. 319-345.

[2] M. E. CHARLES AND P. J. REDBERGER, *The reduction of pressure gradients in oil pipelines by the addition of water: numerical analysis of stratified flow*, Canadian J. Chem. Eng. (1962), pp. 70-75.

[3] D. HASSON, U. MANN AND A. NIR, *Annular flow of two immiscible liquids I. mechanisms*, Canadian J. Chem. Eng., 48 (1970), pp. 514 - 520.

[4] A. E. EVERAGE, Jr., *Theory of bicomponent flow of polymer melts, I. Equilibrium Newtonian tube flow*, Trans. Soc. Rheol., 17 (1973), pp. 629-646.

[5] J. L. WHITE AND B.-L. LEE, *Theory of interface distortion in stratified two-phase flow*, Trans. Soc. Rheol., 19 (1975), pp. 457-479.

[6] J. H. SOUTHERN AND R. L. BALLMAN, *Stratified bicomponent flow of polymer melts in a tube*, Appl. Polymer Symp., 20 (1973), pp. 175-189.

[7] D. L. MACLEAN, *A theoretical analysis of bicomponent flow and the problem of interface shape*, Trans. Soc. Rheol., 17 (1983), pp. 385-399.

[8] N. MINAGAWA AND J. L. WHITE, *Co-Extrusion of unfilled and TiO_2-filled polyethylene: influence of viscosity and die cross-section on interface shape*, Polymer Eng. and Science, 15 (1975), pp. 825 - 830.

[9] C. H. HERTZ AND B. HERMANRUD, *A liquid compound jet*, J. Fluid Mech., 131 (1983), pp. 271 - 287.

[10] K. WALTERS, *Overview of macroscopic viscoelastic flow*, Viscoelasticity and Rheology, ed. Arthur S. Lodge, Michael Renardy, John A. Nohel, Acad. Press (1985), pp. 47 - 80.

[11] V. A. ROMANOV, *Stability of plane-parallel Couette flow*, Func. Anal. & Its Applic., 7 (1973), pp. 137-146.

[12] C-S. YIH, *Instability due to viscosity stratification*, J. Fluid Mech., 27 (1967), pp. 337 - 352.

[13] C.-H. LI, *Instability of three-layer viscous stratified flow*, Phys. Fluids, 12 (1969), pp. 2473 - 2481.

[14] C.-H. LI, *Stability of two superposed elasticoviscous liquids in plane Couette flow*, Phys. Fluids, 12 (1969), pp. 531 - 538.

[15] C.-H. LI, *Role of elasticity on the stability of stratified flow of viscoelastic fluids*, Phys. Fluids, 13 (1970), pp. 1701-1707.

[16] C. E. HICKOX, *Instability due to viscosity and density stratification in axisymmetric pipe flow*, Phys. Fluids, 14 (1971), pp. 251-262.

[17] C. K. WANG, J. J. SEABORG, AND S. P. LIN, *Instability of multi-layered liquid films*, Phys. Fluids, 21 (1978), pp. 1669-1673.

[18] T. W. KAO, *Role of viscosity stratification in the stability of two-layer flow down an incline*, J. Fluid Mech., 33 (1968), pp. 561 - 572.

[19] P. T. THAN, D. D. JOSEPH AND F. ROSSO, *Instability of Poiseuille flow of two immiscible liquids with different viscosities in a channel,* to appear.

[20] A. P. HOOPER, *Long-wave instability at the interface between two viscous fluids: Thin layer effects*, Phys. Fluids, 28 (1985), pp. 1613 - 1618.

[21] A. P. HOOPER AND W. G. C. BOYD, *Shear-flow instability at the interface between two viscous fluids*, J. Fluid Mech., 128 (1983), pp. 507-528.

[22] Y. RENARDY, *Instability at the interface between two shearing fluids in a channel*, Phys. Fluids, 28 (1985), pp. 3441 - 3443.

[23] S. ORSZAG, *Accurate solutions of the Orr-Sommerfeld stability equation*, J. Fluid Mech., 50 (1971), pp. 689 - 703.

[24] D. GOTTLIEB AND S. A. ORSZAG, *Numerical Analysis of Spectral Methods: Theory and Applications*, CBMS-NSF Regional Conference Series in Applied Mathematics, SIAM (1977).

[25] A. P. HOOPER, *Shear-flow instability due to a wall and a viscosity discontinuity at the interface*, to appear, J. Fluid Mech.

[26] YURIKO RENARDY, *The thin-layer effect and interfacial stability in a two-layer Couette flow with similar liquids*, submitted.

[27] D. D. JOSEPH, M. RENARDY AND Y. RENARDY, *Instability of the flow of immiscible liquids with different viscosities in a pipe*, J. Fluid Mech., 141 (1984), pp. 309-317.

[28] YURIKO RENARDY AND DANIEL D. JOSEPH, *Two-fluid Couette flow between concentric cylinders*, J. Fluid Mech., 150 (1985), pp. 381 - 394.

[29] D. D. JOSEPH, Y. RENARDY, M. RENARDY AND K. NGUYEN, *Stability of rigid motions and rollers in bicomponent flows of immiscible liquids*, J. Fluid Mech., 153 (1985), pp. 151 - 165.

[30] Y. RENARDY, *Viscosity and density stratification in vertical Poiseuille flow*, in preparation.

[31] A. P. HOOPER AND R. GRIMSHAW, *Nonlinear instability at the interface between two viscous fluids*, Phys. Fluids, 28 (1985), pp. 37 - 45.

[32] M. RENARDY AND D. D. JOSEPH, *Hopf bifurcation in two-component flow*, SIAM J. Math. Anal. (1986).

[33] M. S. ENGELMAN, *FIDAP – A fluid dynamics analysis package*, Adv. Eng. Software, 4 (1982), pp. 163.

[34] H. SAITO AND L. E. SCRIVEN, *Study of coating flow by the finite element method*, J. Comp. Phys., 42 (1981), pp. 53-76.

[35] C. KOBAYASHI AND L. E. SCRIVEN, *Stability of film flow with variable viscosity*, Paper No. 20d, A.I.Ch.E. Annual Meeting, New Orleans, Louisiana (1981).

[36] H. S. KHESHGI AND L. E. SCRIVEN, *Penalty finite element analysis of unsteady free surface flows*, Finite Elements in fluids - Volume 5, ed. R. H. Gallagher, J. T. Oden, O, C, Zienkiewicz, T. Kawai, and M. Kawahara, John Wiley and Sons Ltd. (1984), pp. 393 -434.

[37] A. DERVIEUX AND F. THOMASSET, *A finite element method for the simulation of a Rayleigh-Taylor instability*, Lecture Notes in Mathematics 771, Springer-Verlag (1980).

[38] G. OOMS, A. SEGAL, S. Y. CHEUNG AND R. V. A. OLIEMANS, *Propagation of long waves of finite amplitude at the interface of two viscous fluids*, Int. J. Multiphase Flow, 11 (1985), pp. 481 - 502.

[39] G. OOMS, A. SEGAL, A. J. VAN DER WEES, R. MEERHOFF AND R. V. A. OLIEMANS, *A theoretical model for core-annular flow of a very viscous oil core and a water annulus through a horizontal pipe*, Int. J. Multiphase Flow, 10 (1984), pp. 41 - 60.

Flow in Hele–Shaw Cells: A Canonical Model for Porous Media (?)

LEONARD SCHWARTZ*

For a number of years, researchers in the oil industry have been concerned with the problem of one fluid pushing another out of a porous medium. Because the problem is one of great complexity, simplified nonlinear models have been formulated including the Hele-Shaw system to be discussed here. We shall begin by defining the problem and presenting the governing equations for the simplest case of flow in a straight channel. The discussion will be extended to consider the role of systemic noise and also the change in behavior in a five-spot flow geometry. Because the problem of interest is "almost ill-posed" at relatively high flooding speeds, low-level noise leads to gross changes in the output parameters of the sytem. Frontal profiles then change from floods with smooth interfaces to jagged patterns with many small fingers. In addition to its relevance to questions of oil production, this problem has recently been adopted by members of the theoretical physics community as a prototype for pattern formation in nonlinear dynamical systems.

It is well known that single-phase flow through porous media is governed by Darcy's law, i.e. fluid velocity is proportional to pressure gradient. Mass conservation then leads to a potential flow problem. Of interest to oil production is the case where one liquid is used to push another out of a porous oil-bearing formation or "reservoir." When one fluid is used to push another a two-phase flow problem arises where, in general, the driving and driven fluids are intimately mixed on the pore level. However, a useful

*Exxon Corporate Research - Science Laboratories
Annandale, New Jersey 08801

idealization is to assume that the fluids are immiscible and that the microscopic displacement process is complete. This leads to two potential-flow domains separated by a moving boundary, the time evolution of which needs to be determined as part of the solution. While formally a viscous creeping flow, it is mathematically analogous to inviscid free-surface and interfacial wave problems and also to flows resulting from the Rayleigh-Taylor instability.

Of greatest interest is the case where the pusher fluid is the less viscous of the two. The fluid interface is then unstable to small perturbations. Often called the Saffman-Taylor instability [1], the resulting flow exhibits viscous "fingers" where the protrusions of the driving liquid grow without limit. When viewed on a macroscopic scale, much of the driven fluid is bypassed by the moving front, leading to reduced values of "sweep efficiency."

A Hele-Shaw cell, consisting of two glass plates separated by a small uniform distance, is an experimental device used to visualize unstable displacements. In a rectilinear channel a single dominant finger forms and, at sufficiently low flow rates, may be observed to propagate without change of form. The standard idealization neglects the wetting layer of the driven fluid left behind on the walls of the device and averages the flow properties in the direction perpendicular to the plates. (This is a highly simplified boundary condition which, while qualitatively correct for Hele-Shaw cells, is not directly applicable to porous media flows.) The resulting potential flow problem is two-dimensional. The viscous contribution to the normal stress balance on the interface is neglected; thus the pressure jump across the interface is given as the product of the interfacial tension and the curvature. Because the differential system is second-order, the tangential stress balance and the side-wall no-slip conditions must be abandoned. For a given initial interface shape, cell geometry and viscosity ratio, the time-dependent problem utilizes the standard kinematic boundary condition to advance the moving boundary. Only a single dimensionless parameter governs the subsequent motion, the ratio of viscous to surface tension forces, i.e. a capillary number. Within the porous-medium flow community, this system, including the above boundary conditions, is usually referred to as the Hele-Shaw equations.

Several flow geometries have been considered [2,3,4]. As an example, we will state the problem that has been treated most extensively, that of fingering in a rectilinear channel when the pusher fluid is taken to be inviscid. Darcy's law takes the form

$$\bar{v} = -M\nabla p, \tag{1a}$$

$$M = b^2/(12\mu) \tag{1b}$$

where b is the plate spacing and μ is the liquid viscosity. Assuming the liquid to be incompressible, the continuity equation is

$$\overline{\nabla}\cdot v = 0 \cdot \tag{2}$$

Combining the two equations, we obtain Laplace's equation for the pressure in the liquid:

$$\nabla^2 p = 0 \quad \cdot \tag{3}$$

On the interface we have

$$p(x,y) = \sigma \frac{d\theta}{ds}, \qquad (x,y) \in \partial D \cdot \tag{4a}$$

where s is the arc length and θ is surface inclination. On the moving interface ∂D the kinematic boundary condition, that a particle on this boundary remain on the boundary for all time, must also be satisfied. Restated, this condition is that the normal component of fluid velocity, with cartesian components (u,v), of a particle occupying a point on the surface, is equal to the normal component of the surface velocity (x_t,y_t) at that point. Thus

$$(u,v)\cdot\overline{n} = (x_t,y_t)\cdot\overline{n}, \qquad (x,y) \in \partial D \tag{4b}$$

where $\bar{n}$ is a normal vector to ∂D and subscripts signify time differentiation. Because Hele-Shaw flow is a subset of creeping motion, inertial contributions to the interface motion are neglected and surface velocity components found by solving (3), at each instant of time, yield the instantaneous surface motion according to (4b).

On the side walls of the cell, the normal component of fluid velocity is zero, i.e.

$$\frac{\partial p}{\partial x} = 0 , \qquad x = \pm L/2 \cdot \tag{5}$$

At upstream infinity, we assume constant velocity,

$$M\nabla p \to -\frac{Q}{bL}\,\overline{y} \qquad y \to +\infty \quad , \tag{6}$$

where Q is the volumetric flow rate into the channel and $\overline{y}$ is a unit vector in the direction of increasing y.

We have solved this system numerically, using a boundary- integral technique to treat the imbedded linear problem. Time integration is done implicitly. Methods involving both distributed sources and distributed vorticity have been developed and each has certain advantages. The scheme is essentially Lagrangian and a key feature involves periodic redistribution of the boundary nodes. For a rectilinear channel, a slightly-perturbed interface evolves into a steady-state propagating finger, in qualitative agreement with experimental results. Previous stability analyses, suggesting that these fingers, if formed, should be unstable, can be understood by recognizing that, while certain small disturbances can be found that will grow initially, ultimately they are "left behind" by the moving finger and will decay.

When the viscosity ratio is unfavorable [1,5], it can be shown that the problem without surface tension is ill-posed in the sense that large wavenumber disturbances grow most rapidly. Because the effective surface tension force is very small in laboratory-scale flows (and even smaller when oil-field dimensions are considered), results are critically sensitive to the ambient level of "noise." Thus we have included a second parameter in the model to represent the noise level. In Hele-Shaw cell experiments this noise may arise via imperfections in the device or vibrations in the environment. Since no actual porous medium is totally homogeneous, noise in porous media flows may be identified with local permeability variations. Computations [3] reveal that, for a given value of capillary number, there is a critical noise level. When this critical level is exceeded, the displacement front becomes jagged, with many side branches. Such flows have recently been observed experimentally [6] in linear Hele-Shaw cells at relatively large capillary number, and are similar in form to results of the simulation.

Of more direct interest to porous media displacements that are relevant to the questions of efficiency when water is used to displace oil, for example, are flows in other geometries. In a linear channel, the fraction of the cell contacted by the less viscous fluid often is not a strong function of the flooding speed or the noise level. Many oil fields are exploited using a pattern of wells where certain of these wells are used to inject water while neighboring wells produce oil. Such a situation is depicted in figure 1a which shows a "five-spot" pattern. The single injection well is idealized as a source of relative strength 4 while the production wells, placed symmetrically as shown, are sinks of strength -1. No steady state is ever achieved here; the moving front begins as a small circle surrounding the injector which expands with time, ultimately breaking-through to a producer. In the simplest case, where the driver fluid viscosity is negligible, breakthrough terminates oil production and the dependence of breakthrough sweep efficiency on flooding speed is the quantity of interest. The single dimensionless control parameter, in the absence of noise, for this geometry is

$$\tau = \sigma M/QL \tag{7}$$

where Q is now the inflow rate to each producer and L is the distance between injector and producer. Taking L as reference length, quadrant symmetry is assumed to reduce computational cost. This is conveniently accomplished by use of the conformal map

$$\zeta = z^{-2}$$

leading to the flow configuration shown in figure 1b.

As a computational example, figure 2a shows successive profiles leading ultimately to breakthrough, for $\tau = 0.001$. The sweep efficiency, defined as the contacted area referenced to the area of the square shown, is 71 %. This may be compared to an identical computation with imposed noise. It is possible to relate the noise magnitude to the permeability variation in a Hele-Shaw cell or porous medium. The permeability k, defined as the product μM, is simply related to the plate spacing for a Hele-Shaw cell by (1b). At each instant of time points on the interface are displaced by small random amounts. It can be demonstrated [4] that the average fractional permeability variation is related to the size of the random displacements according to

$$\langle \delta k/k \rangle = \delta/(2\Delta t) \tag{8a}$$

where

$$x = x_0 + \delta A v. \tag{8b}$$

Here v is the instantaneous surface velocity, Δt is the time interval for the introduction of noise, A is a random number in the interval (-1,1), and $x-x_0$ is the magnitude of the random displacement. The noise level δ is considered to be defined by (8b). In figure 2b we show the same computation but with a noise level equivalent to a 0.5 % permeability variation. The flow pattern is changed markedly and the sweep efficiency is reduced to 69 %. When the flooding speed is increased by a factor of twenty, corresponding to reducing τ to 5×10^{-5}, the resulting flood is as shown in figure (2c). Only the profiles prior to breakthrough are shown for clarity. The sweep efficiency for this flood is 49 % and the multiple-fingering pattern is reminiscent of the experimental results of Habermann [7] who performed experiments in cells filled with packed sand.

It is possible, by perturbation methods [4], to estimate the critical value of noise for transition from smooth profiles, such as shown in figure 2a, to the noise-induced multiple fingering that results in reduced sweep efficiency. The mechanism for stabilization of the front is the suction of the remote wells, while the rate of growth of small disturbances can be obtained from linear stability theory. In the present case, the four-lobed suction pattern will tend to draw the interface away from the noisy perturbations. Only if the order of noise growth exceeds the rate of development of the stable four-lobed pattern, can unstable fingering be expected. Just as in the linear channel, we believe that patterns of the type shown in figure 2 are sta-

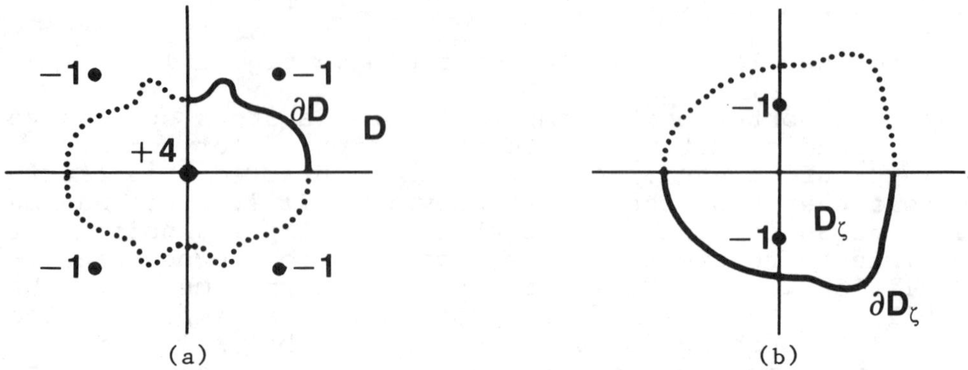

FIG. 1. The five-spot pattern in the physical and transformed planes; (a) z-plane, (b) ζ-plane.

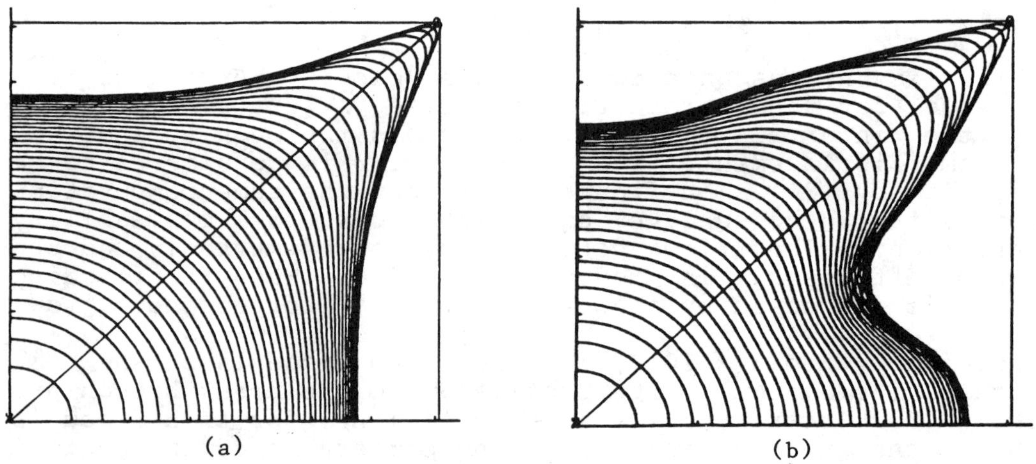

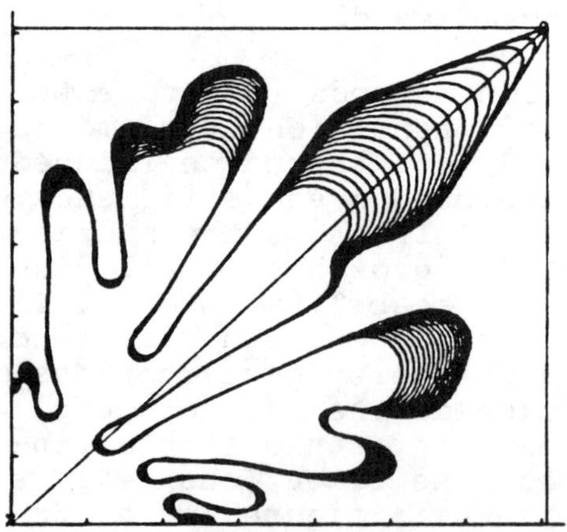

FIG. 2. Progression of the displacement front for one-quarter of a five-spot pattern.
(a) $\tau = 0.001$, no noise;
(b) $\tau = 0.001$, small noise;
(c) $\tau = 5 \times 10^{-5}$, small noise.

ble to low level noise and nonlinearly unstable to noise of sufficient amplitude. In figure 3 we show the dependence of sweep efficiency η on τ as flooding speed is increased. The solid curve is drawn from a number of runs made without introduced noise. Also shown, as a dashed line in the figure, are results obtained when the initial profile contains a small bump of amplitude 0.001. The effect of this disturbance on η is quite pronounced for small values of τ. In all cases tested, with this bump, efficiencies are lower than for the stable patterns; because the imposed disturbance is very small, perhaps comparable to imperfections in a laboratory Hele-Shaw cell, this dashed curve is expected to be more representative of experimental results at low τ values. For relatively large τ, on the other hand, the modes excited by this initial disturbance quickly decay. Thus the bump is seen to have negligible effect for $\tau \geq 0.005$.

Results of runs made with noise equivalent to a two per cent permeability variation for $\tau = 0.001$ are also shown in figure 3. They are presented as an average value with standard deviation error bars using a 10-run sample. Note that the magnitude of the error bars is consistent with the difference between the stable and perturbed results for this value of τ.

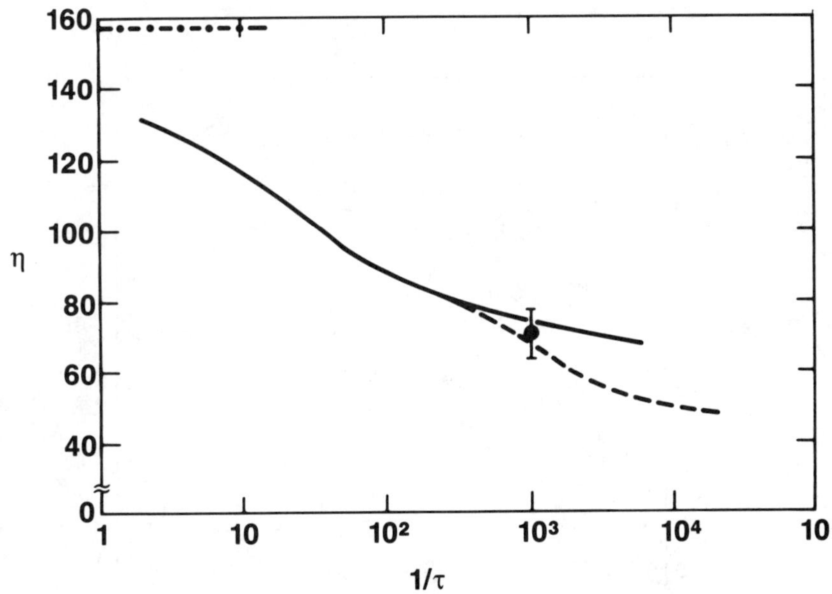

FIG. 3. Sweep efficiency η versus τ^{-1}. unperturbed solutions; - - - results with slightly perturbed initial profile; for $\tau = 0.001$ mean value and standard deviation for 10 runs at noise level equivalent to 2 % permeability variation is shown.

Recently we have extended these calculations to finite viscosity ratio cases [8]. The model now includes 3 parameters. First is the dimensionless surface tension $\tau = \sigma b^2/(12\mu_1 QL)$ where μ_1 is the viscosity of the driven phase. the second parameter is the viscosity ratio $m = \mu_2/\mu_1$ which is less than one when the driver is less viscous. Finally we again use, as a measure of the noise, the amplitude of a small bump, placed on the initial profile. A number of runs were made, for a variety of bump heights. The objective was to determine the critical disturbance amplitude as a function of τ and m. Results are shown in figure 4. For three different combinations of τ and m, we calculate breakthrough sweep efficiency η as a function of A, the magnitude of the imposed disturbance. For the cases corresponding to the smallest values of τ and m, a disturbance amplitude less than 10^{-7} is sufficient to produce a significant change in η. (The fact that the initial alteration is an increase rather than a decrease in η is just a peculiarity of the shape of the disturbance at these values of τ and m.)
Reducing the flooding speed by a factor of five, i. e. $\tau = 0.001$, increases the "tolerable" noise amplitude by about 3 orders of magnitude. Similarly, increases the driver viscosity, i. e. m = 0.2, also increases the stability as expected. For stable displacements, note that η increases as flooding speed is reduced or driver viscosity is increased. We feel that a more important effect, however, is the large increase in critical noise amplitude. For giv-

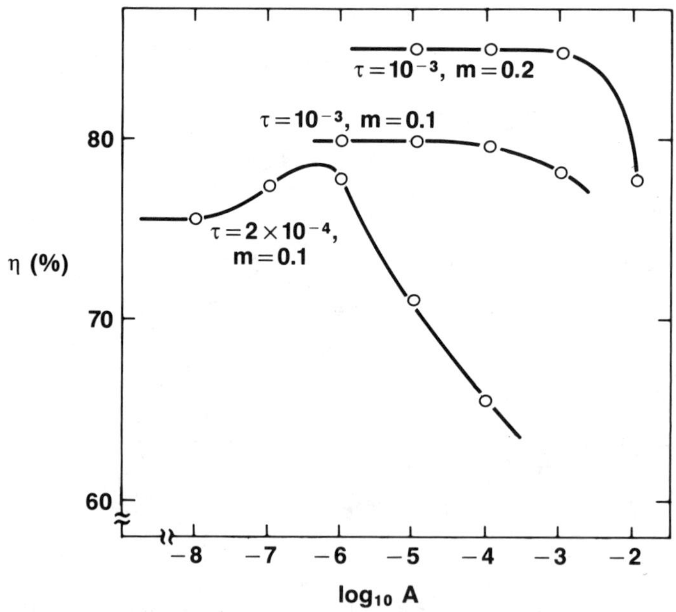

FIG. 4. Variation in sweep efficiency with disturbance amplitude for three combinations of τ and viscosity ratio m.

en values of spatial permeability variation, i. e. noise, and injection rate, the viscosity ratio will determine whether the displacement front will be stable or will break up into a tree-like structure.

Care must be taken in attempting to apply these results to predicting flows in actual porous media. While we believe we now understand the roles of flooding speed, surface tension and permeability variation in Hele-Shaw cells, we expect both quantitative and qualitative differences in porous-media flows. For example, at very low flooding speeds, capillarity can be destabilizing, as shown experimentally by Lenormand et al [9] using lucite network models. Conversely, at somewhat higher speeds, their experiments exhibit true viscous fingering, implying that capillarity has a stabilizing effect. At this stage we view the Hele-Shaw equations as providing a model nonlinear system, the mathematical solution to which exhibits many of the phenomena observed in both Hele-Shaw and porous media experiments. Even when no noise is introduced into the computations, the specific details of the fingering patterns are strongly sensitive to the initial conditions; the large subsequent magnification of disturbances arises because the problem is "almost ill-posed" when the capillary number is large.

While the time-dependent solutions of the Hele-Shaw equations, with noise, reproduce the qualitative features observed in experiments, the numerical value of the width of a steadily propagating finger in a straight channel is consistently underestimated. This has been conjectured to be due to the neglect of the additional interface pressure jump associated with the residual layer of displaced liquid left behind on the cell walls. An asymptotic result of Bretherton [10] predicts the magnitude of this effect. It can be shown that the basic Saffman-Taylor instability of a slightly-perturbed steady-propagating plane interface is significantly altered by wetting-layer phenomen [5]. By altering the numerical algorithm so that this correction can be applied locally, as an "inner solution", an improved match with the available data is obtained. Because this added pressure jump depends on the local value of speed at each point on the front, the imbedded potential problem becomes nonlinear; it can, however, be treated successfully using Newton iteration [11]. It happens that, for virtually all Hele-Shaw experiments to date, this "correction" is the dominant contribution to the interface pressure jump. Yet the fingering phenomena and the stability properties of the fingers, once formed, remain similar to that produced by use of the standard Hele-Shaw equations. This suggests that a class of nonlinear problems of this type may exhibit the same canonical behavior.

REFERENCES

[1] P. G. SAFFMAN and G. I. TAYLOR, The penetration of a fluid into a porous medium or Hele-Shaw cell containing a more viscous liquid, Proc. R. Soc. (London) A 245, (1958) 312-329.

[2] A. J. DEGREGORIA and L. W. SCHWARTZ, A boundary integral method for two-phase displacement in Hele-Shaw cells, J. Fluid Mech. 164 (1986), 383 -400.

[3] A. J. DEGREGORIA and L. W. SCHWARTZ, Finger breakup in Hele-Shaw cells, Physics of Fluids 28, (1985) 2313 - 2314.

[4] L. SCHWARTZ and A. DEGREGORIA, A Hele-Shaw model of sweep efficiency, submitted to Physics of Fluids 1986.

[5] L. SCHWARTZ, Stability of Hele-Shaw Flows; the Wetting-Layer Effect, to appear Phys. Fluids, Sept. 1986.

[6] C. W. PARK and G. M. HOMSY, The instability of long fingers in Hele-Shaw flows, Phys. Fluids 28 (1985) pp. 1583-1585.

[7] B. HABERMANN, The efficiency of miscible displacements as a function of mobity ratio, Trans. AIME 219 (1960), pp. 264-272.

[8] L. SCHWARTZ and A. DEGREGORIA, Mobility ratio effects on fingering and sweep efficiency in Hele-Shaw cells, submitted to Phys. Fluids, 1986.

[9] R. LENORMAND, C. ZARCONE, and A. SAAR, Mechanisms of the displacement of one fluid by another in a network of capillary ducts, J. Fluid Mech. 135 (1983), pp. 337-353.

[10] F. P. BRETHERTON, The motion of long bubbles in tubes, J. Fluid Mech. 10 (1961), 166.

[11] L. SCHWARTZ and A. DEGREGORIA, Simulation of Hele-Shaw fingering with wetting-layer effects included, submitted to Phys. Rev. Lett., 1986.

Computational Fluid Dynamics for Multiphase Flow Systems

WILLIAM T. SHA*

Abstract. Multiphase flow is receiving ever increasing attention, and has become one of the focal points in the engineering community. The driving force behind this trend is the fact that many industrial and commercial applications are multiphase flow systems. In addition, recent advances in computer technology, as well as numerical analysis, have allowed the use of computational fluid dynamics as an essential tool for analyzing and optimizing multiphase flow systems. In spite of both the attention and importance placed on multiphase flows, the current status of knowledge and understanding of multiphase flow systems in general, and computational fluid dynamics in particular, are far from satisfactory. To accelerate the acquisition of urgently needed know-how in multiphase flow, it is necessary to prioritize our efforts and resources. The purpose of this paper is to identify research and development needs in the area of computational fluid dynamics for multiphase flow systems.

Two areas in computational fluid dynamics for multiphase flow systems have been identified--development of governing equations and closure relations, and development of efficient numerical solution procedures. The first area requires critical evaluation of various averaging processes and development of a rational set of multiphase flow governing equations that is sufficiently general and widely applicable. Furthermore, it is important to develop a corresponding set of closure relations based on fundamental understanding of physics, and these closure relations should be validated by carefully planned separate and integral experiments. The second area needs development of new, efficient solution procedures for a set of

*Argonne National Laboratory, 9700 South Cass Avenue, Argonne, Illinois 60439. This work sponsored by the U. S. Nuclear Regulatory Commission under Interagency Agreement DOE 40-550-75

multiphase governing equations with optimum utilization of supercomputers with parallel processors. Some of the guidelines for developing these solution procedures are outlined in this paper.

I. Introduction. In the past two or three decades, interest in multiphase flow has grown rapidly, reflecting the importance of such flows in many industrial and commercial applications. Recent advances in computer technology and numerical analysis have made computational fluid dynamics an essential tool for analyzing multiphase flow systems. Despite the fact that there are now tens of thousands of publications in the area, there still remain major uncertanties in the interpretation and extrapolation of available information. This state of affairs reflects the great complexity of multiphase flow. It should be noted that even single-phase turbulent flows still defy predictions from a fundamental point of view.

II. Objective. The objective of this paper is to identify research and development areas that are urgently needed in computational fluid dynamics of multiphase flow.

III. Research and Development Needs. Two areas of research and development are urgently needed for computational fluid dynamics for multiphase flow systems: governing equations and closure relations, and efficient numerical solution procedures.

A. Governing Equations and Closure Relations. Multiphase flows consist of interacting phases that are dispersed randomly in space and time. In principle, the intraphase conservation equations for mass, momentum, and energy and their associated initial and boundary conditions can be written, and the problem is too complicated to permit detailed solutions. In fact, they are seldom needed in engineering applications. A possible approach is to express the essential dynamics and thermodynamics of such a system in terms of local, global quantities. This can be achieved by applying some sort of averaging process.

Indeed, the connection between single-phase and multiphase flow governing equations, i.e., conservation of mass, momentum, and energy equations, is through averaging processes. These averaging processes can be spatial averaging[1-5], time averaging[6], time-volume averaging[7-9], volume-time averaging[7], Boltzmann statistical averaging[10,11], Lagrangian averaging[12,13], etc.

In the literature, we find several formulations of the governing conservation equations for multiphase flow. Differences exist among these formulation procedures because of the different assumptions and averaging processes employed in the formulations. To show these differences, selected widely used time-averaged and local-volume-averaged multiphase flow governing equations are presented below.

1. Time Averaging. A set of time-averaged multiphase flow governing equations is taken from Ishii[6]:

a. Conservation of Mass

$$\frac{\partial\left(\alpha_k \bar{\bar{\rho}}_k\right)}{\partial t} + \nabla \cdot \left(\alpha_k \bar{\bar{\rho}}_k \bar{V}_k\right) = \Gamma_k .$$

b. Conservation of Momentum

$$\frac{\partial\left(\alpha_k \bar{\bar{\rho}}_k \bar{V}_k\right)}{\partial t} + \nabla \cdot \left(\alpha_k \bar{\bar{\rho}}_k \bar{V}_k \bar{V}_k\right) = -\nabla\left(\alpha_k \bar{\bar{P}}_k\right) + \nabla \cdot \left[\alpha_k \left(\bar{\bar{\tau}}_k + \tau_k^T\right)\right]$$

$$+ \alpha_k \bar{\bar{\rho}}_k \bar{g}_k + \left(\bar{V}_{ki}\Gamma_k + \bar{\bar{P}}_{kl}\nabla\alpha_k + M_k^d\right).$$

c. Conservation of Energy $\left(\text{in terms of enthalpy, } \bar{h}_k\right)$

$$\frac{\partial\left(\alpha_k \bar{\bar{\rho}}_k \bar{h}_k\right)}{\partial t} + \nabla \cdot \left(\alpha_k \bar{\bar{\rho}}_k \bar{h}_k \bar{V}_k\right)$$

$$= \nabla \cdot \alpha_k \quad \bar{\bar{q}}_k + q_k^T + \frac{D_k\left(\alpha_k \bar{\bar{P}}_k\right)}{Dt} - \bar{V}_k \cdot \nabla \cdot \left(\alpha_k \tau_k^T\right) + \alpha_k \bar{\bar{\tau}}_k : \nabla \bar{V}_k$$

$$+ \left(\Gamma_k \bar{h}_{ki} + \frac{\bar{\bar{q}}_k''}{L_s}\right) - \bar{\bar{P}}_{ki}\frac{D_k \alpha_k}{Dt} + M_k^d \cdot \left(\bar{V}_{ki} - \bar{V}_k\right) + W_{ki}^T ,$$

where α_k = time fraction of phase k,

$\bar{\bar{\rho}}_k$ = mean density of phase k,

∇ = divergence operator,

$\bar{V}_k$ = weighted mean velocity of phase k,

Γ_k = mass generation of phase k,

t = time,

$\bar{\bar{P}}_k$ = mean pressure of phase k,

$\bar{\bar{\tau}}_k, \tau_k^T$ = average viscous and turbulent stress tensors of phase k,

g_k = body force field of phase k,

$\bar{V}_{ki}$ = weighted mean velocity of phase k at interface,

$\bar{\bar{P}}_{ki}$ = interfacial mean pressure of phase k,

M_k^d = total drag force,

$\bar{h}_k$ = weighted mean enthalpy of phase k,

$\overline{\overline{q}}_k, q_k^T$ = mean conduction and turbulent heat fluxes,

$\frac{D_k}{Dt} = \frac{\partial}{\partial t} + V_k \cdot \nabla$,

$\overline{h}_{ki}$ = weighted mean enthalpy of phase k at interfaces,

$\overline{\overline{q}}_k''$ = average heat transfer per interfacial area,

$1/L_s$ = total area concentration, and

W_{ki}^T = work due to fluctuation in drag forces.

2. **Local Volume Averaging.** A set of local-volume-averaged multiphase flow governing equations with stationary and dispersed solid structures is taken directly from Sha, Chao, and Soo[5]:

a. Conservation of Mass

$$\gamma_v \frac{\partial}{\partial t} \alpha_k \, {}^{3i}\langle \rho_k \rangle + \nabla \cdot \gamma_A \, \alpha_k \, {}^{2i}\langle \rho_k \underline{U}_k \rangle = \gamma_v \, \Gamma_k \ ,$$

where

$$\gamma_v \Gamma_k = - v^{-1} \int_{A_k} \rho_k (\underline{U}_k - \underline{U}_{ki}) \cdot \underline{n}_k \, dA \ .$$

b. Conservation of Momentum

$$\gamma_v \frac{\partial}{\partial t} \alpha_k \, {}^{3i}\langle \rho_k \underline{U}_k \rangle + \nabla \cdot \gamma_A \, \alpha_k \, {}^{2i}\langle \rho_k \underline{U}_k \underline{U}_k \rangle$$

$$= - \gamma_v \nabla \alpha_k \, {}^{3i}\langle P_k \rangle + \nabla \cdot \gamma_A \, \alpha_k \, {}^{2i}\langle \underline{\underline{\tau}}_k \rangle + \gamma_v \, \alpha_k \left({}^{3i}\langle \rho_k \rangle \, \underline{f} \right.$$

$$\left. + \, {}^{3i}\langle \underline{R}_k \rangle \right) + v^{-1} \int_{A_k} \left(-P_k \underline{\underline{I}}_k + \underline{\underline{\tau}}_k \right) \cdot \underline{n}_k \, dA$$

$$- \, v^{-1} \int_{A_k} \rho_k \underline{U}_k (\underline{U}_k - \underline{U}_{ki}) \cdot \underline{n}_k \, dA \ ,$$

where

$$\gamma_v \, \alpha_k \, {}^{3i}\langle \underline{R}_k \rangle = - v^{-1} \int_{A_{wk}} \left(-P_k \underline{n}_k + \underline{\underline{\tau}}_k \cdot \underline{n}_k \right) dA \ .$$

c. Conservation of Energy (in terms of enthalpy)

$$\gamma_v \frac{\partial}{\partial t} \left(\alpha_k \, {}^{3i}\langle \rho_k h_k \rangle \right) + \nabla \cdot \gamma_A \, \alpha_k \, {}^{2i}\langle \rho_k \underline{U}_k h_k \rangle$$

$$= -\gamma_v \frac{\partial}{\partial t}\left(\alpha_k \, {}^{3i}\langle P_k\rangle\right) + \nabla \cdot \gamma_A \, \alpha_k \, {}^{2i}\langle \rho_k \underline{U}_k\rangle - \gamma_v \alpha_k \, {}^{3i}\langle P_k \, \nabla \cdot \underline{U}_k\rangle$$

$$- \nabla \cdot \gamma_v \, \alpha_k \, {}^{2i}\langle \underline{J}_{qk}\rangle + \gamma_v \, \alpha_k \left({}^{3i}\langle J_{Ek}\rangle + {}^{3i}\langle \Phi_k\rangle + {}^{3i}\langle \dot{Q}_{wk}\rangle\right)$$

$$+ v^{-1} \int_{A_k} P_k(\underline{U}_k - \underline{U}_{ki}) \cdot \underline{n}_k \, dA - v^{-1} \int_{A_k} \underline{J}_{qk} \cdot \underline{n}_k \, dA$$

$$- v^{-1} \int_{A_k} \rho_f h_k(\underline{U}_k - \underline{U}_{ki}) \cdot \underline{n}_k \, dA \, ,$$

where γ_v = local volume porosity,

γ_A = directional surface porosity,

v = local averaging volume,

A = area,

α_k = local volume fraction of phase k,

$\underline{U}_k$ = velocity vector of phase k,

$\underline{U}_{ki}$ = interfacial velocity vector of phase k,

$\underline{n}_k$ = unit outward normal vector of phase k,

A_k = interfacial area of phase k,

A_{wk} = interfacial area between phase k and wall (or solid structure),

$\underline{f}$ = field force per unit mass,

$\underline{\underline{I}}_k$ = unitary tensor,

$\underline{\underline{\tau}}_k$ = viscous stress tensor,

$\underline{R}_k$ = flow resistance of solid structure,

$\underline{J}_{qk}$ = heat flux,

J_{Ek} = heat source,

ϕ_k = dissipation rate per unit volume,

$\dot{Q}_{wk}$ = heat transfer rate at wall,

${}^{2i}\langle\ \rangle$ = intrinsic area average,

${}^{3i}\langle\ \rangle$ = intrinsic volume average,

and the remaining symbols have conventional meanings.

It is important to note that α_k used in the time averaging denotes α_k^t and is defined as fractional residence time of phase k, whereas α_k used in the local volume averaging denotes α_k^v, which represents local volume fraction of phase k. Both α_k^t and α_k^v have different physical meanings and are equivalent only if flow is one-dimensional with a uniform velocity field[9].

The first task, therefore, is to critically evaluate various averaging formulations from the points of view of merit, generality, limitations, physical realism, etc., and to develop a rational set of multiphase flow governing equations that are sufficiently general and widely applicable. In the evaluation process, careful consideration must be given to relating averaging procedures and data obtained through various measurement techniques.

As one would expect, information is lost in any kind of averaging process and it is up to the engineers and scientists to devise means to effectively recover this lost information. This is a closure problem that we must address. At present, most closure relations are developed through empirical correlations based on experimental data. These correlations are generally valid within the data base; extrapolation beyond the data base is often not valid. Thus, it is important to devise a set of appropriate relations for closure based on fundamental understanding of physics. This can be achieved either by coordinating and integrating the phenomenological modeling effort supported by carefully planned experiments, or by theoretical studies.

B. Efficient Numerical Solution Procedures. Once a set of multiphase flow governing equations and its appropriate closure relations have been respectively derived and devised, this set of equations and closure relations, coupled with initial and boundary conditions, can be solved numerically as a boundary value problem in space and an initial value problem in time.

Multiphase flow problems can be solved numerically in two distinct ways[14]. One type of numerical solution procedure applies to situations where one phase is continuous while the other phases are dilute and dispersed. Examples are solid particles in fluid or liquid droplets in a gas. The other numerical solution procedure is applicable where all the phases form interpenetrating continua. This occurs in steam/water mixtures, fluidized beds, liquid-gas reactors, etc.

1. Dilute dispersed-phase model. For the cases in which one continuous phase is present, the computational approach is to solve for the flow field of that phase in a Eulerian framework and to track representative particles of the other phases in a Lagrangian manner. This determination of the particle trajectories is based on the dynamics of the individual particle as influenced by the drag exerted by the surrounding continuous phase. Different types of particles can be tracked; they are characterized by their origin, size, chemical composition, and other parameters. The forces exerted by the particles on the continuous phase are reintroduced in the Eulerian flow calculation. However, the interactions between the dispersed phases are neglected. The final solution is obtained by successive repetition of the continuum flow and particle trajectory calculations.

2. **Multifluid model.** The other class of problems involves a number of interpenetrating continua. The computational procedure then consists of calculating, at a given point, separate velocities and other dependent variables for the different phases. Thus, momentum, continuity, and energy equations are solved, in a Eulerian manner, for each phase. The interlinkage between the velocity fields for different phases is introduced primarily through pressure and the interfacial shear. Correspondingly, both density and temperature (internal energy) fields for different phases are closely related through primary interfacial mass and energy transfer and heat source or sink in the system.

A number of numerical solution procedures for multiphase flow systems have been devised[15-18]. Two arbitraily selected solution procedures for the multi-fluid model are presented below.

a. **RELAP5/MOD2 Model[15].** The solution procedure used in the Reactor Excursion and Leak Analysis Program (RELAP5/MOD2) is based on a semi-implicit numerical scheme in which convective terms in both the continuity and energy equations, pressure term in the momentum equations, and all interfacial interaction terms in the continuity, energy and momentum equations are treated as implicit. All remaining terms are treated as explicit. RELAP5/MOD2 is a one-dimensional, time-dependent two-phase flow code; the step-by-step solution procedure is presented below.

1. Taylor series expansions of $\tilde{\rho}_g^{n+1}$, $\tilde{\rho}_f^{n+1}$, $\tilde{T}_g^{n+1}$, $\tilde{T}_f^{n+1}$, and $\tilde{T}^{s,n+1}$ (see Appendix A-a), are substituted into the expanded form (see Appendix A-b) of continuity and energy equations.

2. The system of equations (3 continuity and 2 energy) has the following form:

$$\underset{\approx}{A}\,\underset{\sim}{X} = \underset{\sim}{b} + \underset{\approx}{g}^1\, v_{g,j+1}^{n+1} + \underset{\approx}{g}^2\, v_{g,j}^{n+1} + \underset{\sim}{f}^1\, v_{f,j+1}^{n+1} + \underset{\sim}{f}^2\, v_{f,j}^{n+1}, \tag{1}$$

where

$$\underset{\approx}{A} = \begin{bmatrix} A_{11} & A_{12} & 0 & A_{14} & A_{15} \\ A_{21} & A_{22} & A_{23} & A_{24} & A_{25} \\ A_{31} & A_{32} & A_{33} & A_{34} & A_{35} \\ A_{41} & A_{42} & A_{43} & A_{44} & A_{45} \\ A_{51} & A_{52} & A_{53} & A_{54} & A_{55} \end{bmatrix} \qquad \underset{\sim}{X} = \begin{bmatrix} \tilde{x}_n^{n+1} - x_n^n \\ \tilde{U}_g^{n+1} - U_g^n \\ \tilde{U}_f^{n+1} - U_f^n \\ \tilde{\alpha}_g^{n+1} - \alpha_g^n \\ P^{n+1} - P^n \end{bmatrix}$$

$$\underset{\sim}{b} = \begin{bmatrix} 0 \\ b_2 \\ b_3 \\ b_4 \\ 0 \end{bmatrix} \quad \underset{\approx}{g}^1 = \begin{bmatrix} g_1^1 \\ g_2^1 \\ 0 \\ g_4^1 \\ g_5^1 \end{bmatrix} \quad \underset{\approx}{g}^2 = \begin{bmatrix} g_1^2 \\ g_2^2 \\ 0 \\ g_4^2 \\ g_5^2 \end{bmatrix} \quad \underset{\sim}{f}^1 = \begin{bmatrix} 0 \\ 0 \\ f_3^1 \\ f_4^1 \\ f_5^1 \end{bmatrix} \quad \underset{\sim}{f}^2 = \begin{bmatrix} 0 \\ 0 \\ f_3^2 \\ f_4^2 \\ f_5^2 \end{bmatrix}$$

3. A Gaussian elimination system solver is used to obtain the bottom row of $\underset{\approx}{A}^{-1}$.

4. By multiplying Eq. (1) by $\underset{\approx}{A}^{-1}$, one can verify that just the bottom row of $\underset{\approx}{A}^{-1}$ is needed to obtain an equation that involves only the unknown variables $(P^{n+1} - P^n)$, $v_{g,j+1}^{n+1}$, $v_{g,j}^{n+1}$, $v_{f,j+1}^{n+1}$, and $v_{f,j}^{n+1}$.

5. Velocity in the momentum equation can be expressed as a function of neighboring pressures (see Appendix A-c).

6. Substituting velocity expressions as shown in Step 5 into Step 4 results in a single equation involving only pressures.

7. A sparse matrix solver is used to obtain $P^{n+1} - P^n$ for each volume.

8. By substituting $P^{n+1} - P^n$ into Step 5, the new time velocity v_g^{n+1} and v_f^{n+1} is obtained for each volume.

9. By substituting P^{n+1}, v_g^{n+1} and v_f^{n+1} into Eq. (1), a single vector on the right side is obtained.

10. The Gaussian elimination system solver is called again to obtain the intermediate time variables, $\tilde{x}_n^{n+1}$, $\tilde{U}_g^{n+1}$, $\tilde{U}_f^{n+1}$, and $\tilde{\alpha}_g^{n+1}$.

11. The mixture density ρ_m^{n+1} is calculated from the mixture continuity equation expressed in a conservative form, and it is compared with $\tilde{\rho}^{n+1}$, which is calculated based on Taylor series expansion knowing P^{n+1}, $\tilde{U}_g^{n+1}$, $\tilde{U}_f^{n+1}$, and $\tilde{x}_n^{n+1}$. The difference between ρ_m^{n+1} and $\tilde{\rho}^{n+1}$ is used to provide a time step control based on mass error.

12. The P^{n+1}, $\tilde{U}_g^{n+1}$, $\tilde{U}_f^{n+1}$, and $\tilde{x}_n^{n+1}$ are used to calculate $\tilde{T}_g^{n+1}$, $\tilde{T}_f^{n+1}$, and $\tilde{T}^{s,n+1}$, and thus Γ_g^{n+1}.

13. A set of nonexpanded equations (see Appendix A-d) of continuity and energy are used to calculate the new time variables: x_n^{n+1}, U_g^{n+1}, U_f^{n+1}, and α_g^{n+1}.

14. Time step control:
    a. Material Courant limit check.
    b. Mass error check.
    c. Material properties out of defined ranges.
    d. Water property errors.
    e. Excessive extrapolation of state variables in the metastable regimes.
    f. Change in structure temperature more than 50°K in one time step.

15. Repeat Steps 1 through 14 until calculational objective is achieved.

b. COMMIX-2 Model[16]. The solution procedures used in the COMMIX (Component Mixing) two-phase code are based on the implicit method. COMMIX-2 is a three-dimensional, time-dependent, two-phase flow computer code developed primarily for liquid metal fast breeder reactor (LMFBR) applications. Since sodium has very high conductivity, it seems reasonable to assume that the liquid and vapor phases of sodium are in thermal equilibrium. Therefore, the five-equation model, (two continuity equations, two momentum equations, and one energy equation) is used in COMMIX-2 and presented in Appendix B. Extension of the five-equation model to the two-fluid model (six-equation model) has been in progress for some time. The step-by-step solution procedure of COMMIX-2 described here is limited to the five-equation model.

1. Store all old time values of dependent variables and thermal physical properties:

$$u_g^n = u_g^{n+1} \qquad u_\ell^n = u_\ell^{n+1}$$
$$v_g^n = v_g^{n+1} \qquad v_\ell^n = v_\ell^{n+1}$$
$$w_g^n = w_g^{n+1} \qquad w_\ell^n = w_\ell^{n+1} \qquad \text{etc.}$$

2. Update pressure boundary condition.

3. Calculate coolant thermal physical properties.

4. Calculate evaporation or condensation rate, Γ_g.

5. Calculate interfacial drag (K) between vapor and liquid phases.

6. Solve mixture energy equation for h_m^{r+1} and compute $\Delta h_m = h_m^{r+1} - h_m^r$, where superscript r denotes iteration within a given time step n.

7. Update coolant thermal physical properties as in Step 3 since new enthalpy has been calculated.

8. Update Γ_g.

9. Update K.

10. Calculate coefficients of both liquid and vapor phase momentum equations in x or r direction.

11. Calculate coefficients of both liquid and vapor phase momentum equations in y or θ direction.

12. Calculate coefficients of both liquid and vapor phase momentum equations in z direction.

13. Calculate coefficients of the liquid phase Poisson equation.

14. Calculate coefficients of the vapor phase Poisson equation.

15. Solve the mixture (sum of liquid and vapor phase) Poisson equation by either SOR or direct matrix inversion method for P (pressure field).

16. Solve for u_ℓ^{r+1}, u_g^{r+1}, v_ℓ^{r+1}, v_g^{r+1}, w_ℓ^{r+1}, and w_g^{r+1} by knowing P using momentum equations.

17. Calculate

$$\Delta u_\ell = u_\ell^{r+1} - u_\ell^r, \qquad \Delta u_g = u_g^{r+1} - u_g^r,$$

$$\Delta v_\ell = v_\ell^{r+1} - v_\ell^r, \qquad \Delta v_g = v_g^{r+1} - v_g^r,$$

$$\Delta w_\ell = w_\ell^{r+1} - w_\ell^r, \qquad \Delta w_g = w_g^{r+1} - w_g^r,$$

18. Calculate mixture velocity u_m^{r+1}, v_m^{r+1}, and w_m^{r+1} .

19. Calculate mixture mass residue (DLMAX) from mixture continuity equation.

20. Check for convergence criteria of velocity, enthalpy, and mass residue:

$$\Delta u_{\ell} < \varepsilon_1$$

$$\Delta u_g < \varepsilon_1$$

$$\Delta v_{\ell} < \varepsilon_1$$

$$\Delta v_g < \varepsilon_1$$

$$\Delta w_{\ell} < \varepsilon_1$$

$$\Delta w_g < \varepsilon_1$$

$$\Delta h_m < \varepsilon_2$$

$$\text{DLMAX} < \varepsilon_3$$

21. Go to Step 3 if Step 20 is not satisfied and continue iteration within a given time step.

22. Go to Step 1 if Step 20 is satisfied and continue the calculation until user-specified termination criteria have been met.

Although numerical solution procedures of multiphase flow systems exist for both the dilute dispersed-phase model and the multifluid model, considerable improvement is needed to make the solution procedures efficient and reliable. Alternative solution procedures suitable for supercomputer applications must be developed and tested.

Based on experience in computational fluid dynamics for multiphase flow systems, guidelines for developing new and efficient solution procedures may be summarized as follows:

(a) The numerical scheme should be consistent and conservative in mass and energy inventories. A greater degree of approximation for momentum effects is considered acceptable. Both mass and energy are convected from the same cell and each is evaluated at the same time level--that is, mass density is evaluated at the old time level, so energy density is also evaluated at the old time level[15].

(b) To achieve fast execution speed for a wide range of engineering problems, it is highly desirable to provide both semi-implicit and implicit numerical schemes as users' options. The former is more suitable for fast transient problems and the latter is particularly suitable for steady-state and slow transients.

(c) Any new and efficient numerical solution procedures must take full advantage of the current supercomputer developments, such as parallel processors, etc.

IV. Conclusions. Many commercial and industrial applications are involved in multiphase flow. Multiphase flow is a complex phenomenon. Present knowledge of multiphase flow is far from satisfactory. Therefore, it is important to channel our resources toward fundamental understanding of the physics of multiphase flow, i.e., to develop a rational set of multiphase flow governing equations and closure relations. Simultaneously, efforts should focus on developing efficient and reliable numerical solution procedures for use in benchmarking computer codes for the analysis of multiphase flow systems. These benchmarking computational tools are used in conjunction with carefully planned experiments iteratively for both ascertaining closure relations and understanding interfacial transport phenomena, etc. Once these computational tools are validated with experimental data, they can be used to optimize design and performance, and to enhance safety of multiphase flow systems.

REFERENCES

[1] T. B. ANDERSON and R. JACKSON, A fluid mechanical description of fluidized beds, Ind. Eng. Chem. Fund. Vol. 6 (1967), pp. 527-539.

[2] J. C. SLATTERY, Flow of viscoelastic fluids through porous media, J. AIChE, Vol. 13 (1967), pp. 1066-1071.

[3] S. WHITAKER, Diffusion and dispersion in porous media, J. AIChE, Vol. 13 (1967), pp. 420-427.

[4] S. WHITAKER, A simple geometric derivation of the spatial averaging theorem, Chem. Eng. Educ. (1985), pp. 18-21, 50-52.

[5] W. T. SHA, B. T. CHAO, and S. L. SOO, Local Volume-Averaged Transport Equations for Multiphase Flow in Regions Containing Distributed Solid Structures, Argonne National Laboratory NUREG/CR-2354, ANL-81-69, December 1981.

[6] M. ISHII, Thermo-Fluid Dynamic Theory of Two Phase Flow, Eyrolles, Paris, 1975.

[7] J. M. DELHAYE and J. L. ACHARD, On the averaging operators introduced in two-phase flow modeling, Proc. of OECD Specialists' Meeting on Transient Two-Phase Flow, Toronto, Pergamon Press (1976).

[8] W. T. SHA and J. C. SLATTERY, Local Volume-Time Averaged Equations of Motion for Dispersed, Turbulent, Multiphase Flows,

Argonne National Laboratory NUREG/CR-1491, ANL-80-51, November 1980.

[9] W. T. SHA, B. T. CHAO, and S. L. SOO, Time- and Volume-Averaged Conservation Equations for Multiphase Flow. Part One: System Without Internal Solid Structures, Argonne National Laboratory NUREG/CR-3989, ANL-84-66, December 1984.

[10] A. V. KALININ, Derivation of fluid mechanics equations for a two-phase medium with phase changes, Heat Transfer Soviet Research, Vol. 2, No. 3 (1970), p. 83.

[11] S. I. PAI, Fundamental equations of a mixture of gas and small spherical solid particles from simple kinetic theory, Int. Sym. on Two-Phase Systems, Haifa, Israel (1971), Paper 6-6.

[12] G. F. CARRIER, Shock waves in dusty gas, J. of Fluid Mechanics, Vol. 4 (1958), p. 276.

[13] N. ZUBER, On the dispersed flow in the laminar flow regime, Chem. Engrg. Science, Vol. 19 (1964a), p. 897.

[14] S. V. PATANKER, Personal Communication, April 1986.

[15] V. K. RAMSON et al., RELAP5/MOD2 Code Manual. Volume 1: Code Structure, Systems Models, and Solution Methods, Argonne National Laboratory NUREG/CR-4312, EGG-2396, August 1985.

[16] W. T. SHA ET AL., COMMIX-2: A Three-Dimensional Transient Computer Program for Thermal Hydraulic Analysis of Two-Phase Flows, Argonne National Laboratory NUREG/CR-4371, ANL-85-47, September 1985.

[17] D. R. LILES and W. H. REED, A semi-implicit method for two-phase flow dynamics, J. Comp. Phys., Vol. 26 (1978), pp. 390-407.

[18] J. H. MCFADDEN ET AL., RETRAN-02 - A Program for Transient Thermal-Hydraulic Analysis of Complex Fluid Flow Systems, Vol. 1, Electric Power Research Institute, Palo Alto, California, EPRI NP-1850 CCM, May 1981.

Appendix A

REACTOR EXCURSION AND LEAK ANALYSIS PROGRAM

RELAP5/MOD2

a. Taylor Series Expansion

$$\tilde{\rho}^{n+1}_{g,L} = \rho^{n}_{g,L} + \left(\frac{\partial \rho_g}{\partial P}\right)^{n}_{L} \left(P^{n+1}_{L} - P^{n}_{L}\right) + \left(\frac{\partial \rho_g}{\partial X_n}\right)^{n}_{L} \left(\tilde{X}^{n+1}_{n,L} - X^{n}_{n,L}\right)$$

$$+\left(\frac{\partial \rho_g}{\partial U_g}\right)_L^n \left(\tilde{U}_{g,L}^{n+1} - U_{g,L}^n\right)$$

$$\tilde{\rho}_{f,L}^{n+1} = \rho_{f,L}^n + \left(\frac{\partial \rho_f}{\partial P}\right)_L^n \left(P_L^{n+1} - P_L^n\right) + \left(\frac{\partial \rho_f}{\partial U_f}\right)_L^n \left(\tilde{U}_{f,L}^{n+1} - U_{f,L}^n\right) .$$

$$\tilde{T}_L^{s,n+1} = T_L^{s,n} + \left(\frac{\partial T^s}{\partial P}\right)_L^n \left(P_L^{n+1} - P_L^n\right) + \left(\frac{\partial T^s}{\partial X_n}\right)_L^n \left(\tilde{X}_{n,L}^{n+1} - X_{n,L}^n\right)$$

$$+\left(\frac{\partial T^s}{\partial U_g}\right)_L^n \left(\tilde{U}_{g,L}^{n+1} - U_{g,L}^n\right)$$

$$\tilde{T}_{g,L}^{n+1} = T_{g,L}^n + \left(\frac{\partial T_g}{\partial P}\right)_L^n \left(P_L^{n+1} - P_L^n\right) + \left(\frac{\partial T_g}{\partial X_n}\right)_L^n \left(\tilde{X}_{n,L}^{n+1} - X_{n,L}^n\right)$$

$$+\left(\frac{\partial T_g}{\partial U_g}\right)_L^n \left(\tilde{U}_{g,L}^{n+1} - U_{g,L}^n\right)$$

$$\tilde{T}_{f,L}^{n+1} = T_{f,L}^n + \left(\frac{\partial T_f}{\partial P}\right)_L^n \left(P_L^{n+1} - P_L^n\right) + \left(\frac{\partial T_f}{\partial U_f}\right)_L^n \left(\tilde{U}_{f,L}^{n+1} - U_{f,L}^n\right)$$

b. Governing Equations in Expanded Form.

The sum of phasic continuity equations:

$$\alpha_g \frac{\partial \rho_g}{\partial t} + \alpha_f \frac{\partial \rho_f}{\partial t} + (\rho_g - \rho_f) \frac{\partial \alpha_g}{\partial t} + \frac{1}{A} \frac{\partial}{\partial x} (\alpha_g \rho_g v_g A + \alpha_f \rho_f v_f A) = 0 .$$

The difference of phasic continuity equations:

$$\alpha_g \frac{\partial \rho_g}{\partial t} - \alpha_f \frac{\partial \rho_f}{\partial t} + (\rho_g + \rho_f) \frac{\partial \alpha_g}{\partial t} + \frac{1}{A} \frac{\partial}{\partial x} (\alpha_g \rho_g v_g A - \alpha_f \rho_f v_f A)$$

$$= - \frac{2[H_{ig}(T^s - T_g) + H_{if}(T^s - T_f)]}{h_g^* - h_f^*} + 2\Gamma_w .$$

The noncondensible gas continuity equation*

$$\rho_g X_n \frac{\partial \alpha_g}{\partial t} + \alpha_g X_n \frac{\partial \rho_g}{\partial t} + \alpha_g \rho_g \frac{\partial X_n}{\partial t} + \frac{1}{A} \frac{\partial}{\partial x} \left(\alpha_g \rho_g X_n v_g A\right) = 0 \ .$$

The vapor energy equation:

$$\left(\rho_g U_g + P\right) \frac{\partial \alpha_g}{\partial t} + \alpha_g U_g \frac{\partial \rho_g}{\partial t} + \alpha_g \rho_g \frac{\partial U_g}{\partial t}$$

$$+ \frac{1}{A} \left[\frac{\partial}{\partial x} \left(\alpha_g \rho_g U_g v_g A\right) + P \frac{\partial}{\partial x} \left(\alpha_g v_g A\right)\right]$$

$$= - \left(\frac{h_f^*}{h_g^* - h_f^*}\right) H_{ig}\left(T^s - T_g\right) - \left(\frac{h_g^*}{h_g^* - h_f^*}\right) H_{if}\left(T^s - T_f\right)$$

$$+ \left[\left(\frac{1 + \varepsilon}{2}\right) h_g^s + \left(\frac{1 - \varepsilon}{2}\right) h_f^s\right] \Gamma_w + Q_{wg} + DISS_g \ .$$

The liquid energy equation:

$$- \left(\rho_f U_f + P\right) \frac{\partial \alpha_g}{\partial t} + \alpha_f U_f \frac{\partial \rho_f}{\partial t} + \alpha_f \rho_f \frac{\partial U_f}{\partial t}$$

$$+ \frac{1}{A} \left[\frac{\partial}{\partial x} \left(\alpha_f \rho_f U_f v_f A\right) + P \frac{\partial}{\partial x} \left(\alpha_f v_f A\right)\right]$$

$$= \left(\frac{h_f^*}{h_g^* - h_f^*}\right) H_{ig}\left(T^s - T_g\right) + \left(\frac{h_g^*}{h_g^* - h_f^*}\right) H_{if}\left(T^s - T_f\right)$$

$$- \left[\left(\frac{1 + \varepsilon}{2}\right) h_g^s + \left(\frac{1 - \varepsilon}{2}\right) h_f^s\right] \Gamma_w + Q_{wf} + DISS_f \ .$$

The sum of phasic momentum equations:

$$\alpha_g \rho_g \frac{\partial v_g}{\partial t} + \alpha_f \rho_f \frac{\partial v_f}{\partial t} + \frac{1}{2} \alpha_g \rho_g \frac{\partial v_g^2}{\partial t} + \frac{1}{2} \alpha_f \rho_f \frac{\partial v_f^2}{\partial t}$$

$$= - \frac{\partial P}{\partial x} + \rho B_x - \alpha_g \rho_g v_g FWG - \alpha_f \rho_f v_f FWF - \Gamma_g\left(v_g - v_f\right) \ .$$

---

*It is assumed that noncondensible gas is very dilute and its presence will not alter the properties of other phases. It is further assumed that noncondensible gas is in equilibrium thermally and mechanically with steam in the system.

The difference of phasic momentum equations:

$$\frac{\partial v_g}{\partial t} - \frac{\partial v_f}{\partial t} + \frac{1}{2}\frac{\partial v_g^2}{\partial t} - \frac{1}{2}\frac{\partial v_f^2}{\partial t}$$

$$= -\left(\frac{1}{\rho_g} - \frac{1}{\rho_f}\right)\frac{\partial P}{\partial x} - v_g FWG + v_f FWF$$

$$+ \Gamma_g[\rho v_I - (\alpha_f\rho_f v_g + \alpha_g\rho_g v_f)]/(\alpha_g\rho_g\alpha_f\rho_f)$$

$$- \rho FI(v_g - v_f) - C[\rho^2/(\rho_g\rho_f)]\,\frac{\partial(v_g - v_f)}{\partial t} ,$$

where

$$\frac{\partial\alpha_f}{\partial t} = -\frac{\partial\alpha_g}{\partial t} , \text{ and}$$

$$\Gamma_g = -\frac{H_{ig}(T^s - T_g) + H_{if}(T^s - T_f)}{h_g^* - h_f^*} + \Gamma_{w_o} .$$

c. **Momentum Equations:** In the semi-implicit scheme, the momentum equations can be expressed as

$$v_{g,j}^{n+1} = A^1\left(P_L^{n+1} - P_K^{n+1}\right) + C^1$$

$$v_{f,j}^{n+1} = B^1\left(P_L^{n+1} - P_K^{n+1}\right) + D^1 ,$$

where A^1, B^1, C^1, and D^1 again contain only n time level variables.

d. Governing Equations in Non-Expanded Form

Phasic continuity equations:

$$\frac{\partial}{\partial t}(\alpha_g\rho_g) + \frac{1}{A}\frac{\partial}{\partial x}(\alpha_g\rho_g v_g A) = \Gamma_g$$

$$\frac{\partial}{\partial t}(\alpha_f\rho_f) + \frac{1}{A}\frac{\partial}{\partial x}(\alpha_f\rho_f v_f A) = -\Gamma_g .$$

Phasic energy equation:

$$\frac{\partial}{\partial t}(\alpha_g\rho_g U_g) + \frac{1}{A}\frac{\partial}{\partial x}(\alpha_g\rho_g U_g v_g A)$$

$$= - P \frac{\partial \alpha_g}{\partial t} - \frac{P}{A}\frac{\partial}{\partial x}(\alpha_g v_g A) + Q_{wg} + Q_{ig} + \Gamma_{ig} h_g^* + \Gamma_w h_g^s + DISS_g$$

$$\frac{\partial}{\partial t}(\alpha_f \rho_f U_f) + \frac{1}{A}\frac{\partial}{\partial x}(\alpha_f \rho_f U_f v_f A)$$

$$= - P \frac{\partial \alpha_f}{\partial t} - \frac{P}{A}\frac{\partial}{\partial x}(\alpha_f v_f A) + Q_{wf} + Q_{if} - \Gamma_{ig} h_f^* - \Gamma_w h_f^s + DISS_f$$

Phasic momentum equation:

$$\alpha_g \rho_g A \frac{\partial v_g}{\partial t} + \frac{1}{2}\alpha_g \rho_g A \frac{\partial v_g^2}{\partial x} = - \alpha_g A \frac{\partial P}{\partial x} + \alpha_g \rho_g B_x A - (\alpha_g \rho_g A)\, FWG(v_g)$$

$$+ \Gamma_g A(v_{gI} - v_g) - (\alpha_g \rho_g A)\, FIG(v_g - v_f)$$

$$- C\alpha_g \alpha_f \rho A \frac{\partial (v_g - v_f)}{\partial t}$$

$$\alpha_f \rho_f A \frac{\partial v_f}{\partial t} + \frac{1}{2}\alpha_f \rho_f A \frac{\partial v_f^2}{\partial x} = - \alpha_f A \frac{\partial P}{\partial x} + \alpha_f \rho_f B_x A - (\alpha_f \rho_f A)\, FWF(v_f)$$

$$- \Gamma_g A(v_{fI} - v_f) - (\alpha_f \rho_f A)\, FIF(v_f - v_g)$$

$$- C\alpha_f \alpha_g \rho A \frac{\partial (v_f - v_g)}{\partial t} ,$$

where

$$\Gamma_f = - \Gamma_g ; \qquad \Gamma_g = \Gamma_{ig} + \Gamma_w ,$$

$$v_{gI} = v_{fI} = v_I ,$$

$$\alpha_g \rho_g \, FIG = \alpha_f \rho_f \, FIF = \alpha_g \alpha_f \rho_g \rho_f \, FI ,$$

$$Q = Q_{wg} + Q_{wf} ,$$

$$DISS_g = \alpha_g \rho_g \, FWG \, v_g^2 ,$$

$$DISS_f = \alpha_f \rho_f \, FWF \, v_f^2 ,$$

$$DISS = DISS_g + DISS_f$$

$$C = \text{coefficient of virtual mass},$$

and the remaining symbols have conventional meaning.

Appendix B

COMMIX-2 COMPUTER PROGRAM

FIVE-EQUATION MODEL

Phasic continuity equations:

$$\frac{\partial(\alpha_g \rho_g)}{\partial t} + \nabla \cdot (\alpha_g \rho_g \underline{u}_g) = \Gamma_g$$

$$\frac{\partial(\alpha_\ell \rho_\ell)}{\partial t} + \nabla \cdot (\alpha_\ell \rho_\ell \underline{u}_\ell) = \Gamma_\ell = -\Gamma_g$$

Phasic momentum equations:

$$\frac{\partial(\alpha_g \rho_g \underline{u}_g)}{\partial t} + \nabla \cdot (\alpha_g \rho_g \underline{u}_g \underline{u}_g)$$

$$= -\alpha_g \nabla P + \alpha_g \rho_g \underline{g} + K(\underline{u}_\ell - \underline{u}_g) - \alpha_g \underline{D}_g + \{[0,\Gamma_g]\underline{u}_\ell - [0,\Gamma_g]\underline{u}_g\}$$

$$\frac{\partial(\alpha_\ell \rho_\ell \underline{u}_\ell)}{\partial t} + \nabla \cdot (\alpha_\ell \rho_\ell \underline{u}_\ell \underline{u}_\ell)$$

$$= -\alpha_\ell \nabla P + \alpha_\ell \rho_\ell \underline{g} + K(\underline{u}_\ell - \underline{u}_g) - \alpha_\ell \underline{D}_\ell + \{[0,\Gamma_g]\underline{u}_\ell - [0,\Gamma_g]\underline{u}_g\}$$

Mixture energy equation:

$$\frac{\partial(\rho_m h_m)}{\partial t} + \nabla \cdot (\rho_m \underline{U}_m h_m) + \nabla \cdot [x(1 - x)\, \rho_m (h_g - h_\ell)\, V_{SL}]$$

$$= \frac{\partial P}{\partial t} + \underline{u}_m \cdot \nabla P - \nabla \cdot \underline{q} + Q + \left[\frac{1}{\rho_m} \alpha_g \alpha_\ell (\rho_\ell - \rho_g)\, \underline{V}_{SL}\right] \cdot \nabla P \,,$$

where

$$\alpha_g + \alpha_\ell = 1,$$

$$[a,b] = \max\,(a,b),$$

$\underline{V}_g, \underline{V}_\ell$ = momentum density sources arising from viscous dissipation of vapor phase and liquid phase, respectively,

K = interface drag coefficient,

$\underline{D}_g, \underline{D}_\ell$ = wall drag forces for vapor and liquid respectively,

X = quality,

$\underline{V}_{SL}$ = slip velocity,

$\underline{q}$ = heat flux, and

Q = volumetric heat source.

Note that the last term on the right side of the mixture energy equation is neglected in COMMIX-2. The magnitude of this term is much smaller than the last term on the left side of the same equation.

Three Phase Flow in Porous Media: Recent Developments in Nonstrictly Hyperbolic Conservation Laws

MICHAEL SHEARER* AND DAVID G. SCHAEFFER**

Abstract. Generalizations to three phase flow of the Buckley-Leverett equation are discussed. Under simplifying constitutive assumptions on the relative permeabilities, the equations are hyperbolic but not strictly hyperbolic. The characteristic speeds coincide at a point in state space known as an <u>umbilic point</u>. Using a normal form depending on two parameters, the basic features of the classification of hyperbolic equations with an umbilic point are presented. The Riemann initial value problem assumes a central role in numerical front tracking methods used to simulate secondary oil recovery. The presence of an umbilic point substantially complicates the solution of the Riemann problem, introducing new phenomena, some of which are discussed here.

1. Introduction. In this paper, we describe recent progress concerning generalizations of the Buckley-Leverett equation to three phase flow in a porous medium. The Buckley-Leverett equation [2] is a scalar conservation law

$$u_t + k(u)_x = 0 \tag{1.1}$$

governing the one-dimensional flow of two fluids (with volume fractions u and 1-u) in a porous medium. The nonlinearity k is given in terms of the relative permeabilities f(u), g(1-u) for the two phases, by

$$k(u) = f(u)/(f(u) + g(1-u)). \tag{1.2}$$

The generalization of (1.1), (1.2) to three phase flow is

---

*Department of Mathematics, North Carolina State University, Raleigh, NC 27695. Research supported by U.S. Army Research Office, grant number DAAL03-86-K-004.

**Department of Mathematics, Duke University, Durham, NC 27706. Research supported by NSF grant number MCS-8401591.

straightforward. There are three volume fractions u, v, 1-u-v, and the law of conservation of mass gives rise to a 2 × 2 system of conservation laws

$$u_t + F(u,v)_x = 0 \tag{1.3}$$

$$v_t + G(u,v)_x = 0.$$

The functions F, G have the special form

$$F = f/(f+g+h);\ G = g/(f+g+h) \tag{1.4}$$

in terms of the relative permeabilities f, g and h of the phases. The volume fractions lie between 0 and 1, so that (u,v) is restricted to the <u>physical triangle</u> $0 \leq u,v,\ 1-u-v \leq 1$.

A major obstacle to studying equations (1.3), (1.4) has been the lack of experimental or theoretical information concerning the dependence of the relative permeabilities f, g, h on (u,v).

Recall that system (1.3) is <u>hyperbolic</u> if the Jacobian matrix $J(u,v) = \partial(F,G)/\partial(u,v)$ has <u>real</u> eigenvalues λ_1, λ_2. <u>Strict</u> hyperbolicity requires $\lambda_1 \neq \lambda_2$. A point (u,v) is an <u>umbilic point</u> if J(u,v) is a multiple of the identity. In a forthcoming paper, we show that for a general class of f, g, h consistent with properties of two-phase flow, equations (1.3), (1.4) cannot be strictly hyperbolic throughout the interior of the physical triangle.

To develop more specific information concerning the failure of strict hyperbolicity, Marchesin and Paes-Leme (cf. [10]) proposed a simplified model in which f, g, h depend only on their respective volume fractions:

$$f = f(u),\ g = g(v),\ h = h(1-u-v). \tag{1.5}$$

We refer to this as the <u>Marchesin model</u>. Under natural assumptions on f, g, h (see condition (A) of §2), equations (1.3) - (1.5) are hyperbolic, with umbilic points at the vertices of the physical triangle, and at one interior point. In §2, we outline the classification of 2 × 2 hyperbolic systems near an isolated umbilic point.

A related model known as <u>Stone's model</u> also takes f = f(u) and g = g(v), but h = h(u,v) assumes a different form. For an example of Stone's model, Bell, Trangenstein and Shubin [1] showed numerically that the eigenvalues of J(u,v) are complex in a region within the physical triangle.

The Riemann problem is an initial value problem for system (1.3) with initial data consisting of two constants forming a single jump discontinuity. Solving the Riemann problem for any choice of the initial constants is central to numerical methods based on Glimm's scheme [3] and Godunov's method [5], as well as in numerical simulations of 2-dimensional flows using front-tracking [4].

The solution of the Riemann problem involves shock waves and smooth waves known as rarefaction waves.. The values of (u,v) on either side of these waves leads to a pictorial representation of the solution in the (u,v) - plane. In studying the Riemann problem near an umbilic point, we find several complicating features that can be understood systematically in the pictorial representation. We present some of these features with examples in §3.

2. Classification of equations near an umbilic point. Properties of hyperbolic equations (1.3) near an umbilic point (u^*, v^*) depend upon the form of the quadratic terms in the Taylor series expansion of $(F(u,v), G(u,v))$ about (u^*, v^*). To focus on these terms, we consider purely quadratic nonlinearities $Q: \mathbb{R}^2 \longrightarrow \mathbb{R}^2$:

$$U_t + Q(U)_x = 0 \tag{2.1}$$

To ensure hyperbolicity of system (2.1), we assume

(H) $dQ(U)$ has real eigenvalues for all U.

Then up to a linear, constant change of variables in U,

$$Q(U) = dC(U), \tag{2.2}$$

where

$$C(u,v) = au^3/3 + bu^2v + uv^2. \tag{2.3}$$

Using this normal form, derived in [10], we can classify properties of equation (2.1) in terms of a,b. We find four substantially different cases. The corresponding regions in the (a,b) - plane are shown in Figure 1. The four cases are distinguished by the patterns of <u>rarefaction curves</u> shown in Figure 2. Rarefaction curves are directed segments of integral curves of the right eigenvectors $r_k(U) (k = 1,2)$ of $dQ(U)$. The direction (indicated by arrows in Figure 2) is that of increasing characteristic speed $\lambda_k(U)$. Rarefaction curves are important for the Riemann problem because they represent the values of U in smooth centered rarefaction wave solutions $U(x/t)$ of (2.1), where $x/t = \lambda_k(U)$.

<u>Remarks 1</u>. The classification is stable to perturbations of Q by cubic and higher order terms, providing the corresponding (a,b) does not lie on one of the boundary curves in Figure 1.

<u>2</u>. The distinction between cases I and II is due to the differing orientations of the rarefaction curves R_k. In fact, in Case I, there are three inflection loci on which $\lambda_k(U)$, when restricted to R_k, has a critical point, whereas in cases II - IV there is exactly one inflection locus. Inflection loci correspond to a loss of genuine nonlinearity (cf. [8, 9, 10]): $r_k(U) \cdot d\lambda_k(U) = 0$.

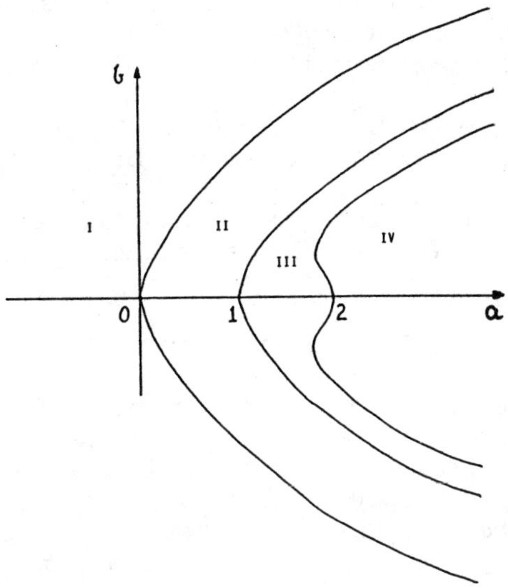

FIGURE 1. The four cases.

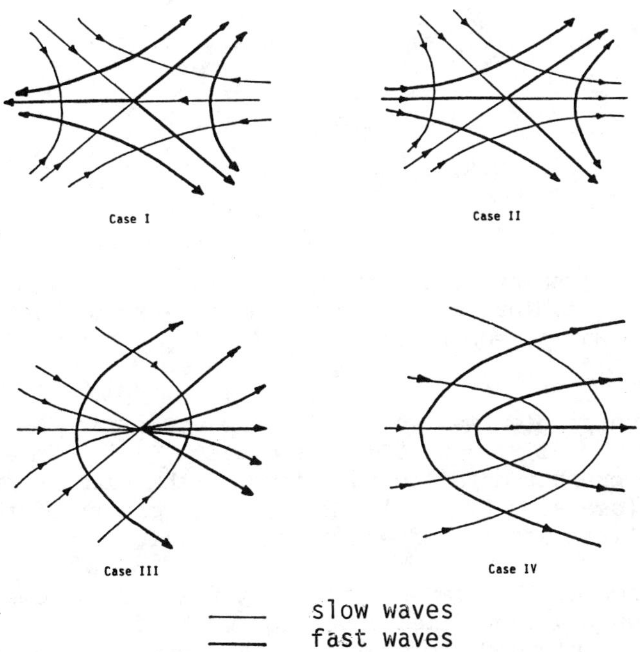

FIGURE 2. Rarefaction curves.

To relate the classification to the Marchesin model (1.3) - (1.5), we make the following assumption, which roughly summarizes known properties of relative permeabilities in two phase flow.

(A) $f(0) = f'(0) = 0$, $f''(u) > 0$ for $0 < u < 1$, and similarly for $g(v)$, $h(w)$.

Theorem. [10] If the relative permeabilities f,g,h in (1.4) have the form (1.5) and satisfy (A), then equations (1.3) have an isolated umbilic point (u^*,v^*) with $0 < u^*, v^*, 1-u^*-v^* < 1$. Moreover, equations (1.3) fall into Case I or Case II at (u^*,v^*).

Remark. Under assumption (A), the corners of the physical triangle are also umbilic points, with $a = 1$, $b = 0$. The corners are, therefore, degenerate. However, preliminary calculations show that the inclusion of cubic terms leads to properties (for instance of the rarefaction curves) similar to those of Case III. Thus, Cases I - III are known to be relevant to three phase flow in a porous medium.

For general equations (1.3), consistent with known two-phase flow properties, the corners are umbilic points that are not necessarily degenerate. A detailed treatment of issues surrounding general models of the form (1.3), (1.4), and the corners of the physical triangle, will be presented in a future paper.

3. Solution of the Riemann problem. In this section we discuss the Riemann problem for equations (1.3) near an umbilic point. Specifically, consider the initial value problem

$$U_t + Q(U)_x = 0 \qquad -\infty < x < \infty, \quad t > 0 \tag{3.1}$$

$$U(x,0) = \begin{cases} U_L & \text{if } x < 0 \\ U_R & \text{if } x > 0, \end{cases} \tag{3.2}$$

with $Q = Q(u,v)$ given by (2.2), (2.3). Centered solutions of (3.1), (3.2) consist of combinations of shock waves (across which $U(x/t)$ jumps) and rarefaction waves, separated in the (x,t)-plane by constant values of U.

The Riemann problem was solved for one symmetric choice of Q in Case I, namely $a = -1$, $b = 0$ [12]. (see also [6].) In a forthcoming paper [11], we present detailed solutions of the Riemann problem for Cases II - IV (see also [7, 13]). Here, we give part of the solution in Case III, to illustrate the effect of the umbilic point.

The solution of the Riemann problem for 2×2 systems generally involves a (possibly composite) wave from each of the two characteristic families, and an intermediate state U_1. For example, in Figure 3, we show a composite slow rarefaction-shock and a fast shock.

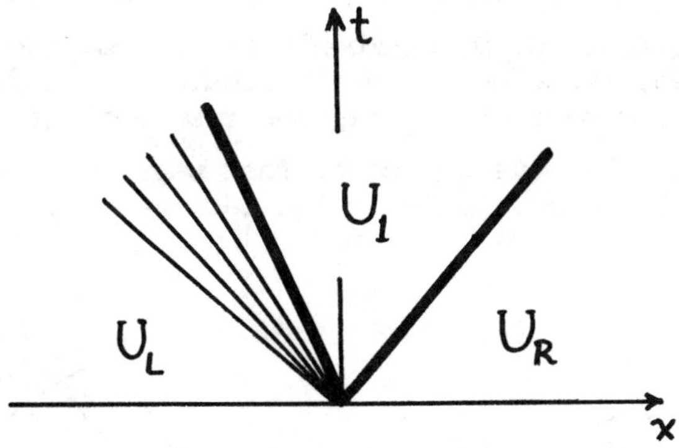

FIG. 3 (RS)S solution of the Riemann problem

We represent all possible (RS)S solutions in the U_R plane, for fixed U_L, as follows. The set of possible U_1 behind a slow (RS) wave forms a curve in the U plane. For each U_1 on this 1-<u>wave curve</u>, there is a new curve of possible U_R ahead of the fast shock. These U_R 2-<u>wave curves</u> sweep out a U_R - region as U_1 varies. We label the region (RS)S. Note that we have set up a coordinate system in this region using the wave curves. As U_L varies the region distorts continuously, and may disappear.

To solve the Riemann problem, we consider all combinations of waves, with the restriction that shocks must satisfy an appropriate admissibility condition. In the neighborhood of an umbilic point, in cases II - IV, the Lax condition [8] is appropriate. We present the solution for Case III in Figures 4 - 6. In Figure 4, we have identified

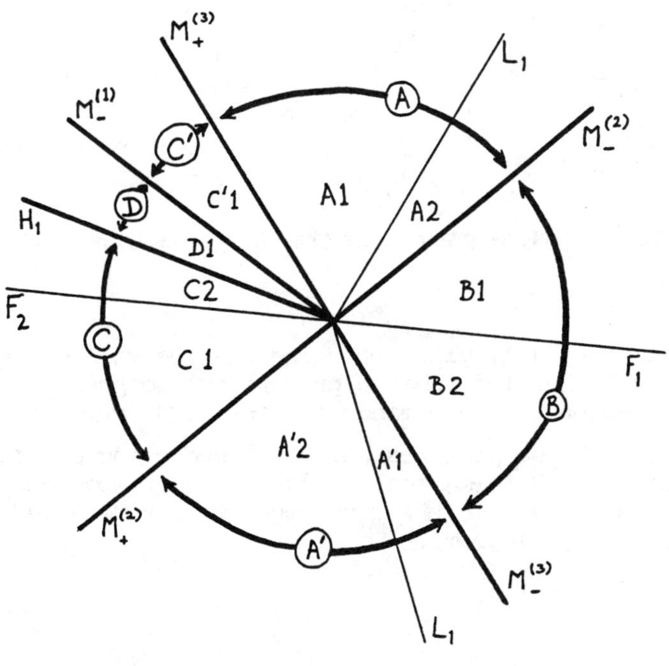

FIG. 4. U_L sectors, Case III

U_L sectors in which the U_k-regions are topologically unchanged. Within each major sector, A, B, C, D, between heavy lines, the pattern of wave curves is unchanged; these are given in Figure 5. In Figure 6 we show representative diagrams of the U_R-regions from each of the major sectors. Note that the wave curves of fast waves may terminate (on the knotted lines). This corresponds to the fast wave beginning to overlap

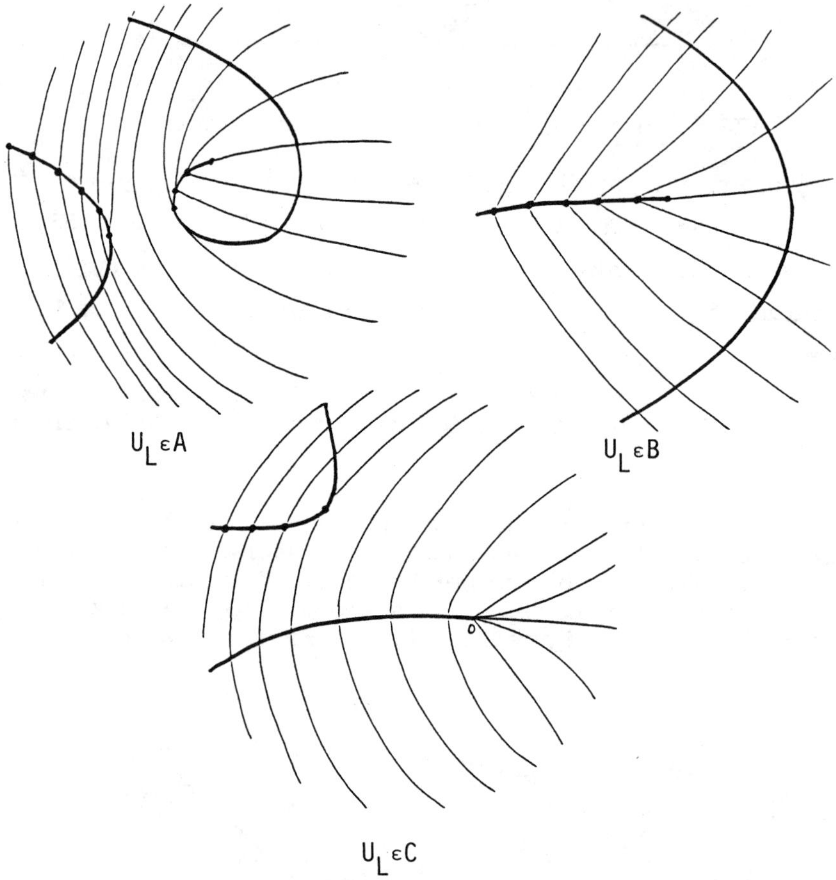

FIG. 5. Wave curves in the U_R plane, Case III

the slow wave in the (x,t) plane, forming what we call an <u>overcompressive wave</u>. The knotted lines represent overcompressive waves. The patterns of U_R regions undergo minor changes as U_L crosses the lighter rays of Figure 4. These pictures were obtained using a combination of computer graphics and Mathematical analysis. Corresponding pictures for Cases II and IV are given in [11], together with a detailed interpretation of the diagrams.

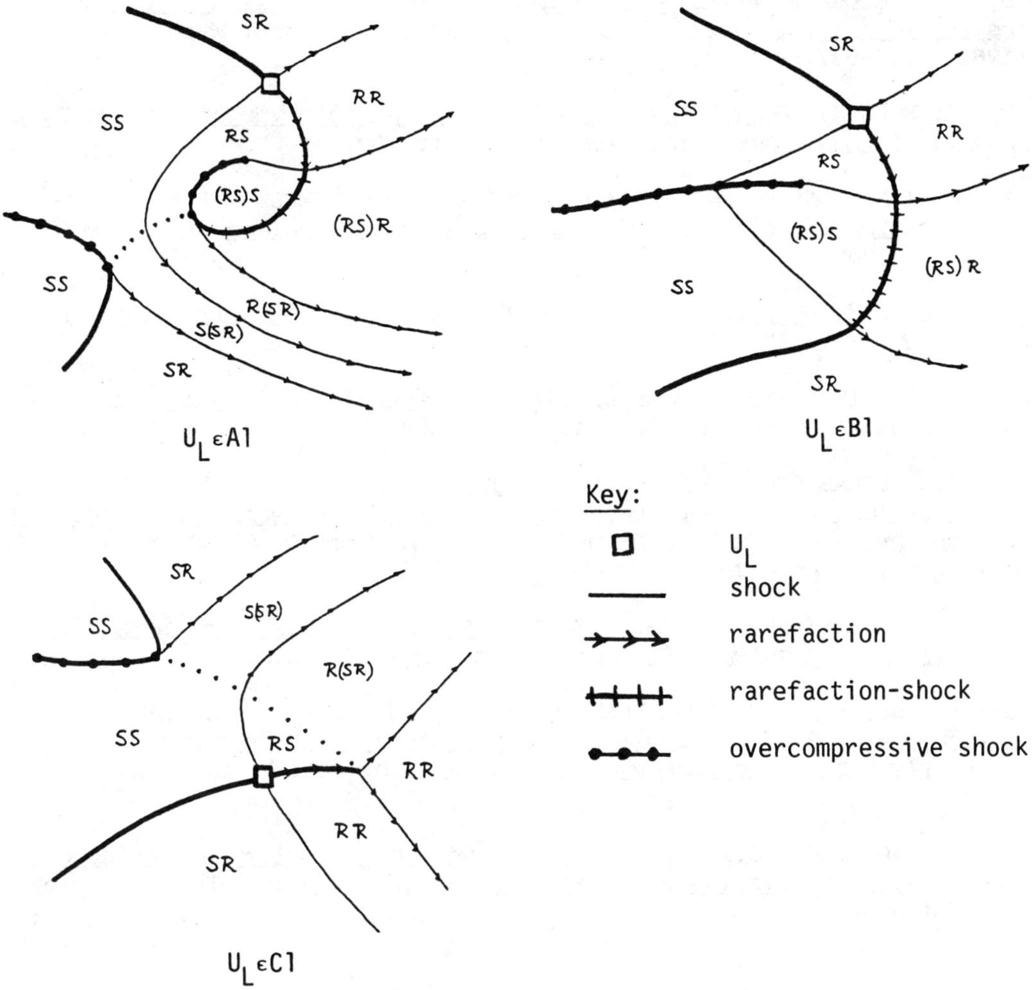

FIG. 6. U_R regions, Case III

REFERENCES

[1] J.B. BELL, J.A. TRANGENSTEIN, and G.R. SHUBIN, Conservation Laws of Mixed Type Describing Three-Phase Flow in Porous Media, Exxon Production Research, preprint, 1985.

[2] S.E. BUCKLEY and M.C. LEVERETT, Mechanism of fluid displacement in sands, Trans AIME, 146(1942), 107-116.

[3] J. GLIMM, Solution in the large for nonlinear hyperbolic systems of equations. Comm. Pure Appl. Math. 18(1965), 697-715.

[4] J. GLIMM, E. ISAACSON, D. MARCHESIN and O. MCBRYAN, Front tracking for hyperbolic systems, Adv. Appl. Math. 2(1981), 91-119.

[5] S.K. GODUNOV, Bounds on the discrepancy of approximate solutions constructed for the equations of gas dynamics, J. Comp. Math. Phys., 1(1961), 623-637. (Russian).

[6] H. HOLDEN, On the Riemann Problem for a Prototype of a Mixed Type Conservation Law, preprint, New York University, 1986.

[7] E. ISAACSON, D. MARCHESIN, B. PLOHR and B. TEMPLE, The Classification of Solutions of Quadratic Riemann Problems, I, II, MRC preprints, 1985.

[8] P.D. LAX, Hyperbolic systems of conservation laws II, Comm. Pure Appl. Math., 19(1957), 537-566.

[9] T.-P. LIU, The Riemann problem for general 2×2 conservation laws, Trans. Amer. Math. Soc., 199(1974), 89-112.

[10] D.G. SCHAEFFER and M. SHEARER, The classification of 2×2 systems of nonstrictly hyperbolic conservation laws, with application to oil recovery; Appendix with D. Marchesin, P.L. Paes-Leme, Comm. Pure Appl. Math., to appear.

[11] D.G. SCHAEFFER and M. SHEARER, Riemann problems for nonstrictly hyperbolic 2×2 systems of conservation laws. To appear.

[12] M. SHEARER, D.G. SCHAEFFER, D. MARCHESIN and P.L. PAES-LEME, Solution of the Riemann problem for a prototype 2×2 system of nonstrictly hyperbolic conservation laws, Arch. Rat. Mech. Anal., to appear.

[13] Z. TANG and T.C.T. TING, Wave curves for the Riemann problem of plane waves in simple isotropic elastic solids, preprint, University of Ill., Chicago, 1985.

Recent Trends in the Continuum Theory of Phase Transitions

M. SLEMROD*

Abstract. This paper surveys some recent results on solvability of the balance laws of momentum, mass, and energy for a material exhibiting first order phase transitions.

The purpose of this note is to outline some recent ideas in the continuum theory of phase transitions. The main issues are simple. Do unique weak solutions exist for the equations of balance of mass, momentum, and energy for van der Waals like fluids. If such solutions do exist, how can we compute them? Are such computations consistent with laboratory experiments?

First let us recall the equations of balance of mass, momentum, energy (neglecting viscous, capillary and body forces). The equations written in Lagrangian coordinates x, t, $-\infty < x < \infty$, $t > 0$,

$$u_t + p_x = 0 , \tag{1}$$

$$w_t - u_x = 0 , \tag{2}$$

$$E_t + (pu)x = 0 , \quad \left(E = \frac{u^2}{2} + \varepsilon\right) . \tag{3}$$

---

*Department of Mathematical Sciences
Rensselaer Polytechnic Institute
Troy, New York 12180-3590
Research sponsored in part by the Air Force Systems Command, USAF, Contract/Grant No. 85-0239. The U. S. Government's right to retain a nonexclusive royalty-free license in and to the copy right covering this paper, for governmental purposes, is acknowledged.

Here $u(x,t)$ denotes the velocity, $w(x,t)$ the specific volume, $\theta(x,t)$ the temperature, $p(w,\theta)$ the pressure $\varepsilon(w,\theta)$ the specific internal energy, and E = specific total energy. Of interest to us will be the case where

$$p(w,\theta) = \frac{R\theta}{w-b} - \frac{a}{w^2} \ .$$

This is the classical van der Waals equation of state. We take

$$\varepsilon(w,\theta) = -\frac{a}{w} + c_v\theta + \text{constant}$$

where c_v is the specific heat at constant volume.

If $\theta < \theta_{crit} = \frac{8a}{27bR}$ the $p(w,\theta)$ has the classical non-monotone isotherms shown in Figure 1. Then for

$$b < w < \alpha \ , \quad w > \beta \ , \quad p_w(w,\theta) < 0$$

and for

$$\alpha < w < \beta \ , \quad p_w(w,\theta) > 0 \ .$$

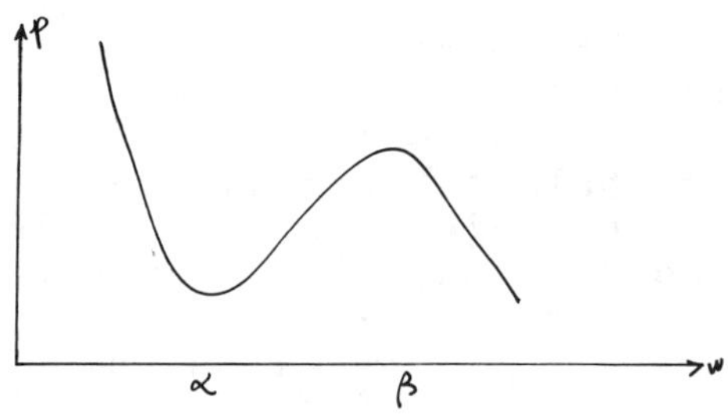

FIGURE 1

For isothermal motions where (3) is ignored becomes a mixed hyperbolic-elliptic system i.e., hyperbolic in the "liquid" $b < w < \alpha$ region and "vapor" $\beta < w < \infty$ region but elliptic in the "spinodal" $\alpha < w < \beta$ region.

Given our usual experience on elliptic initial value problems and the well known Hadamard instability one might at first glance think (1)-(3) to be ill-posed. The point I wish to stress is that I believe (1)-(3) to be well posed in a sense I will now make precise. By well-posedness I shall mean (1)-(3) possesses piecewise continuous solutions (u,w,Θ) satisfying (1)-(3) in the sense of distributions. <u>Furthermore, w takes values only in the hyperbolic regimes $b < w < \alpha$, $w > \beta$.</u> In the hyperbolic regimes shocks and contact discontinuities may occur. These waves satisfy the Rankine-Huginot jump conditions

$$-s\,[u] + [p] = 0\ ,$$

$$-s\,[w] - [u] = 0\ , \qquad \text{(R-H)}$$

$$-s\,[E] + [pu] = 0\ .$$

If w possesses a jump across hyperbolic regimes (R-H) must still be satisfied. In this case the shock wave is called a _phase_ _boundary_.

The idea for this point of view was first raised by R. James [1] in his fundemental paper on propagating phase boundaries. In that paper the question of _admissibility_ of phase boundaries was raised but left _unresolved_. I considered the issue of admissibility from the perspective of _shock_ _structure_ in several papers [2], [3], [4], (jointly _with R. Hagan_) as did M. Shearer [5], [6] and J. Serrin and R. Hagan [7].

The main point of the shock structure theory is to incorporate effects of viscosity, capillarity, and heat conduction into (1)-(3). This can be done as follows. Write the total stress τ as consisting of two parts τ_1, τ_2 where τ_1 contains the elastic and viscous contributions and τ_2 contains the capillarity contribution. Specifically we write

$$\tau = \tau_1 + \tau_2$$

$$\tau_1 = -p(w,\theta) + \mu u_x$$

$$\tau_2 = -Aw_{xx}$$

where A,μ are positive (small) constants. Now the balance of mass and momentum are

$$u_t - \tau_x = 0\ , \tag{4}$$

$$w_t - u_x = 0\ . \tag{5}$$

The compararitively difficult equation to write down is one for balance of energy. The main idea is to follow Felderhof [8] (see also Dunn and Serrin [9]) and postuate that the rate of change in internal energy depends only on the elastic and viscous contributions to the stress and _not_ on the capillarity contribution, i.e., we write

$$\varepsilon_t = \tau_1 u_x + h_x \tag{6}$$

for the balance of energy equation. Here h is the heat flux.

A motivation for (6) can be found in the following computation. Let $\psi(w,\theta) = \varepsilon(w,\theta) - \theta\eta(w,\theta)$ devote the Helmholtz free energy when $\eta(w,\theta)$ is the specific entropy. The thermodynamic field relations are

$$(7) \qquad \eta(w,\theta) = -\frac{\partial\psi}{\partial\theta}, \qquad p(w,\theta) = -\frac{\partial\psi}{\partial w} .$$

Now differentiate ψ with respect to t:

$$(8) \qquad \psi_t = \varepsilon_t - \theta_t\eta - \theta\eta_t$$

and by the chain rule

$$(9) \qquad \psi_t = \psi_w w_t + \psi_\theta\, \theta_t = -pw_t - \eta\theta_t \text{ (by(7))}.$$

Now combine (8),(9) to see

$$\varepsilon_t - \theta_t\eta - \theta\eta_t = -pw_t - \eta\theta_t$$

or

$$\theta\eta_t = \varepsilon_t + pu_x$$

where we have used (5). But now our postulate (6) gives

$$\theta\eta_t = \tau_1 u_x + pu_x + h_x ,$$

i.e.,

$$\theta\eta_t = -pu_x + \mu u_x^2 + pu_x + h_x,$$

$$(10) \qquad \theta\eta_t = \mu u_x^2 + h_x .$$

If now constitute h by Fourier's law (for simplicity)

$$(11) \qquad h = K\theta_x$$

where K is a positive constant the (10) becomes

$$\eta_t = \mu\frac{u_x{}^2}{\theta} + K\frac{\theta_{xx}}{\theta}$$

$$= \mu\frac{u_x{}^2}{\theta} + K\left(\frac{\theta_x}{\theta}\right)_x + K\frac{\theta_x{}^2}{\theta^2}$$

and hence we have the desired entropy production (Clausius-Plank) inequality

(12) $$\eta_t - K\left(\frac{\theta_x}{\theta}\right)_x = \mu \frac{u_x{}^2}{\theta} + K \frac{\theta_x{}^2}{\theta} > 0 .$$

On the other hand if we had attempted to use the full stress τ in (6) we would be led to an indeterminate form for $\eta_t - K\left(\frac{\theta_x}{\theta}\right)_x$ violating the Clausius-Plank) inequality

Finally to write (6) in conservation form we put

$$\tilde{E}(w,u,\theta) = \frac{u^2}{2} + \varepsilon(w,\theta) + A\frac{w_x{}^2}{2}$$

(the total specific energy = kinetic energy + internal energy + interfacial energy). Then

$$\tilde{E}_t = uu_t + \varepsilon_t + Aw_x w_{xt} ,$$

$$= u\tau_x + \tau_1 u_x + Aw_x u_{xx} + K\theta_{xx} ,$$

$$= u\tau_{1x} + u\tau_{2x} + \tau_1 u_x + Aw_x u_{xx} + K\theta_{xx} .$$

$$\tilde{E}_t = (u\tau)_x - u_x\tau_2 + Aw_x u_{xx} + K\theta_{xx} ,$$

$$= (u\tau)_x + Aw_{xx}u_x + Aw_x u_{xx} + K\theta_{xx} ,$$

(using the definition τ_2)

$$\tilde{E}_t = (u\tau)_x + A(u_x w_x)_x + K\theta_{xx} .$$

Hence we can record the balance of momentum, mass, and energy as

(13) $$u_t = -p(w,\Theta)_x + \mu u_{xx} - Aw_{xxx} ,$$

(14) $$w_t = u_x ,$$

(15) $$\tilde{E}_t = \left(u[-p + \mu u_x - Aw_{xx}]\right)_x + A(u_x w_x)_x + K\theta_{xx} .$$

The goal is then to solve (13),(14),(15) with initial data

$$u_0(x), \qquad w_0(x), \qquad \theta_0(x)$$

and hopefully pass to the limit in such solutions as μ, A, $k \to 0$ to recover a distributional solution of (1). This program has not yet been carried out. The only successes so far are in the isothermal case where (15) is ignored and the data is consistent with Rankine-Hugoniot jump conditions (shock structure problem) [2],[4],[5], [6] and work in [3] and [9] for the full system (13), (14), (15). An important new result has been obtained by M. Grinfeld [10] who generalized and improved earlier results for both the isothermal and full systems by use of the Conley index technique. Again, this work requires data consistent with the Rankine-Hugoniot jump conditions.

For arbitrary data the results are more sparse. The main result so far is M. Shearer's [6] proof showing solvebility of the Riemann problem (1),(2) for data close to the Maxwell line equilibrium solution of (1),(2). Shearer also shows each shock wave and phase boundary in his construction is admissible with respect to structured shock theory based on (13),(14).

Finally, we note that for (1),(2) with p replaced by a van der Waals plus Maxwell type construction Roytburd and Slemrod [11] have obtained some partial results on solvbility of (1),(2) as limits of solutions of (13),(14).

With regard to computation I note the papers [11], [12] where the Lax-Friedrichs method has been proposed and used for numerical intergration of (1),(2). The reason for the choice of Lax-Friedrichs is that it provides the desired simulation of viscosity and capillarity which proved useful i.e., the shock structure analyses.

To my knowledge a serious computational attempt to compare theory with experimenthas not yet been done.

REFERENCES

[1] JAMES, R. (1980), The Propagation of Phase Boundaries in Elastic Bars. Archive for Rational Mechanics and Analysis 73, 125-158.

[2] SLEMROD, M. (1983), Admissibility Criteria for Propagating Phase Boundaries in a van der Waals Fluid. Archive for Rational Mechanics and Analysis 81, 301-315.

[3] SLEMROD, M. (1984), Dynamic Phase Transitions in a van der Waals Fluid. J. Differential Equations 52, 1-23.

[4] HAGAN, R. and M. SLEMROD (1984), The Viscosity-Capillarity Criterion for Shocks and Phase Transitions. Archive for Rational Mechanics and Analysis 83, 333-361.

[5] SHEARER, M. (1983), Admissibility Criteria for Shock Wave Solutions of a System of Conservation Laws of Mixed Type. Proc. Royal Soc. Edinburgh 93A, 233-244.

[6] SHEARER, M., Nonuniqueness of Admissible Solutions of Riemann Initial Value Problems for a System of Conservation Laws of Mixed Type, to appear J. Differential Equations.

[7] HAGAN, R. and J. SERRIN (1984), One Dimensional Shock Layers in Korteweg Fluids, in Phase Transformations and Material Instabilities in Solids, Ed. M.E. Gurtin New York: Academic Press 113-128.

[8] FELDERHOF, B.U. (1970), Dynamics of the Diffuse Gas-Liquid Interface near the Critical Point. Physica 48, 541-560.

[9] DUNN, J.E., and J. SERRIN (1983), On the Thermodynamics of Interstitial Working. Institute for Mathematics and its Application, University of Minnesota, Minneapolis, MN, preprint #24, to appear Archive for Rational Mechanics and Analysis.

[10] M. GRINFELD, (1986), Topological Techniques in Dynamic Phase Transitions. Ph.D. Thesis, Department of Mathematical Sciences, Rensselaer Polytechnic Institute.

[11] M. SLEMROD (1985), Lax-Friedrichs and the Viscosity-Capillarity Criterion, in Physical Mathematics and Nonlinear Partial Differential Equations, ed. J. Lightbourne and S. Rankin, Marcel-Dekker, 75-84.

[12] M. SLEMROD and J. FLAHERTY (1986), Numerical Integration of the Riemann Problem for a van der Waals fluid, to appear Conference on Phase Transitions, eds. Gittus, Aifantis.

Continuum Equations of Multiphase Flow and Physical Systems

S. L. SOO*

Abstract. The basis of volume averaging of a mixture divided by interfaces followed by time averaging has been treated. Consideration of barycentric frame of reference suggests the significance of phase configurations and the requirement of averaging of volumetric quantities, fluxes and stresses only. Local deviations from averages are shown to be negligible for a dispersed system. Closure calls for relating the transfer integrals to constitutive relations. The relations are simplified for stratified flow systems and dispersed systems. Physical examples of phase in pure stratified flow and averaged motions of phases in a fluidized bed provide the physical meaning and show the usefulness of continuum equations in treating certain cases of multiphase flow.

1. Introduction. Multiphase systems have been defined in the dynamic sense in that elements of an inhomogeneous fluid separated by interfaces which possess similar dynamic response characteristics may be treated as a dynamic phase. For example, in the simple case of a particulate suspension with a distribution in particle sizes but of similar particle material, as the suspending fluid is accelerated, particles of a given size range will be similarly accelerated, thus constituting a dynamic phase [1]. Similar considerations for identifying dynamic phases may arise due to different thermal responses, as well as similar particles carrying different electrostatic charges. The concept of dynamic phases is based on different phases occupying the same volume in space; volume is therefore a basic measure. Extension can be made to multiphase systems with general phase configurations. Especially when considering dispersed systems, some averaging procedures

*Department of Mechanical and Industrial Engineering, University of Illinois at Urbana-Champaign, 1206 West Green Street, Urbana, IL 61801.

are needed in formulating their equations of conservation. Statistical or ensemble average is not directly applicable because of the basic need of taking into account the interfaces.

2. Significance of phase configurations. Before getting into detailed mathematical procedures, the nature and significance of the interface can be seen from considering the Bernoulli equations of phases and their mixture. When it is written for the steady flow of a phase k of incompressible fluids in the well known form, the mixture (m) then has the relation:

$$(1) \qquad (1/2)\, \rho_m U_m^2 + \Sigma_k (1/2)\, \rho_k (U_k - U_m)^2 + P_m + \rho_m gz = \text{constant}$$

where, for the mixture, $P_m = \Sigma_k P_k$, and $\rho_m = \Sigma_k \rho_k$, $\rho_m U_m = \Sigma_k \rho_k U_k$, according to the barycentric frame of reference and $\rho_m U_m^2 = \Sigma_k \rho_k U_k^2 + \Sigma_k \rho_k (U_k - U_m)^2$; U denotes velocities, ρ the densities based on the volume of the mixture, P is the pressure, z is the elevation, and g is the gravitational acceleration. This formulation represents the case ignoring transfer of inertial force across the interface, that is, the case of pure stratified flow. In this case the one-dimensional dynamic equation is ill-posed, since a common set of characteristics does not exist. The drift flux formulation [2] of two phase flow belong to this type. This separated flow formulation is applicable to, for instance, batch settling [3] where separation of phases occur in the nature of exponential decay and the equations are parabolic, giving imaginary characteristics.

However, if one considers a highly dispersed system that can be approximated by a macroscopically homogeneous mixture, the Bernoulli equation which accounts for conservation based on the mixture will give, for phase k:

$$(2) \qquad (1/2)\, \rho_k U_k^2 - (1/2)\, \rho_k (U_k - U_m)^2 + P_k + \rho_k gz = \text{constant}$$

Eq. (2) is exact for highly dispersed flow [4]. Physically a highly dispersed system is one in which the characteristic dimension of the dispersed phase is smaller than the characteristic path length (mean free path of a gas or scale of turbulence of a fluid, for instance) of the suspending fluid. The close coupling of phases provides for a common set of real characteristics for one-dimensional motion of the mixture. Other flow regimes, depending on configuration of phases, will have to be represented by relations intermediate between these two limiting cases. As long as some inertia forces is transferred across the interface, the formulation will be well-posed [4].

3. Volume and time averaging. Averaging of conservation relations was first suggested by Birkhoff [5]. The volume averaging theorems of continuum mechanics of Slattery [6] and Whitaker [7], originally introduced to deal with flow through porous media, make possible averaging within a control volume including moving interfaces. The averaging relations are valid for:

(3) (Characteristic length of pores and phases)
<< (Characteristic length of averaging volume)
<< (Characteristic length of the physical system)

Physical quantities of such a volume fraction α_k of each phase are readily identified with the dynamic phases; $\rho_k = \alpha_k \bar{\rho}_k$, $\bar{\rho}_k$ being the density of material constituting phase k.

A different approach was suggested by Delhaye and Achard [8] to apply time averaging to the equations of conservation by considering the passage of interfaces at time t over an interval T. For an accumulated residence time T_k of phase k, they gave, from applying the Leibnitz rule for the differentiation of an integral:

$$(\partial/\partial t) \int_{T_k} f_k \, dt = f_k \left[t + (T/2)\right] - f_k \left[t - (T/2)\right] \tag{4}$$

While a limiting procedure can be applied in the mathematical sense, a physically meaningful averaging time must be large enough to include a large number of interfaces of frequency of passage ν_s, yet it should be small when compared to the characteristic transit time of the mixture through the system. Hence T cannot be taken to the limiting value of zero and the time of passage of an interface is not independent of time coordinate t for Eq. (4) to be valid. One also notes that the corresponding quantity of volume fraction of a phase in the above is fraction residence time of that phase; the latter is identical to the the volume fraction only in the case of one-dimensional steady uniform flow.

In dealing with a multiphase system, the cumulative nature of volume fraction is highly desirable; for instance, $\rho_m = \Sigma_k \alpha_k \bar{\rho}_k$ in the above. Therefore, volume averaging is preferred to time averaging to facilitate relating the formulation to physical quantities and measurements.

The interface disappears after the volume averaging. Time averaging after volume averaging was performed by extending the Reynolds analysis via averaging over a time interval T such that [9]

(5) (Characteristics time of high frequency fluctuations) << T
<< (Characteristic time of low frequency fluctuations)

The high frequency fluctuations may be represented by that of turbulence while the low frequency change may be that of, say, a blowdown process, and not necessarily periodic.

4. Averaged continuum equations. The volume averaging theorems [7] are applied to the continuum equations of phase k in volume v_k encompassed by control volume v with a moving interface A_k between phase k and another phase f, say, with outward normal $\underline{n}_k$ and at velocity $\underline{W}_k$; inside v_k the velocity is $\underline{U}_k$ and the density is ρ_k. Other properties are pressure P_k, internal energy u_k, etc. Volume averaging gives terms from the phasic equations including volume averages of density, momentum per unit volume, momentum flux, internal energy per unit volume, internal energy flux, gradients of pressure, shear stress, heat flux, and dissipation, etc., as well as interface transfer integrals of rate of change of momentum, stress $\underline{\underline{\tau}}_k$, energy flux, heat flux $\underline{J}_{qk}$, and heat source J_{Ek}.

One notes that volume averaging, when the barycentric frame of reference is satisfied, is applicable only to quantities per unit volume or area. While the general representation may take the form

(5) $$^{i}\langle f_k \rangle = v_k^{-1} \int_{v_k} f_k \, dv_k$$

for the intrinsic average, and

(6) $$\langle f_k \rangle = v^{-1} \int_{v_k} f_k \, dv_k = \alpha_k \, {}^{i}\langle f_k \rangle$$

for the control volume average or extensive average, the averaging operation has to be linked to physical quantities according to:

(7) $$\langle \rho_k \rangle = v^{-1} \int_{v_k} \rho_k \, dv_k = \alpha_k \, {}^{i}\langle \rho_k \rangle$$

(8) $$^{i}\langle \underline{U}_k \rangle = v_k^{-1} \int_{v_k}^{v_k} \rho_k \, \underline{U}_k \, dv_k \, / \, {}^{i}\langle \rho_k \rangle$$

since $\langle \underline{U}_k \rangle$ over the control volume has no useful physical meaning. In addition, we have

(9) $$^{i}\langle \underline{U}_k \, u_k \rangle = v_k^{-1} \int_{v_k} \rho_k \, \underline{U}_k \, u_k \, dv_k / \, {}^{i}\langle \rho_k \rangle$$

$\langle u_k \rangle$ is also meaningless; u_k can be replaced by $\underline{U}_k$ for momentum flux, etc. Further, the average temperature of phase k is given by:

(10) $$^{i}\langle T_k \rangle = v_k^{-1} \int_{v_k} \rho_k \, c_{vk} \, T_k \, dv / \, {}^{i}\langle \rho_k c_{vk} \rangle$$

$^{i}\langle \rho_k c_{vk} \rangle$ is the average heat capacity of phase k per unit volume. $\langle P_k \rangle$, $\langle \underline{\underline{\tau}}_k \rangle$, etc are readily visualized.

Volume averaging gives the mass conservation equation for phase k in the form:

(11) $$(\partial/\partial t) \langle \rho_k \rangle + \nabla \cdot \langle \rho_k \rangle \, {}^{i}\langle \underline{U}_k \rangle = -v^{-1} \int_{A_k} \rho_k (\underline{U}_k - \underline{W}_k) \cdot \underline{n}_k \, dA = \Gamma_k .$$

The momentum equation takes the form:

(12) $$(\partial/\partial t) \langle \rho_k \rangle \, {}^{i}\langle \underline{U}_k \rangle + \nabla \cdot \langle \rho_k \rangle \, {}^{i}\langle \underline{U}_k \, \underline{U}_k \rangle = -\nabla \langle P_k \rangle + \nabla \cdot \langle \underline{\underline{\tau}}_k \rangle + \rho_k \underline{f}$$
$$- v^{-1} \int_{A_k} \rho_k \underline{U}_k \, (\underline{U}_k - \underline{W}_k) \cdot \underline{n}_k \, dA - v^{-1} \int_{A_k} (P_k \underline{\underline{I}} - \underline{\underline{\tau}}_k) \cdot \underline{n}_k \, dA.$$

The energy equation in terms of internal energy is:

(13) $$(\partial/\partial t) \langle \rho_k \rangle \, {}^{i}\langle u_k \rangle + \nabla \cdot \langle \rho_k \rangle \, {}^{i}\langle \underline{U}_k u_k \rangle = - \langle P_k \nabla \cdot \underline{U}_k \rangle - \nabla \cdot \langle \underline{J}_{qk} \rangle$$

$$+ J_{Ek} + \langle \phi_k \rangle - v^{-1} \int_{A_k} \rho_k u_k (\underline{U}_k - \underline{W}_k) \cdot \underline{n}_k \, dA$$

$$+ v^{-1} \int_{A_k} \underline{J}_{qk} \cdot \underline{n}_k \, dA$$

where $\phi_k = \underline{\underline{\tau}}_k : \nabla \underline{U}_k$ is the dissipation function. In addition, there are interface balance equations for mass, momentum, and energy; Γ_k is further related to the physical process like $\underline{\underline{\tau}}_k$, $\underline{J}_{qk}$, J_{Ek}. Note that all the transfer integrals consist of local quantities.

Interface balances further give, for mass balance: $(\underline{W}_f = \underline{W}_k, \underline{n}_f = -\underline{n}_k)$

(14) $$\Gamma_k = - v^{-1} \int_{A_k} \rho_k (\underline{U}_k - \underline{W}_k) \cdot \underline{n}_k \, dA = v^{-1} \int_{A_k} \rho_f (\underline{U}_f - \underline{W}_f) \cdot \underline{n}_f \, dA;$$

for momentum balance:

(15) $$\int_{A_k} \rho_k \underline{U}_k (\underline{U}_k - \underline{W}_k) \cdot \underline{n}_k \, dA - \int_{A_k} \rho_f \underline{U}_f (\underline{U}_f - \underline{W}_f) \cdot \underline{n}_f \, dA$$

$$+ \int_{A_k} (P_k \underline{\underline{I}} - \underline{\underline{\tau}}_k) \cdot \underline{n}_k \, dA + \int_{A_k} (P_f \underline{\underline{I}} - \underline{\underline{\tau}}_f) \cdot \underline{n}_f \, dA = 0;$$

where the effect of surface tension is neglected for the present, and for the energy balance:

(16) $$\int_{A_k} \rho_k [u_k + (1/2) \underline{U}_k^2] (\underline{U}_k - \underline{W}_k) \cdot \underline{n}_k \, dA$$

$$+ \int_{A_k} \rho_f [u_f + (1/2) \underline{U}_f^2] (\underline{U}_f - \underline{W}_f) \cdot \underline{n}_f \, dA$$

$$+ \int_{A_k} \underline{J}_{qk} \cdot \underline{n}_k \, dA + \int_{A_k} \underline{J}_{qf} \cdot \underline{n}_f \, dA$$

$$+ \int_{A_k} \underline{U}_k \cdot (P_k \underline{\underline{I}} - \underline{\underline{\tau}}_k) \cdot \underline{n}_k dA + \int_{A_k} \underline{U}_f \cdot (P_f \underline{\underline{I}} - \underline{\underline{\tau}}_f) \cdot \underline{n}_f \, dA = 0;$$

all averaged over the control volume v. Note that for each interface, Eq. (14) constitutes two independent equations since Γ_k, like $\underline{\underline{\tau}}_k$ and $\underline{J}_{qk}$, and J_{Ek}, is also related to the physical process. Note also that the momentum transfer integral is not necessarily zero when the mass transfer integral is zero, that is, when there is no net mass transfer [4].

Time averaging, besides reducing averages of products to products of averages, introduces what corresponds to eddy diffusivities D_{km}^T, Reynolds stresses, and eddy conductivities κ_{km}^T according to : (super t for time averages over T according to Eq. (5), LF for low and (') for high frequencies)

(17) $$^t\langle\langle \rho_k \rangle \, ^i\langle \underline{U}_k \rangle\rangle = \langle \rho_k \rangle_{LF} \, ^i\langle \underline{U}_k \rangle_{LF} + \, ^t\langle\langle \rho_k' \underline{U}_k' \rangle\rangle$$

$$= \langle \rho_k \rangle_{LF} \, ^i\langle \underline{U}_k \rangle_{LF} - D_{km}^T \nabla \langle \rho_k \rangle_{LF}.$$

and

$$(18)\qquad {}^{t}\langle\langle\rho_k\rangle\ {}^{i}\langle \underline{U}_k\ u_k\rangle\rangle = \langle\rho_k\rangle_{LF}\ {}^{i}\langle\underline{U}_k\rangle_{LF}\ {}^{i}\langle u_k\rangle_{LF} + \langle\rho_k\rangle_{LF}\ {}^{t}\langle{}^{i}\langle\underline{U}_k' u_k'\rangle\rangle + {}^{i}\langle u_k\rangle_{LF}\ {}^{t}\langle\langle\rho_k' \underline{U}_k'\rangle\rangle + {}^{i}\langle\underline{U}_k\rangle_{LF}\ {}^{t}\langle\langle\rho_k' u_k'\rangle\rangle .$$

where the eddy conductivity is given by:

$$(19)\quad -\langle\rho_k\rangle_{LF}\ {}^{t}\langle{}^{i}\langle\underline{U}_k' u_k'\rangle\rangle = \langle\rho_k\rangle_{LF}\ D_{uk}^{T}\ \nabla\ {}^{i}\langle u_k\rangle_{LF} \equiv -\underline{J}_{qk}^{T} = \kappa_{km}^{T}\ \nabla\ {}^{i}\langle T_k\rangle$$

Replacing u_k with $\underline{U}_k$ in Eq. (18) give Reynolds stress in the form $\langle\rho_k\rangle\ {}^{t}\langle{}^{i}\langle U_k' U_k'\rangle\rangle$. In addition, there are other transport parameters in the form of additional transfer integrals as those in Eqs. (11) to (13).

5. Closure. For the sake of simplicity, closure can be considered based on the volume averaged equations. The latter, when high frequency fluctuations are absent and averaging volume is small, will have averages of products ${}^{i}\langle\underline{U}_k u_k\rangle \sim {}^{i}\langle\underline{U}_k\rangle\ {}^{i}\langle u_k\rangle$, for instance, although deviations [10] need to be accounted for when the averaging volume is small compared to that of a phase (a limiting case is illustrated in Example 1 of the next section). For a dispersed system of spherical solid particles (phase k) in a fluid (f), it is readily shown that the error of neglecting local deviations in the fluid velocity $\underline{U}_f$ is nearly: $({}^{i}\langle\underline{U}_f\rangle - {}^{i}\langle\underline{U}_k\rangle)\ \alpha_k/(1-\alpha_k)$, a small quantity for small spheres or a dilute suspension.

For n phases, there are n(n-1)/2 interfaces, and the interface balances contribute (n-1) + 3n(n-1)/2 independent equations and Eqs. (11) to (13) give 3 n equations, or a total of 4n - 1 + 3n(n-1)/2 independent vector equations. The dependent variables include n sets of P_k's, $\underline{U}_k$'s, and u_k's each, (n-1) of volume fraction α_k's, n(n-1)/2 of (A_k/v)'s, $\underline{n}_k$'s, and normal velocities of interfaces $(\underline{W}_k \cdot \underline{n}_k)$'s each. Closure is assured by the constitutive equations of interface transfer integrals such as $\Gamma_k = -n_k\ dm_k/dt$ for n_k droplets per unit volume evaporating at the rate of dm_k/dt for mass m_k of each droplet (the limiting case is by solving local phasic equations).

When reduced to a two-phase dispersed system of spherical particles, we note that the above reduced to 6 dependent variables of ${}^{i}\langle\rho_k\rangle$, ${}^{i}\langle\underline{U}_k\rangle$, ${}^{i}\langle u_k\rangle$ (k = 1, 2) plus α_k, dR_k/dt and R_k, the radius of the particulate phase. The 6 averaged equations plus 3 constitutive equations for the transfer integrals also result from simplification in the case of a dispersed system. A case is also illustrated next as Example 2.

6. Physical systems. The limitations and usefulness of the averaged equations are illustrated in the following examples of pure stratified flow and dispersed flow:

Example 1: A pure stratified steady incompressible laminar channel flow system is shown in Fig. 1. In this case, Eq. (3) is put to its limit and phasic equations have to be solved for closure as the constitutive equations are not known beforehand. Since the continuity equation is already satisfied, we only need to account for the x-component of the momentum equation where closure of the averaged equation calls for the determination of the transfer integral of shear stress:

$$v^{-1} \int_{A_k} \underline{\underline{\tau}}_k \cdot \underline{n}_k dA_k \equiv \rho_k F_{kf} ({}^i\langle \underline{U}_f \rangle - {}^i\langle \underline{U}_k \rangle) \tag{20}$$

at the interface where F_{kf} is the relaxation time constant for momentum transfer from phase f to phase k and F_{kw} for that from the wall. For pressure drop $d\langle P\rangle/dx = -C$, and $P_k = P_f$

$$v^{-1} \int_{A_k} P_k \underline{n}_k \, dA = P_k \, d\alpha_k/dx \tag{21}$$

Solution for laminar flow gives ($\langle P_k \rangle = \alpha_k {}^i\langle P_k \rangle$)

$$0 = -\frac{d}{dx} {}^i\langle P_k \rangle + \bar{\rho}_k F_{kw} (-{}^i\langle U_k \rangle) + \bar{\rho}_k F_{kf} ({}^i\langle U_f \rangle - {}^i\langle U_k \rangle) \tag{22}$$

where the constitutive relations are given by solving the local equations with

$$F_{kw} = \frac{6 \mu_k}{\bar{\rho}_k y_{ok}^2} \frac{[1 + (B \mu_k / y_{ok}^2)]}{[1 + 3(B \mu_k / y_{ok}^2)]},$$

$$F_{kf} = \frac{6 \mu_k \mu_f}{\bar{\rho}_k y_{ok}^2 y_{of}} \left[\frac{\mu_k}{y_{ok}} + \frac{\mu_f}{y_{of}}\right]^{-1},$$

$$B = (y_{ok} + y_{of}) \left[\frac{\mu_k}{y_{ok}} + \frac{\mu_f}{y_{of}}\right]^{-1};$$

We get

$${}^i\langle U_k \rangle = \frac{C y_{ok}^2}{12 \mu_k} \left(1 + \frac{3 B \mu_k}{y_{ok}^2}\right) \tag{23}$$

and the interface velocity is $W_k = CB/2$.

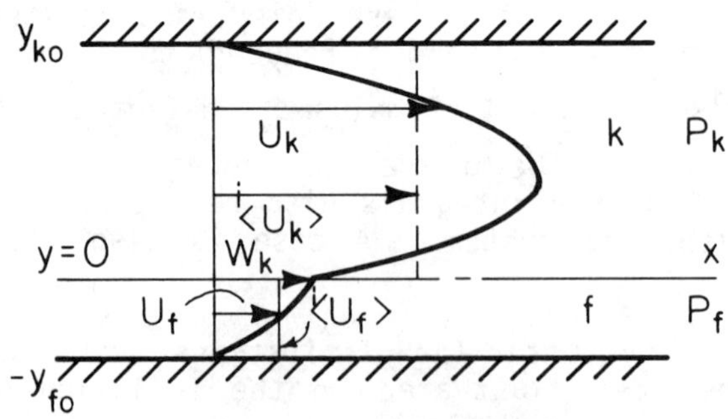

FIG. 1. Incompressible fully developed laminar flow-pure stratified flow.

Example 2. A dispersed system with strong phase interactions is represented by a fluidized bed. A knowledge of the averaged motion of a fluidized bed is important to the transport process of mass and energy, and erosion. Extending from an earlier study [11], a computation of the circulatory motion in a cylindrical fluidized bed is illustrated. This average circulatory motion of particles was experimentally determined by using a radioactive tracer [12] to be ascending at the center and descending near the wall (ACDW) or in the reverse direction (AWDC) and may have multiple cell patterns depending on the gas velocity and bed geometry. For a cylindrical bed of radius r_o and expanded bed height h (Fig. 2) of a single species of solid particles in an isothermal gas. The radial and axial coordinates of the system are r and z with the origin at the bottom center of the bed, and z is aligned in the opposite direction of gravity of acceleration g. Gas enters at the bottom through a porous plate at an average velocity of W_o. The conjugate velocity components in the r,z directions are U, W of the gas and U_p, W_p of the particle cloud of density $\rho_p = \alpha_p \bar{\rho}_p$ which is taken to be uniform in this primary solution of one toroidal cell, although the solution can be extended to multiple cells. The average equations are now reduced to: (averaging signs are dropped).

(24) $$(1/r)\ (\partial/\partial r)\ (rU) + (\partial W/\partial z) = 0$$

(25) $$(1/r)\ (\partial/\partial r)\ (r\ U_p) + (\partial W_p/\partial z) = 0$$

(26) $$0 = \rho_p\ F(W - W_p) + \mu_p[(1/r)\ (\partial/\partial r)\ r(\partial\ W_p/\partial r) + (\partial^2\ W_p/\partial z^2)] - \rho_p\ g$$

since the viscosity of particle-particle interaction μ_p predominates, and

(27) $$0 = -(\partial P/\partial z) - \rho_p\ F\ W_o - (\rho_p + \rho)\ g$$

The latter momentum equation is for the mixture. The boundary conditions are: At r = 0, $U = U_p = 0$, $\partial W/\partial r = \partial W_p/\partial r = 0$; at $r = r_o - a$, a being the particle radius ($r_o >> a$), $W_p = W_{pw}(z)$, $W = W_o + W_w(z)$, $U = U_p = 0$; and for the slip motion at the wall [11], $-\lambda_p(\partial W_p/\partial r)_w = W_{pw}(z) - W_w(z)$, where $\lambda_p \sim a/3(2)^{1/2}\alpha_p$ is the mean free path of particle-particle interaction; and at z = 0 and z = h, $W_p = 0$, $\dot{m}_p/\alpha_p\ \bar{\rho}_p = 2\pi \int_0^{r_o} W_p\ r\ dr = 0$.

Eqs. (24) to (27) are satisfied by the solutions:

(28) $$W = W_o + \sin(\pi\ z/h) \sum_n c_n(r/r_o)^n$$

(29) $$U = -(\pi\ r_o/h)\ \cos(\pi\ z/h) \sum_n c_n/(n+2)\ (r/r_o)^{n+1}$$

(30) $$W_p = \sin(\pi\ z/h) \sum_n a_n(r/r_o)^n$$

(31) $$U_p = -(\pi\ r_o/h)\ \cos(\pi\ z/h) \sum_n [a_n/(n+2)]\ (r/r_o)^{n+1}$$

for $c_o - a_o \ll W_o$, which is physically valid; and the boundary conditions reduce the above series to polynomials

with a_n's amd c_n's given by:

$a_0 = -w_{pw}[16/5) + (1/20)(\gamma/\kappa)] + (13/140)(w_w/\kappa)$, $a_1 = 0$,

$a_2 = w_{pw}[14 + (1/3)(\gamma/\kappa)] - (26/63)(w_w/\kappa)$, $a_3 = 0$,

$a_4 = -w_{pw}[21 + (3/4)(\gamma/\kappa)] + (13/28)(w_w/\kappa)$,

$a_5 = w_{pw}[56/5) + (7/15)(\gamma/\kappa)] - (13/90)(w_w/\kappa)$; and

$c_o = \gamma\, a_o - 4\,\kappa\, a_2$, $c_1 = 0$, $c_2 = \gamma\, a_2 - 16\,\kappa\, a_4$, $c_3 = -25\,\kappa\, a_5$,

$c_4 = \gamma\, a_4$, $c_5 = \gamma\, a_5$; where $\gamma = 1 + [(\pi^2\mu_p)/(h^2\rho_p F]$, $\kappa = [\mu_p/(r_o^2\rho_p F]$.

The slip velocity at the wall $w_w = \Sigma_n c_n$, $w_{pw} = \Sigma_n a_n$ are given by:

$$w_{wp}[(16/5) + (1/20)(\gamma/\kappa) - (1/20)(\gamma^2/\kappa) - 56\,\kappa - (68/15)\,\gamma] + (13/20)\,w_w[(1/7)(\gamma/\kappa) - (1/7)(1/\kappa) + (160/63)] = (g/F) - W_o \tag{32}$$

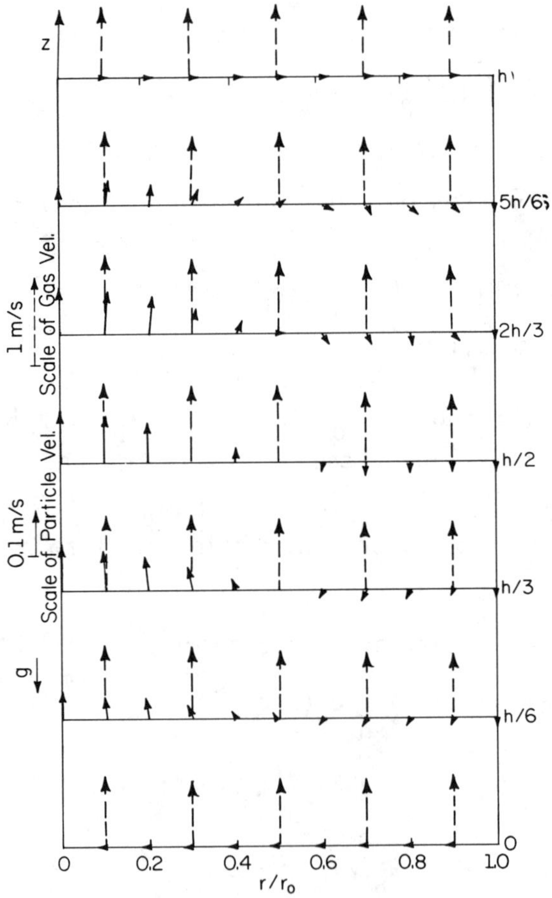

FIG. 2. Motion of phases in the fluidized bed in the generic example as shown by velocity vectors (solids ACDW).

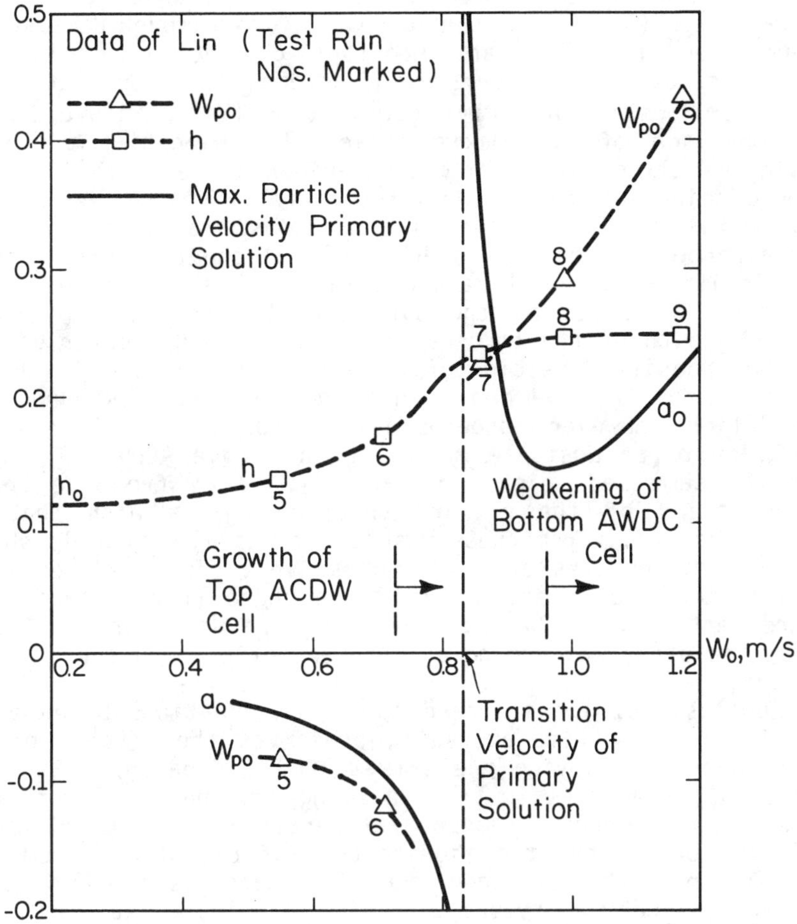

FIG. 3. Comparison of primary solution to experimental results of Lin [12] and the trends of transition from AWDC to ACDW in the experimental system and the single cell solution.

and

$$w_{wp} = w_w[1 - (13/42)\ (\lambda_p/r_o)\ (1/\kappa)] \tag{33}$$

These latter two relations determine w_w and w_{pw} at $r = r_o - a$.

A generic case is illustrated in Fig. 2 for the case of physical parameters given by: $r_o = 0.1$ m, $h = 0.22$ m, $\alpha_p = 0.30$, $W_o = 0.8$ m/s, $\bar{\rho}_p = 2500$ kg/m^3, $\mu_p = 0.1822$ kg/m-s, $F = 12$ s^{-1}, $\gamma = 1.0124$, $\kappa = 0.006073$, with $w_{pw} = -0.01905$ m/s and $w_w = 0.01$. In this case we get $a_0 = 0.1263$, $a_2 = -0.7618$, $a_4 = 1.599$, $a_5 = 0.9744$; $c_0 = 0.1464$, $c_2 = -0.9266$, $c_3 = 0.1479$, $c_4 = 1.619$, $c_5 = -0.9865$. This is a case of ACDW with velocity vectors as shown. The trend conforms to the observations [12]. One notes that the maximum velocities of circulation W_{po} is given by values of a_0 along the z-axis of these cases which check with the experimental results within the order of magnitude because of the simplifying specifications, an important one being that of uniform density; and the available accuracy in the empirical correlations of viscosity μ_p and relaxation time constant F [3]. Figure 3 shows an overall comparison of the maximum particle velocities given by the primary single cell solution to the experimental results [12] including expanded bed heights h for various interstitial velocities of the gas. The experimental results include a continuous transition from AWDC to ACDW over a range of growth of a ACDW cell at the top. Starting from a low air velocity a AWDC cell appears first at a low maximum velocity downward which increases as the air velocity increases. Instability leads to the growth of the toroidal cell at the upper outer edge of run No. 6 in the opposite direction of the prevailing AWDC mode, leading to the transition to the ACDW modes in Runs 7 and 8. The curves of the primary solution, however, show a discontinuous transition between a single AWDC cell (an unstable mode) to a single ACDW cell (a stable mode) for the same bed heights as in [12]. Away from the transition region, the trend of the primary solution agrees with that of the experimental results. From these results it was also readily shown that the motion of the bubbles of dilute phase in a fluidized bed is influenced by the motion of the dense particle phase (for instance, bubbles rise toward center for ACDW mode and along the wall for the AWDC mode) rather than the other way around as has been often conjectured.

7. Conclusions. The averaged continuum equations based on volume averaging followed by time averaging preserves the concept of dynamic phases with phases occupying the same volume in space and accounts for the presence of interfaces in the averaging procedure. A condition for preserving the barycentric frame of reference has been introduced. The scale relation determines the significance of deviations from averages and suggests that the averaged equations are especially useful in dealing with dispersed systems. The physical examples of pure stratified flow and fluidized bed demonstrate the feasibility and usefulness in applications.

Acknowledgements. Study was supported in part by the National Science Foundation and the ALCOA Foundation.

REFERENCES

[1] S. L. SOO, Dynamics of Multiphase Flow Systems, Ind. Eng. Chem Fund., 4 (1965), pp. 426-433.

[2] N. ZUBER, Flow excursions and oscillations in boiling two-phase flow systems with heat additions, Proc. Symp. Two-phase Flow Dynamics, Eindhoven, 1 (1967), pp. 1071-1089.

[3] S. L. SOO, Fluid dynamics of multiphase systems, Blaisdell Pub. Co., (1967), pp. 367-370.

[4] S. L. SOO, Effects of configuration of phases on dynamic relations, AIChE Symp. Ser. 208, Heat Transfer (R. F. Stein, Ed.), 77 (1981), pp. 152-160.

[5] G. BIRKHOFF, Averaged conservation laws in pipe, J. of Math. Analysis and Applications, 1 (1964), pp. 66-77.

[6] J. C. SLATTERY, Flow of viscoelastic fluids through porous media, AIChE J., 13 (1967), pp. 1066-1071.

[7] S. WHITAKER, Advances in Theory of Fluid Motion in Porous Media, Ind. Eng. Chem., 61 (1969), pp. 14-28.

[8] J. M. DELHAYE and J. L. ACHARD, On the averaging operators introduced in two-phase flow modeling, Proc. of OECD/NEA Specialists' Meeting on Transient Two-phase Flow, Toronto, Pergamon Press (S. Banerjee and K. R. Weaver, Eds.), 1 (1976), pp. 5-84.

[9] W. T. SHA, B. T. CHAO, and S. L. Soo, Averaging procedures of multiphase conservation equations, First Proc. of Nuclear Thermal Hydraulics, ANS (1983), pp. 3-12.

[10] T. B. ANDERSON and R. JACKSON, A fluid mechanical description of fluidized beds, Ind. Chem. Fund., 6 (1967), pp. 527-539.

[11] S. L. SOO, A note on motions of phases in a fluidized bed, J. Powder Tech., 45 (1986), pp. 169-172.

[12] J. S. LIN, M. M. CHEN, and B. T. CHAO, A novel radioactive particle tracking facility for measurement of solids motion in gas fluidized beds, AIChE J., 31 (1985), pp. 465-473.

Effective Transport Properties of Multiphase Media from the Microstructure

S. TORQUATO*

Abstract. We derive a series representation of a general n-point distribution function H_n, that characterizes the microstructure of multiphase media, which enables us to systematically compute H_n for a certain class of models. It is shown that any of the statistical correlation functions which have arisen in expressions for the effective conductivity σ_e of composite media and the fluid permeability k of porous media can be obtained from this single expression for H_n. Using the series representation of the H_n, we evaluate H_n for a system of randomly centered spheres, for any n. We also introduce and briefly describe a new set of statistical correlation functions which are expected to reflect information about percolating clusters in the system. This is followed by a description of a new highly accurate expression for the conductivity of dispersions of inclusions. Some preliminary calculations of the Weissberg-Prager upper bound on k for beds of penetrable spheres, through second-order in the sphere volume fraction, is discussed.

1. Introduction. In order to exactly predict the effective transport property of random multiphase media, an infinite set of correlation functions, which statistically characterize the microstructure, must be known [1,2]. Except in a few special cases, the infinite set of correlation functions is never known and hence an exact

---

* Departments of Mechanical and Aerospace Engineering and of Chemical Engineering, North Carolina State University, Raleigh, North Carolina 27695-7910.

calculation of the bulk property is generally out of the question. Rigorous bounding methods provide a means of estimating the effective property given a limited amount of microstructural information on the heterogeneous material. Rigorous bounds are useful because: (i) they enable one to test the merits of a theory, (ii) one of the bounds can typically provide a relatively accurate estimate of the property, and (iii) as successively more microstructural information is included the bounds become progressively tighter. Here we shall study and evaluate bounds on the effective electrical conductivity σ_e of composites and on the fluid permeability k of porous media that depend upon the details of the microstructure through various sets of statistical correlation functions.

The two-phase conductivity bounds of Beran [3] depend upon the correlation functions $S_1(\underset{\sim}{x}_1)$, $S_2(\underset{\sim}{x}_1,\underset{\sim}{x}_2)$, and $S_3(\underset{\sim}{x}_1,\underset{\sim}{x}_2,\underset{\sim}{x}_3)$. The $S_n(\underset{\sim}{x}^n)$ ($\underset{\sim}{x}^n \equiv \underset{\sim}{x}_1,\ldots,\underset{\sim}{x}_n$) give the probability of finding n points at positions $\underset{\sim}{x}^n$, respectively, all in one of the two phases, say phase 1. (For statistically homogeneous media, S_1 is simply equal to ϕ_1, the volume fraction of phase 1.) Milton [4] has formally derived general bounds on σ_e that depend upon the set $S_1,S_2,\ldots,S_n$, for arbitrary n. Torquato [5] has recently obtained bounds on σ_e for systems composed of spheres of variable penetrability distributed throughout a matrix. These bounds depend upon a different set of correlation functions, namely, $G_0,G_1,\ldots,G_n$. $G_n(\underset{\sim}{x};\underset{\sim}{r}^n)$ characterizes the correlation associated with finding a point at $\underset{\sim}{x}$ in the matrix phase and a configuration of spheres with centers at $\underset{\sim}{r}^n$. The permeability bounds of Weissberg and Prager [6] and of Doi [7] depend upon, among other quantities, certain two-point correlation functions that involve information about the two-phase interface. (The Weissberg-Prager bound also incorporates G_0,G_1, and G_2.) Moreover, Prager [8] derived bounds on k that involve S_1,S_2, and S_3.

Historically, practical application of all but one of the bounds described above (i.e., the bounds due to Torquato [5]) has been very limited because of the difficulty involved in obtaining even lower-order correlation functions either theoretically or experimentally. Recent advances in the characterization of the microstructure of random media suggests that this impasse is in the process of being overcome. For a statistically inhomogeneous two-phase medium consisting of identical particles dispersed throughout a "matrix" phase (phase 1) such that the location of each inclusion is fully specified by a position vector (e.g., spheres, disks, and oriented cubes, squares, and ellipsoids), the S_n can be systematically represented as an infinite series in terms of the n-particle probability density functions ρ_n [9,10]; functions which are, in principle, known quantities for the ensemble being considered. (The quantity $\rho_n(\underset{\sim}{r}^n)$ characterizes the probability of finding any n particles with configuration $\underset{\sim}{r}^n$.) This series representation of the S_n has led to the evaluation of lower-order S_n (i.e., S_1,S_2, and S_3) [11-15] and hence of conductivity bounds [16-19] for nontrivial models of composite and porous media. Recently, a highly accurate expression for σ_e of dispersions, which depends upon S_1, S_2, and S_3 was derived [2]. (The results of this study will be briefly described in Section 3.) On the experimental side, Berryman [20] has shown that image processing techniques can be employed to efficiently measure the S_n from samples of the material. Using series representations of the G_n, the conductivity bounds due to Torquato [5] have also been evaluated for various models [5,21].

There has been comparatively much less progress in the calculation of permeability bounds. Bounds on k computed to date are only useful for very high porosities [6,22]. A stumbling block in the evaluation of such bounds has been a lack of a systematic means of computing the correlation functions that arise here, i.e., quantities involving two-phase interface information.

We have recently [23] derived a series expression for a very general n-point distribution function H_n, for <u>inhomogeneous</u> media composed of particles distributed throughout matrix or void. The H_n $(\underset{\sim}{x}^m; x^{p-m}; \underset{\sim}{r}^q)$ gives the correlation associated with finding: m points at $\underset{\sim}{x}^m$ on certain surfaces within the system, p-m points at $\underset{\sim}{x}^{p-m}$ $(\equiv \underset{\sim}{x}_{m+1}, \ldots, \underset{\sim}{x}_p)$ in certain portions of the matrix (or void) phase, and the centroids of the particles at $\underset{\sim}{r}^q$, where $n = p + q$. Using this single expression for H_n, we may now derive and compute any of the various sets of correlation functions that have been described in the literature, i.e., S_n, G_n, and the surface correlation functions, for a wide class of models of isotropic, anisotropic, and inhomogeneous media.

In Sections 2 and 3, we describe some recent advances in the characterization of the microstructure and in the calculation of bounds on σ_e and k, respectively. In Section 2 we present a series representation of the general n-point distribution function H_n. It is shown how any of the correlation functions described above can be obtained from this single expression for H_n. The nature of this series enables us to obtain successive rigorous upper and lower bounds on the H_n. Using the series representation of the H_n, we evaluate H_n for a system of randomly centered particles, for arbitrary n. We also introduce and briefly discuss a new set of statistical correlation functions which are expected to contain information about the percolation regime of the medium. In Section 3 we describe a new highly accurate expression for the conductivity of dispersions of inclusions. We also present some preliminary calculations of the Weissberg-Prager upper bound on k for spheres of variable penetrability through second order in the sphere volume fraction ϕ_2. This low-density result for k is compared to an analogous upper bound on σ_e for spheres which are insulating relative to the matrix.

2. Characterization of the Microstructure

2.1 n-Point Distribution Function H_n. Consider adding p spherical "test" or "solute" particles of radii $b_1,\ldots,b_p$, respectively, centered at positions $\underset{\sim}{x}^p$, respectively, to a system of N equi-sized spherical ("solvent") particles of radius R centered at $\underset{\sim}{r}^N$, respectively, where N is sufficiently large to justify a statistical treatment. It is important to emphasize that the solvent particles of radius R represent the actual inclusions in the composite. The p solute particles are capable of excluding the centers of the solvent particles from spheres of radius $a_1,\ldots,a_p$, respectively, surrounding the solute particles. For $b_i > 0$, $a_i = R + b_i$, and for $b_i = 0$, we shall allow the solute particles to penetrate the solvent particles so that $a_i = R - c_i$ $(0 \le c_i \le R)$, where a_i is the minimum separation distance between solute and solvent centers. For simplicity, we let the solute particles be perfectly penetrable to one another, i.e., they are mutually spatially uncorrelated. It is natural to associate with each solute particle i a subdivision of space into two regions: the space available to the ith solute particle (i.e., the space outside N spheres of radius a_i centered at $\underset{\sim}{r}^N$), D_i, and the complement space D_i^*. The entire system is a domain of space D ($= D_i + D_i^*$) of volume V. Let $\mathcal{S}_i$ denote the surface between D_i and D_i^*. Note it is only when $b_i = 0$ and $a_i = R$ that D_i and $\mathcal{S}_i$ refer to the matrix (or void) phase space (i.e., the space outside the solvent particles) and the two-phase interface, respectively.

The solvent particles are in general distributed with arbitrary degree of impenetrability. The degree of impenetrability is characterized by some parameter λ whose value varies between zero (in the case where the spheres are randomly centered and thus completely uncorrelated, i.e., "fully penetrable spheres") and unity (in the instance of totally impenetrable spheres). Examples of sphere distributions involving intermediate values of λ include the permeable-sphere model [24] and the penetrable-concentric-shell (PCS) model [25]. In the PCS model (depicted in Fig. 1) spheres (disks) of radius R are statistically distributed throughout space subject only to the condition of a mutually impenetrable core of radius λR, $0 \le \lambda \le 1$. Such a model enables one to study the effect of the degree of connectivity on the effective transport property. The degree of the connectivity of the individual phases may greatly influence the transport properties of disordered multiphase media, particularly when the phase properties differ significantly. For example, for λ nearly equal to one, the PCS model may be employed to mimic the particle contacts present in a bed of rigid spheres near the close-packing volume fraction. The PCS model may also serve as a useful model for consolidated media, such as sandstones and other rocks, and sintered materials.

The n-point distribution function $H_n(\underset{\sim}{x}^m;\underset{\sim}{x}^{p-m};\underset{\sim}{r}^q)$, which characterizes the special mixture of solute and

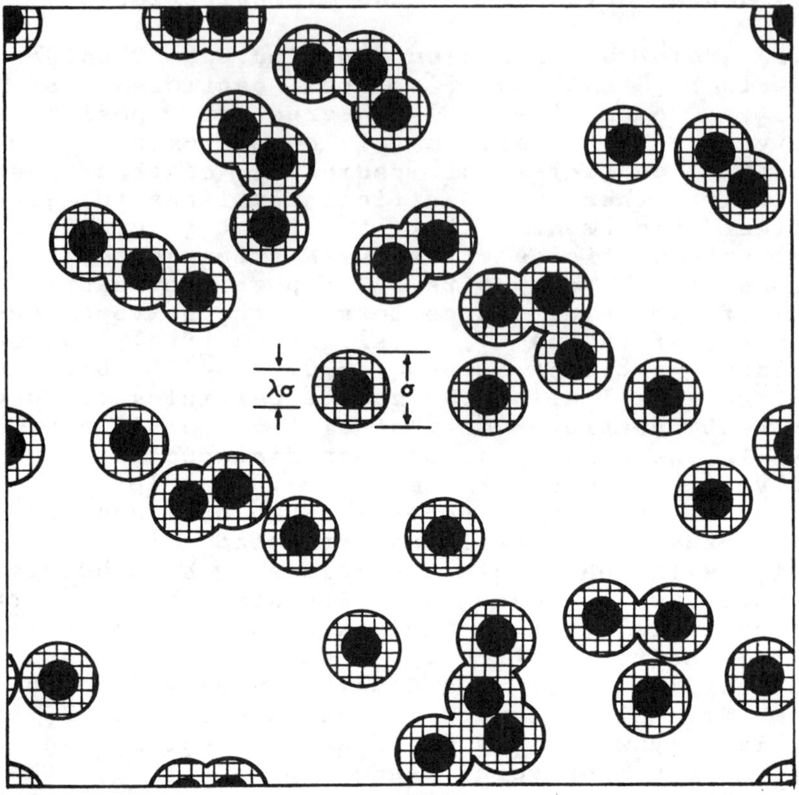

FIG. 1. A computer-generated realization of a distribution of disks of radius $R = \sigma/2$ (shaded region) in a matrix (unshaded region) in the PCS model [25]. The disks have an impenetrable core of diameter $\lambda\sigma$ indicated by the smaller, black circular region. Here $\lambda = 0.5$ and the particle volume fraction is 0.3.

solvent particles described above, gives the correlation associated with finding: the center of a solute particle of radius b_1 at $\mathbf{x}_1$ on $\mathcal{S}_1, \ldots,$ and the center of a solute particle of radius b_m at $\mathbf{x}_m$ on $\mathcal{S}_m$, and the center of a solute particle of radius b_{m+1} at $\mathbf{x}_{m+1}$ in $D_{m+1}, \ldots,$ and the center of a solute particle of radius b_p at $\mathbf{x}_p$ in D_p, and that any q of the solvent particles have configuration $\mathbf{r}^q$, where $n=p+q$ [23]. From this single function one can obtain all of the various sets of correlation functions described in the Introduction by letting the radii of all the solute particles shrink to zero ($b_i=0$) and setting $a_i=R$, $i=1,\ldots p$. For example, in this limit, the n-point matrix probability function $S_n(\mathbf{x}^n)=H_n(\emptyset;\mathbf{x}^n;\emptyset)$ and $G_q(\mathbf{x}_1;\mathbf{r}^q)=H_n(\emptyset;\mathbf{x}_1;\mathbf{r}^q)$, where $\emptyset$ denotes the null set. Similarly, in this limit $H_2(\mathbf{x}_1;\mathbf{x}_2;\emptyset)$ and $H_2(\mathbf{x}_1,\mathbf{x}_2;\emptyset;\emptyset)$ are the surface-matrix and surface-surface correlation functions defined by Doi [7], and $H_2(\mathbf{x}_1;\emptyset;\mathbf{r}_1)$ is a two-point surface correlation function closely related to the one defined by Weissberg and

Prager [6].

It can be shown that [23]

$$(1)\quad H_n(\underset{\sim}{x}^m;\underset{\sim}{x}^{p-m};\underset{\sim}{r}^q) = \sum_{s=0}^{\infty} (-1)^s H_n^{(s)}(\underset{\sim}{x}^m;\underset{\sim}{x}^{p-m};\underset{\sim}{r}^q),\quad n=p+q,$$

where

$$(2)\quad H_n^{(s)}(\underset{\sim}{x}^m;\underset{\sim}{x}^{p-m};\underset{\sim}{r}^q) = (-1)^m \frac{\partial}{\partial a_1}\cdots\frac{\partial}{\partial a_m} G_n^{(s)}(\underset{\sim}{x}^p;\underset{\sim}{r}^q),$$

$$(3)\quad G_n^{(s)}(\underset{\sim}{x}^p;\underset{\sim}{r}^q) = \prod_{\ell=1}^{q}\prod_{k=1}^{p} e(y_{k\ell};a_k)$$

$$\times \int \rho_{q+s}(\underset{\sim}{r}^{q+s}) \prod_{j=q+1}^{q+s} m^{(p)}(\underset{\sim}{x}^p;\underset{\sim}{r}_j)d\underset{\sim}{r}_j,$$

and

$$(4)\quad G_n^{(0)}(\underset{\sim}{x}^p;\underset{\sim}{r}^q) \equiv \rho_q(\underset{\sim}{r}^q) \prod_{\ell=1}^{q}\prod_{k=1}^{p} e(y_{k\ell};a_k)\quad .$$

Here ρ_n is the n-particle probability density described in the Introduction and

$$(5)\quad m^{(p)}(\underset{\sim}{x}^p;\underset{\sim}{r}_j) = 1 - \prod_{i=1}^{p} [1-m(y_{ij};a_i)],$$

$$(6)\quad m(y_{ij};a_i) = \begin{cases} 1 & ,\ y_{ij} < a_i \\ 0 & ,\ \text{otherwise} \end{cases},$$

$$(7)\quad e(y_{ij};a_i) = 1 - m(y_{ij};a_i)\quad ,$$

and $y_{ij} = |\underset{\sim}{x}_i - \underset{\sim}{r}_j|$. Given ρ_n for the ensemble, we may now compute H_n from Eqs. (1)-(7). Note that Eqs. (1)-(4) apply not only to inhomogeneous sphere (disk) distributions but to inhomogeneous distributions of particles such that the location of each particle is fully specified by a position vector (i.e., oriented cubes, squares, ellipsoids, ellipses, etc.). Of course when considering particles which have a nonspherical shape, Eq. (6) must be appropriately generalized.

The nature of the infinite series (1) enables us to obtain successive upper and lower bounds on H_n. We find [23] for even m

$$(8)\quad H_n \begin{matrix} \ge \\ \le \end{matrix} \begin{cases} W_n^{(\ell)} & \text{for } \ell \text{ odd} \\ & \text{for } \ell \text{ even} \end{cases}$$

and for odd m

$$(9)\quad H_n \; \begin{matrix}\geq \\ \leq\end{matrix} \; W_n^{(\ell)} \quad \begin{cases} \text{for } \ell \text{ even} \\ \text{for } \ell \text{ odd} \end{cases}$$

where

$$(10)\quad W_n^{\ell} = \sum_{s=0}^{\ell} (-1)^s H_n^{(s)}$$

is the partial sum and $H_n^{(s)}$ is given by Eq. (2).

Consider the evaluation of H_n for a mixture of p solute particles and N solvent particles which are randomly centered (i.e., fully penetrable spheres). For such a system $\rho_n(\underset{\sim}{r}^n) \to \rho^n$, where $\rho = N/V$ is the number density of spheres. Upon use of Eqs. (1)-(7) we find [23]

$$(11)\quad H_n(\underset{\sim}{x}^m;\underset{\sim}{x}^{p-m};\underset{\sim}{r}^q)=(-1)^m\rho^q\exp[-\rho V_p(\underset{\sim}{x}^p)]$$

$$\times \frac{\partial}{\partial a_1}\cdots\frac{\partial}{\partial a_m}\left[\prod_{\ell=1}^{q}\prod_{k=1}^{p} e(y_{k\ell};a_k)\right]$$

$$+(-1)^m\rho^q\left[\prod_{\ell=1}^{q}\prod_{k=q}^{p} e(y_{k\ell};a_k)\right]\frac{\partial}{\partial a_1}\cdots\frac{\partial}{\partial a_m}\exp[-\rho V_p(\underset{\sim}{x}^p)].$$

Here the quantity $V_p(\underset{\sim}{x}^p)$ is the union volume of p spheres of radius $a_1,\ldots,a_p$, respectively, centered at $\underset{\sim}{x}_1,\ldots,\underset{\sim}{x}_p$, respectively.

Using Eqs. (1)-(7) we have also evaluated [26], for beds of impenetrable spheres, the surface-void and surface-surface correlation functions defined by Doi [7] and a surface-particle center correlation function closely related to one defined by Weissberg and Prager [6].

2.n-Point Cluster Correlation Functions. The lower-order correlation functions described above do not reflect information about percolating clusters within the system. For example, the two-point matrix probability function $S_2(x)$ for fully penetrable spheres [11] at the percolation-threshold sphere volume fraction, equal to about 0.3 [27], is not singularly different than $S_2(x)$ at other volume fractions (see Fig.2). In the study of disordered multiphase media, it would be highly desireable to define and employ correlation functions, analogous to the ones described in Section 2.1, which contain percolation-sensitive information.

Here, for the first time, we introduce and briefly describe such quantitites for media composed of particles dispersed throughout a matrix or void phase. This is

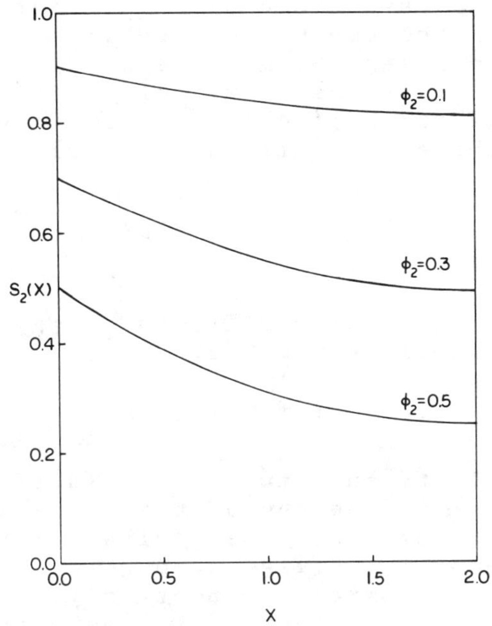

FIG. 2. The two-point matrix probability function $S_2(x)$ for a system of fully penetrable spheres of unit radius at $\phi_2=0.1$, 0.3, and 0.5 [11]. For $r \geq 2$, $S_2=(1-\phi_2)^2$.

accomplished by first defining the intermediate quantity $P_n(\underset{\sim}{x}^n)$, which gives the probability of finding each of the n points at $\underset{\sim}{x}^n$ in n different particles, regardless of whether any point is simultaneously in more than one sphere. $P_n(\underset{\sim}{x}^n)$ in turn is related to the probability of finding n points anywhere in the particle phase, $\bar{S}_n(\underset{\sim}{x}^n)$. The latter quantity can be expressed in terms of the n-point matrix probability functions via the relation [9]

$$(12)\quad \bar{S}_n(\underset{\sim}{x}^n) = 1 - \sum_{i=1}^{n} S_1(\underset{\sim}{x}_i) + \sum_{i<j}^{n} S_2(\underset{\sim}{x}_i,\underset{\sim}{x}_j) - \sum_{i<j<k}^{n} S_3(\underset{\sim}{x}_i,\underset{\sim}{x}_j,\underset{\sim}{x}_k) + \ldots$$

Let $\{\gamma\} = \{\gamma_1 \mid \gamma_2 \mid \ldots \mid \gamma_k\}$ denote the unordered partitioning of the set $\{\underset{\sim}{x}_1,\ldots,\underset{\sim}{x}_n\}$ into k disjoint, unordered subsets. Let the n-point cluster function $C_n(\gamma_1 | \gamma_2 | \ldots | \gamma_k)$ be the probability of finding each point of γ_1 in N_1 different particles which are members of the same cluster, each point of γ_2 in N_2 different particles which are members of the same cluster, etc., regardless of whether any point is simultaneously in more than one particle, such that the clusters associated with each subset are different. N_i is the number of points in the subset γ_i. A cluster is defined as that part of the particle phase which can be reached from a particle-phase point without passing through the matrix or void phase. Clusters may form as the result of overlap between the particles or if the particles come in contact with one another. From these definitions it is clear that

$$(13)\quad P_n(\underset{\sim}{x}^n) = \sum_{\{\gamma\}} C_n(\gamma_1 \mid \gamma_2 \mid \cdots \mid \gamma_k),$$

where the sum is over all partitions of $\{\gamma\}$. For the case n = 2, Eq. (13) simply gives that

$$(14)\quad P_2(\underset{\sim}{x}_1,\underset{\sim}{x}_2) = C_2(\underset{\sim}{x}_1,\underset{\sim}{x}_2) + C_2(\underset{\sim}{x}_1 \mid \underset{\sim}{x}_2) ,$$

i.e., the probability that two points fall in two different particles is equal to the sum of the probabilities that the points fall in two different particles which belong to the same cluster and two different particles which do not belong to the same cluster, respectively.

We still, however, have not described how n-point cluster functions can be calculated. For simplicity we focus our attention on the calculation of the two-point cluster functions for isotropic dispersions of equi-sized spheres. Using Eq. (12) in conjunction with the results of Ref. 14, it is easily shown that

(15) $\bar{S}_2(r) =$

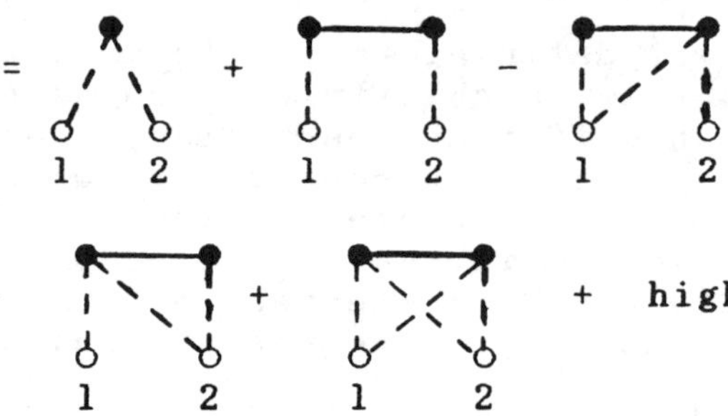

+ higher-order graphs,

where - - - is an m bond, ——— is a $g_2(r)$ bond, $g_2(r)=\rho_2(r)/\rho^2$ is the radial distribution function, and $r = |\underset{\sim}{x}_1 - \underset{\sim}{x}_2|$. Here we have used graphical representation of integrals described in Ref. 14. For example,

(16)

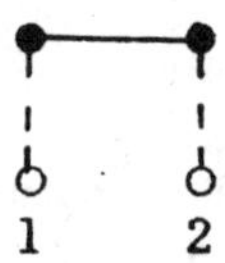

$$= \rho^2 \int\int g_2(r_{34}) m(|\underset{\sim}{x}_1 - \underset{\sim}{r}_3|)\, m(|\underset{\sim}{x}_2 - \underset{\sim}{r}_4|)\, d\underset{\sim}{r}_3 d\underset{\sim}{r}_4,$$

where $r_{34} = |\underset{\sim}{r}_3 - \underset{\sim}{r}_4|$. The first graph in Eq. (15) is simply the contribution to $\bar{S}_2(r)$ when both points lie in the same sphere. The next graph gives the additional contribution of $\bar{S}_2$ regardless of whether point 1 and/or 2 lie in more than one sphere. Hence, all other graphs are extraneous in determining P_2 since they specifically involve cases where point 1 and/or 2 are constrained to lie in more than one sphere. Therefore, by definition of the P_n and by Eq. (15), we have exactly

(17) $P_2(r) =$ 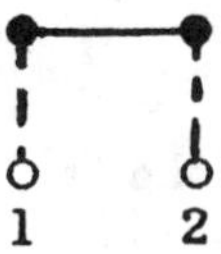 .

It remains now to employ Eq. (14) to obtain the two-point cluster functions. This is done by noting that the $g_2(=\rho_2/\rho^2)$ involved in Eq. (17) can be decomposed into a sum of terms involving pairs of particles which are part of the same cluster g_2^+ and pairs of particles which are not part of the same cluster g_2^*:

(18) $g_2(r) = g_2^+(r) + g_2^*(r)$.

Here, for example, $g_2^+(r)$ is the pair-connectedness function and is defined such that $\rho^2 \, g_2^+(r) \, d\underset{\sim}{r}_1 d\underset{\sim}{r}_2$ is the probability of finding the center of a particle in volume element $d\underset{\sim}{r}_1$ about $\underset{\sim}{r}_1$ and another particle, of the same cluster, in the volume $d\underset{\sim}{r}_2$ about $\underset{\sim}{r}_2$, $r = |\underset{\sim}{r}_1 - \underset{\sim}{r}_2|$ [28]. The quantity $g_2^*(r)$ is the pair-blocking function and is defined such that $\rho^2 \, g_2^*(r) \, d\underset{\sim}{r}_1 d\underset{\sim}{r}_2$ is the probability of finding the center of a particle in volume $d\underset{\sim}{r}_1$ about $\underset{\sim}{r}_1$ and another particle, not belonging to the same cluster, in the volume $d\underset{\sim}{r}_2$ about $\underset{\sim}{r}_2$ [28]. Consequently, from Eqs. (14) and (18), we have that $C(r) \equiv C_2(\underset{\sim}{x}_1, \underset{\sim}{x}_2)$ is given by the contribution to $P_2(r)$ involving the $g_2^+(r)$ only and $\bar{C}(r) \equiv C_2(\underset{\sim}{x}_1 | \underset{\sim}{x}_2)$ is given by the contribution to $P_2(r)$ involving the $g_2^*(r)$ only.

To illustrate an application of this formalism, we consider a percolating system of hard spheres and a percolating system of overlapping spheres. Examples of the former include systems of hard spheres at the close-packing limit [29] and the adhesive- or sticky-sphere model of Baxter [30]. In both cases the radial distribution function possesses a Dirac-delta contribution when the particles are in contact with one another. The Baxter model is characterized by a "stickiness" parameter μ, with $\mu \to 0$ corresponding to non-sticky hard spheres. In the sticky-sphere model, the percolation-threshold sphere volume fraction ϕ_2 decreases from random close-packing limit of 0.64 ($\mu = 0$) as μ increases. Examples of the latter include the permeable-sphere [24] and PCS [25] models. For percolating systems of equi-sized spheres, Eqs. (14), (17), and (18) the two-point cluster functions are given by

(19) $C(r) =$ 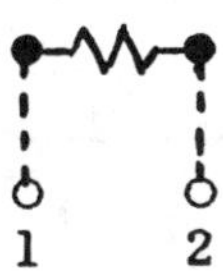

and

(20) $\bar{C}(r)$ = 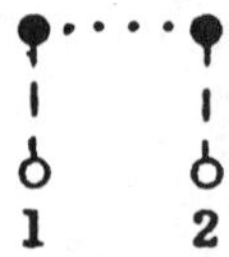 ,

where —/\/\— is a $g^+(r)$ bond and $\cdots\cdot$ is a $g^*(r)$ bond. Hence, given g^+ and g^* we may calculate $C(r)$ and $\bar{C}(r)$, respectively. For the sticky-sphere model or the permeable-sphere model, one can calculate the convolution integral (19) using, for example, the Percus-Yevick solution for g^+ and g^* derived by Chiew and Glandt [31] in the same manner described in Ref. 14. In general, since $g^+(r)$ is short-ranged for volume fractions sufficiently less than the percolation volume fraction and becomes long-ranged near the percolation transition, $C(r)$ itself will behave in a similar fashion. $C(r)$ for general models is also expected to become long- ranged as the percolation point is approached from below. Therefore, $C(r)$ should serve as a very useful signature of the microstructure.

3. Conductivity and Permeability Expressions. Here we shall discuss some recent progress we have made in the evaluation of conductivity and permeability bounds which incorporate some of the statistical quantities described in Section 2. We first describe a new result for the conductivity problem. This is followed by a presentation of some new results for the permeability problem.

3.1 Conductivity Results. Recently, Torquato [2] has derived a perturbation expansion of σ_e for general two-phase isotropic media of arbitrary dimensionality d in powers of

$$\beta_{ij} = \frac{\sigma_i - \sigma_j}{\sigma_i + (d-1)\,\sigma_j}, \quad i \neq j . \tag{21}$$

It was also shown that the [m/m] Pade approximant of this series yielded nth-order bounds, for any n=2m(m≥1), derived by Milton [4]. The reason why Pade approximants provide rigorous bounds is that the effective conductivity, as noted first by Milton and Golden [32], can be written as a Stieltjes series. Certain Pade approximants of Stieltjes functions are known to form converging upper and lower bounds on the function [33]. The [2,2] Pade approximant of the expansion of σ_e/σ_j in powers of β_{ij} is [2]

$$\frac{\sigma_e}{\sigma_j} = \frac{1+[(d-1)\phi_i-\gamma_i/\zeta_i]\beta_{ij}+(1-d)[\phi_j\zeta_i+\phi_i\gamma_i/\zeta_i]\beta_{ij}^2}{1-[\phi_i+\gamma_i/\zeta_i]\beta_{ij}+[\phi_j(1-d)\zeta_i+\phi_i\gamma_i/\zeta_i]\beta_{ij}^2} \tag{22}$$

where $i \neq j$. For $\sigma_2 \geq \sigma_1$, Eq. (22) is a lower bound and an upper bound for j=1 and j=2, respectively. The details of the microstructure enter through d, ϕ_i, ζ_i, and γ_i. The three-point parameter ζ_i is an integral that depends upon the n-point probability functions S_1, S_2 and S_3. The four-point parameter γ_i has been shown [2] to be identically zero for d = 2. In general, $\gamma_i \neq 0$ for d = 3.

Torquato [2] found that the expression for σ_e/σ_1 which results from Eq. (22) for three-dimensional composites with $\gamma_2 = 0$

$$(23) \quad \frac{\sigma_e}{\sigma_1} = \frac{1 + 2\phi_2\beta_{21} - 2\phi_1\zeta_2\,\beta_{21}^2}{1 - \phi_2\beta_{21} - 2\phi_1\zeta_2\,\beta_{21}^2}$$

should yield a highly accurate expression for dispersions provided that the characteristic length of a cluster of the dispersed phase (phase 2), on average, is much smaller than the characteristic length of the macroscopic sample. For such a dispersion, Eq. (23) will behave as an approximate but very accurate lower bound on σ_e/σ_1 when $\alpha = \sigma_2/\sigma_1 \geq 1$ and as an approximate but very accurate upper bound on σ_e/σ_1 for $\alpha \leq 1$.

To test relation (23) we have compared predicted values of σ_e/σ_1 obtained from it for periodic lattices of spheres (using the tabulation of ζ_2 for such system obtained by McPhedran and Milton [34]) to exact numerical solutions of σ_e/σ_1 [35,36]. The agreement is found to be excellent for a wide range of α and sphere volume fraction ϕ_2. Fig. 3 compares exact numerical results for σ_e/σ_1 of three cubic lattices, at the extreme condition $\alpha = \infty$ (up to 96% of their respective close-packing values), to corresponding results predicted by relation (23). Equation (23) is seen to provide an excellent estimate of σ_e/σ_1 for these models, even up to ϕ_2 approaching the percolation or close-packing volume fraction, implying that it incorporates the salient multipolar effects that are especially important when ϕ_2 is large.

Moreover, we have compared predicted values of σ_e/σ_1 obtained from Eq. (23) for random distributions of impenetrable spheres (using the results of Torquato and Lado [19]) to Turner's measurements of the electrical conductivity of a <u>fluidized</u> bed of impenetrable spheres [37]. Fig. 4 gives such a comparison for the extreme case $\alpha = \infty$. Note that Turner's data gives values of σ_e/σ_1, up to $\phi_2 = 0.6$, i.e., about 94% of the random close-packing value. The agreement here is not as good as in the case of cubic lattices of spheres. There are two basic reasons for this which actually are unrelated to the accuracy of Eq. (23). Firstly, the model assumes an equilibrium distribution of impenetrable spheres. At moderate to high densities, a fluidized bed is probably not accurately modelled by an equilibrium dispersion of spheres. Secondly, the values of ζ_2 used to generate the

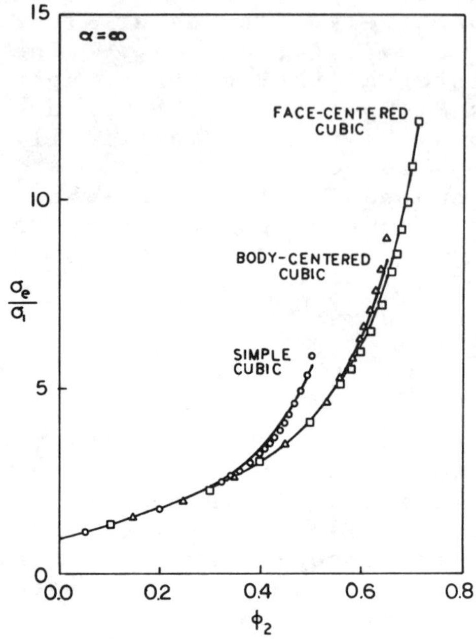

Fig. 3.

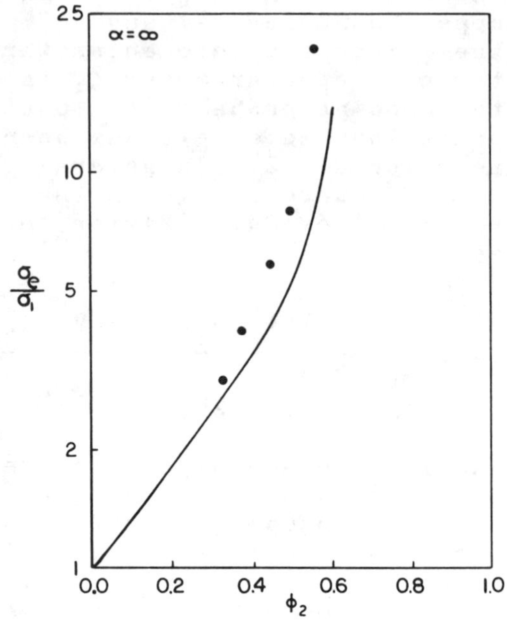

Fig. 4.

FIG. 3. The reduced effective conductivity σ_e/σ_1, for $\alpha=\sigma_2/\sigma_1=\infty$ as a function of the sphere volume fraction ϕ_2, for three cubic lattices of spheres. Exact values for simple, body-centered, and face-centered cubic [35,36] are denoted by O, Δ, and □, respectively. Solid lines represent predicted values from Eq. (23).

FIG. 4. The reduced effective conductivity σ_e/σ_1, for $\alpha=\sigma_2/\sigma_1=\infty$ as a function of the sphere volume fraction ϕ_2, for random arrays of impenetrable spheres. Filled circles are Turner's data [37] for a fluidized bed of impenetrable spheres at α=14,400. Solid line represents predicted values from Eq. (23).

predicted values of σ_e/σ_1 were obtained using the superposition approximation [19]. Recently, Beasley and Torquato [21] have shown that the third-order coefficient of an expansion of ζ_2 in powers of ϕ_2 is larger than the corresponding coefficient obtained using the superposition approximation. (The approximation gives the zeroth- , first- , and second-order coefficients exactly.) Therefore, through third-order in ϕ_2, the exact value of ζ_2 will always be higher than the value of ζ_2 calculated using the superposition approximation. It is expected that this will be true in general. An underestimation of ζ_2 translates to an underestimation of σ_e/σ_1. The error introduced as a result of employing the superposition approximation, however, is not expected to eliminate the discrepancy between the predicted values of σ_e/σ_1 for an equilibrium distribution of impenetrable spheres and Turner's measured values of σ_e/σ_1 for a fluidized bed of impenetrable spheres.

3.2 Permeability Results. Using the results for the general n-point distribution function H_n described in Section II, we have begun to evaluate the Weissberg-Prager bound on the permeability for various random distributions of equi-sized spheres of radius R in void. The Weissberg-Prager bound on k has been, until recently, evaluated only for the case of fully penetrable spheres. For such a geometry the correlation functions involved are trivial.

Torquato and Beasley [38] have recently evaluated the Weissberg-Prager upper bound on k for equi-sized spheres distributed with arbitrary degree of impenetrability λ, up through second order in ϕ_2, in the PS [24] and PCS [25] models. Here we report results for the two extremes, i.e., fully penetrable spheres($\lambda = 0$) and totally impenetrable spheres ($\lambda = 1$). The Weissberg-Prager lower bound on the inverse permeability k^{-1} is given by

$$(24)\quad \frac{k_s}{k_{wp}} = 1 + \frac{3}{2}\phi_2 + \ldots$$

for $\lambda = 0$ and by

$$(25)\quad \frac{k_s}{k_{wp}} = 1 + \left(\frac{193}{81} + 3\ \ln 3\right)\phi_2 + \ldots$$

$$\approx 1 + 5.68\phi_2 + \ldots$$

for $\lambda = 1$, where

$$(26)\quad k_s = \frac{2\ R^2}{9\phi_2}$$

is the exact permeability result for Stokes flow through an infinitely dilute assemblage of equi-sized spheres. These results are to be contrasted with exact low-density expansion for the permeability of an assemblage of hard spheres first derived by Childress [39], Howells [40], and Hinch [41]:

$$(27)\quad \frac{k_s}{k} = 1 + \frac{3}{\sqrt{2}}\phi_2^{1/2} + \frac{135}{64}\phi_2 \ln\phi_2 + 16.5\ \phi_2 + \ldots$$

In Table 1 we compare Eqs. (24), (25) and (27) for various $\phi_2 \le 0.1$.

TABLE 1

Normalized permeabilty k_s/k from, respectively, Eqs. (24), (25) and (27), as a function of the sphere volume fraction ϕ_2 for $\phi_2 \le 0.1$.

ϕ_2	k_s/k_{wp} (λ=0)	k_s/k_{wp} (λ=1)	k_s/k (λ=1)
0.0001	1.000	1.001	1.021
0.001	1.002	1.006	1.069
0.06	1.090	1.262	2.154
0.1	1.150	1.568	2.835

There are several important comments we would like to note regarding this table. Firstly, at the same ϕ_2, k_s/k_{wp} for fully penetrable spheres is smaller than the corresponding value for totally impenetrable spheres. This supports our intuition that, at the same ϕ_2, an assembly of overlapping spheres has a higher permeability than a nonoverlapping one. Secondly, using the general formalism of Sec. 2.1, we have evaluated the Doi bound on k_s/k, through first order in ϕ_2 for λ=1, and found that the first-order coefficient is exactly equal to 5.0 (which is to be contrasted with the value 5.68 of Eq. (25)). This implies that the Weissberg-Prager bound on k_s/k is sharper than the Doi bound through first order in ϕ_2. This is not surprising since the former depends upon one-, two-, and three-point distribution functions and the latter involves only one- and two-point functions. Lastly, for this reason, the Weissberg-Prager bound is expected to yield the best available bound on the permeability of nonoverlapping spheres through all orders in ϕ_2. We are currently in the process of carrying out such a calculation.

It is of interest to compare our calculations of the Weissberg-Prager bound to conductivity bounds which, like the former, are based on the solution to the one-sphere boundary-value problem. Such conductivity bounds have been derived by Torquato [5]. For a matrix or void phase which is conducting relative to the sphere phase, it has been shown that [5]

$$(28) \quad \frac{\sigma_e}{\sigma_1} \le 1 - \frac{3}{2}\phi_2 + 0.644\,\phi_2^2 + \ldots$$

for nonoverlapping spheres. The corresponding exact result obtained by Jeffrey [42] is given by

$$(29) \quad \frac{\sigma_e}{\sigma_1} = 1 - \frac{3}{2}\phi_2 + 0.588\,\phi_2^2 + \ldots$$

It is seen that at $\phi_2 = 0.1$, Eqs. (28) and (29) both yield $\sigma_e/\sigma_1 = 0.856$. This is to be contrasted with the larger discrepancies between Eqs. (25) and (27) at the same value of ϕ_2. This is an interesting result since both bounds contain the same level of information. Consequently, permeability bounds must incorporate more morphological information than conductivity bounds in order for the former to yield results as accurate as those obtained from the latter.

Acknowledgements

This work was in part supported by the Office of Basic Energy Sciences, U.S. Department of Energy, under Grant No. DE-FG05-86ER13482 and by the National Science Foundation under Grant No. CBT-8514841. The author is grateful to J. D. Beasley for a critical reading of this manuscript.

REFERENCES

[1] W. F. BROWN, Solid Mixture Permittivities, J. Chem. Phys., 23 (1955), pp. 1514-1517.

[2] S. TORQUATO, Effective Electrical Conductivity of Two-Phase Disordered Composite Media, J. Appl. Phys., 58 (1985), pp. 3790-3797.

[3] M. BERAN, Use of Variational Approach to Determine Bounds for the Effective Permittivity in Random Media, Nuovo Cimento, 38 (1965), pp. 771-782.

[4] G. W. MILTON, Bounds on the Transport and Optical Properties of a Two-Component Composite Material, J. Appl. Phys., 52 (1981), pp. 5294-5304.

[5] S. TORQUATO, Bulk Properties of Two-Phase Disordered Media. III. New Bounds on the Effective Conductivity of Dispersions of Penetrable Spheres, J. Chem. Phys., 84 (1986), pp. 6345-6359.

[6] H. L. WEISSBERG and S. PRAGER, Viscous Flow Through Porous Media. III. Upper Bounds on the Permeability for a Simple Geometry, Phys. Fluids, 13 (1970), pp. 2958-2965.

[7] M. DOI, A New Variational Approach to the Diffusion and the Flow Problem in Porous Media, J. Phys. Soc. Japan, 40 (1976), pp. 567-572.

[8] S. PRAGER, Viscous Flow Through Porous Media, Phys. Fluids, 4 (1961), pp. 1477-1482.

[9] S. TORQUATO and G. STELL, Microstructure of Two-Phase Random Media. I. The n-Point Probability Functions, J. Chem. Phys., 77 (1982), pp. 2071-2077.

[10] S. TORQUATO and G. STELL, Microstructure of Two-Phase Random Media. II. The Mayer-Montroll and Kirkwood-Salsburg Hierarchies, J. Chem. Phys., 78 (1983), pp. 3262-3272.

[11] S. TORQUATO and G. STELL, Microstructure of Two-Phase Random Media. III. The n-Point Matrix Probability Functions for Fully Penetrable Spheres, J. Chem. Phys. 79 (1983), pp. 1505-1510.

[12] S. TORQUATO and G. STELL, Microstructure of Two-Phase Random Media. IV. Expected Surface Area of a Dispersion of Penetrable Spheres and Its Characteristic Function, J. Chem. Phys., 80 (1984), pp. 878-880.

[13] S. TORQUATO and F. LADO, Characterization of the Microstructure of Distributions of Rigid Rods and Disks in a Matrix, J. Phys. A: Math. Gen., 18 (1985), pp. 141-148.

[14] S. TORQUATO and G. STELL, Microstructure of Two-Phase Random Media. V. The n-Point Matrix Probability Functions for Impenetrable Spheres, J. Chem. Phys., 82 (1985), pp. 980-987.

[15] J. M. HAILE, C. MASSOBRIO, and S. TORQUATO, Two-Point Matrix Probability Function for Two-Phase Random Media: Computer Simulation Results for Impenetrable Spheres, 83 (1985), pp. 4075-4078.

[16] S. TORQUATO and G. STELL, Bounds on the Effective Thermal Conductivity of a Dispersion of Fully Penetrable Spheres, Lett. Appl. Eng. Sci., 23 (1985), pp. 375-383.

[17] S. TORQUATO and J. D. BEASLEY, Effective Properties of Fiber-Reinforced Materials. I. Bounds on the Thermal Conductivity of Dispersions of Fully Penetrable Cylinders, Int. J. Eng. Sci., 24 (1986), pp. 415-434.

[18] F. LADO and S. TORQUATO, Effective Properties of Two-Phase Disordered Composite Media I. Simplification of Bounds on the Conductivity and Bulk Modulus of Dispersions of Impenetrable Spheres, Phys. Rev. B, 33 (1986), pp. 3370-3378.

[19] S. TORQUATO and F. LADO, Effective Properties of Two-Phase Disordered Composite Media. II. Evaluation of Bounds on the Conductivity and Bulk Modulus of Dispersions of Impenetrable Spheres, Phy. Rev. B., 33 (1986), pp. 6428-6435.

[20] J. G. BERRYMAN, Measurement of Spatial Correlation Functions Using Image Processing Techniques, J. Appl. Phys., 57 (1985), pp. 2374-2384.

[21] J. D. BEASLEY and S. TORQUATO, Bounds on the Conductivity of Suspensions of Impenetrable Spheres, J. Appl. Phys., in press..

[22] J. G. BERRYMAN and G. W. MILTON, Normalization Contraint for Variational Bounds on Fluid Permeability, J. Chem. Phys., 82 (1985), pp. 754-760.

[23] S. TORQUATO, Characterization of the Microstructure of Disordered Two-Phase Media, to be published.

[24] J. SALACUSE and G. STELL, Polydisperse Systems: Statistical Thermodynamics, with Applications to Several Models Including Hard and Permeable Spheres, J. Chem. Phys., 77 (1982), pp. 3714-3725.

[25] S. TORQUATO, Bulk Properties of Two-Phase Disordered Media. I. Cluster Expansion for the Effective Dielectric Constant of Dispersions of Penetrable Spheres, J. Chem. Phys., 81 (1984), pp. 5079-5088.

[26] S. TORQUATO, Interfacial Surface Statistics Arising in Diffusion and Flow Problems in Porous Media, J. Chem. Phys., in press.

[27] S. W. HAAN and R. ZWANZIG, Series Expansions in a Continuum Percolation Problem, J. Phys. A: Math. Gen., 10 (1977), pp. 1547-1555.

[28] A. CONIGLIO, U. DE ANGELIS, and A. FORLANI, Pair Connectedness and Cluster Size, J. Phys. A: Math. Gen., 10 (1977), pp. 1123-1139.

[29] J. G. BERRYMAN, Random Close Packing of Hard Spheres and Disks, Phys. Rev. A, 27 (1983), pp. 1053-1061.

[30] R. J. BAXTER, Percus-Yevick Equation for Hard Spheres with Surface Adhesion, J. Chem. Phys., 49 (1968), pp. 2770-2774.

[31] Y. C. CHIEW and E. D. GLANDT, Percolation Behavior of Permeable and of Adhesive Spheres, J. Phys. A, 16 (1983), pp. 2599-2608.

[32] G. W. MILTON and K. GOLDEN, Thermal Conduction in Composites, Proceedings of the 18th International Thermal Conductivity Conference, Rapid City, South Dakota, 1983.

[33] G. A. BAKER, Essentials of Pade Approximants, Academic, New York, 1975.

[34] R. C. McPHEDRAN and G. W. MILTON, Bounds and Exact Theories for the Transport Properties of Inhomogeneous Media, Appl. Phys., A 26, (1981), pp. 207-220.

[35] R. C. McPHEDRAN and D. R. McKENZIE, The Conductivity of Lattices of Spheres I. The Simple Cubic Lattice, Proc. R. Soc. London Ser. A, 359 (1978), pp. 45-63.

[36] D. R. McKENZIE, R. C. McPHEDRAN, and G. H. DERRICK, The Conductivity of Lattices of Spheres II. The Body-Centered and Face-Centered Cubic Lattices, Proc. R. Soc. London Ser. A, 362 (1978), pp. 211-232.

[37] J. C. R. TURNER, Two-Phase Conductivity: The Electrical Conductance of Liquid-Fluidized Bed of Spheres, Chem. Eng. Sci., 31 (1976), pp. 487-492.

[38] S. TORQUATO and J. D. BEASLEY, to be published.

[39] S. CHILDRESS, Viscous Flow Past a Random Array of Spheres, J. Chem. Phys., 56 (1972), pp. 2527-2539.

[40] I. D. HOWELLS, Drag Due to the Motion of a Newtonian Fluid Through A Sparse Random Array Of Small Fixed Rigid Objects, J. Fluid Mechanics, 64 (1974), pp. 449-475.

[41] E. J. HINCH, An Averaged-Equation Apporach to Particle Interactions in a Fluid Suspension, J. Fluid Mechanics, 83 (1977), pp. 695-720.

[42] D. J. JEFFREY, Conduction Through a Random Suspension of Spheres, Proc. R. Soc. London Ser. A, 335 (1973), pp. 355-367.

Numerical Studies of Large Amplitude Instabilities of Fluid Interfaces

GRÉTAR TRYGGVASON*

Abstract. Numerical studies of two "fingering" instabilities are discussed. One is the Taylor-Saffman instability between two fluids in a Hele-Shaw cell, the other is the Rayleigh-Taylor instability between two inviscid fluids of different density. The physics involved is quite different; one is a viscous, dissipative system, the other is a inviscid, conservative system. However, in both cases the interface between the fluids may be modelled as a vortex sheet, and by using a hybrid Eulerian-Lagrangian numerical technique their motion may be simulated very efficiently. For many systems the interactions and competition of many initial waves are responsible for the statistical properties of large-scale mixing. Examples of large simulations are presented, and it is discussed how simulations of only one wave may give an incomplete picture of the large amplitude evolution and the role of the controlling parameters.

1. Introduction. Large amplitude instabilities of fluid interfaces generally form complex unsteady and highly nonlinear systems. Only in exceptional cases are analytical solutions obtainable, and then only for steady, or quasi-steady, states. To capture the full evolution, experimental and numerical investigations are necessary. The need to supplement experiments with numerical simulations is, of course, akin to the need to supplement experiments with analytical investigations. Experimentally it is difficult to achieve the same control over the involved physics, the initial conditions, and external forcing as when working with an analytical model. Changes of parameters and elimination of terms representing forces thought to be unimportant are easily done in the equations but are often difficult, or even impossible, in the laboratory. The analytical models are also often applicable to situations where experiments are difficult (such as for low gravity phenomena) or impossible (such as for astrophysical or geophysical phenomena). Unfortunately, the ability to simulate numerically arbitrary analytical models is quite limited. Besides being almost exclusively confined to two-dimensional models, the systems generally have to be quite small. For many systems of interest the most important quantities are the average, or bulk, properties. Only for a very dilute system is the behavior of a single "structure" (e.g., bubble, solid particle, etc.) representative of the bulk flow. The collective interactions of the "structures" often have a profound influence on the bulk properties and make "large" simulations following the evolution of many "structures" essential.

While numerical integration of the Navier-Stokes equations for a fluid containing several blobs of different fluids must wait for more powerful generations of supercomputers

* Department of Mechanical Engineering and Applied Mechanics, University of Michigan

or better numerical methods or both, "large" simulations of some relatively simple models are feasible at the present time. These models, although much simpler than the full problem, often exhibit quite rich behavior and frequently account well for experimentally observed phenomena. Simulations of two such simplified systems are discussed here. One is the Taylor-Saffman instability between two fluids in a Hele-Shaw cell, the other is the Rayleigh-Taylor instability between two inviscid fluids of different density. Both instabilities are due to a heavy fluid initially overlying a lighter one in a gravitational field, but acceleration toward the heavier fluid can also drive a Rayleigh-Taylor instability, and velocity in the direction of the more viscous fluid can generate a Taylor-Saffman instability. A planar interface represents an unstable solution, and once perturbed, the fluids will mix and "bubbles," "spikes," or "fingers" of one fluid penetrate the region initially occupied by the other fluid. The initial evolution of small perturbations, in particular the wavelength of the most unstable wave, is well predicted by linear analyses. These waves quickly outgrow the linear phase but are relatively independent of each other to begin with. Studies of one such wave, or "structure," (with periodic boundary conditions) will therefore capture the evolution for the initial stage. However, for larger amplitudes these structures may interact in such a manner as to change the character of the mixed region considerably. Often these interactions manifest themselves in competitions (and possibly mergings) of the primary structures, leading to an increase in the mean length scales of the mixed region.

The physics of these systems is quite different; the Taylor-Saffman instability is a viscous, dissipative system, but the Rayleigh-Taylor instability is an inviscid, conservative system. However, in both cases the flow within each fluid is a potential flow, and the interface may be described as a (generalized) vortex sheet. The possibility of formulating these systems in terms of vortex sheets allows the use of numerical techniques particularly well suited for flows where a discontinuity in material properties must be accurately tracked. The evolution equation for the vortex sheet strength depends on the particular dynamics of each problem ranging from a simple algebraic equation for fluids in a Hele-Shaw cell with similar viscosities to an integro-differential equation involving the interface acceleration for a strongly stratified Rayleigh-Taylor instability. Once the vortex sheet strength is found, the velocities can be determined in the same way for both problems. The general kinematic relation between vorticity (ω) and velocity ($\underline{u}$) for incompressible, two-dimensional flow is

$$\nabla^2\psi = -\omega; \qquad \underline{u} = \nabla \times \psi\hat{k} \tag{1}$$

where ψ is the stream function. When the vorticity is confined to a sheet it may be expressed as

$$\omega(\underline{x}) = \int \delta(\underline{x}(s,t) - \underline{x})\gamma(s)\,ds \tag{2}$$

where γ is the vortex sheet strength, $\underline{x}$ is a position vector, s the archlength, and $\delta(\underline{x})$ is a two-dimensional delta function. The solution of (1) can be written as an integral over the vorticity containing part of the fluid, and in the case of a vortex sheet we have

$$\underline{U}(s,t) = \frac{1}{2\pi} P \int \frac{\hat{k} \times (\underline{x}(s,t) - \underline{x}(s',t))}{|\underline{x}(s,t) - \underline{x}(s',t)|^2}\gamma(s',t)\,ds' \tag{3}$$

for the velocity ($\underline{U}$) of the vortex sheet itself. Here P stands for the principal value, s for the archlength, and $\hat{k}$ is a unit vector perpendicular to the plane. Traditional vortex methods replace the continuous sheet by a row of points and the integral (3) by summation. This has the drawback that it is necessary to sum over all interface points to find the velocity of each interface point. Hence, if N is the number of interface points, these methods need at least $O(N^2)$ operations per time step. For the simulations of a single wave this is quite reasonable and accurate results can be obtained at modest cost. We have used such traditional methods (with, however some essential modifications) in our study of the single wave Rayleigh-Taylor instability (see Fig. 4). However, for large simulations of many competing modes the $O(N^2)$ operation count becomes prohibitively expensive. To circumvent these difficulties, we have coupled the Lagrangian interface (the vortex sheet) with a Eulerian grid such that the velocities are found by solving the Poisson equation (1) on the fixed grid. The Poisson equation may be solved by $O(M^2 \log M)$ operations, where M is the number of mesh blocks in each direction, and for a highly contorted interface, we may assume approximately one interface point per mesh block, so $N \approx M^2$ and the operations needed to find the interface velocity are $O(N \log N)$. This method, the so-called Vortex-In-Cell (VIC) technique, has been used to follow the evolution of vorticity for inviscid, homogeneous fluids for many years [6]. Several improvements to the basic method have been suggested over the years, some of which we have implemented, although in many cases we have found the original version to be sufficiently accurate. Meng and Thomson [20] were the first to use this method for weakly stratified flow. We have extended it to arbitrary stratification and refined it somewhat. In addition to its speed and relative accuracy, the method is quite robust in the sense that a very coarse resolution will produce solutions having all the general features of the fully converged solution. Where we have been able to compare this method with either traditional vortex methods or other front tracking methods, it has been 10-100 times faster for comparable resolution.

Although the motivation to use the Vortex-In-Cell technique is the desire to investigate statistical properties of interfacial mixing, we have quite successfully applied the method to interfaces perturbed by only one wave. Indeed, we were among the first to simulate the evolution of the Taylor-Saffman instability, and we have overcome problems associated with the Rayleigh-Taylor instability of fluids of similar densities where traditional methods failed.

2. Flow in a Hele-Shaw Cell, the Taylor-Saffman Instability. The motivation for studies of viscous fingering in Hele-Shaw cells has traditionally been the analogy of the model with two-phase flow in porous medium. In recent years, however, this problem is being viewed more and more as a canonical model for large amplitude instabilities and pattern formation in dissipative systems. Many recent investigations have also been motivated by the analogy of the Hele-Shaw model with solidification. (When conductivity in the solid is assumed to vanish, and the interface moves slowly compared with the characteristic diffusion time in the melt, the governing equations reduce exactly to a special case of the Hele-Shaw equations.) The early examination of this problem was done by Saffman and Taylor [25], and Chuoke, van Meurs and van der Poel [7]. One of the main findings of the Saffman and Taylor study was that for large viscosity ratios (where one of the fluids has negligible viscosity; this is often referred to as the single fluid case) the interactions of the initial fingers lead to a single large finger which then propagated down the cell without change in shape. They found an analytic expression for the shape of this finger, assuming no surface tension. The width of the finger was contained as a undetermined parameter in this solution, and the question of uniqueness occupied investigators for a long time. Important progress was made by McLean and Saffman [17] who showed that the inclusion of surface tension lead to a steady-state finger whose width depended on the surface tension. However, subsequent studies [33] showed that although the continuous family of solution was gone, infinitely many other solutions still exist for non-zero surface tension.

The formulation of this problem as a vortex sheet goes back to Jossalin de Jong [10], but one of the first numerical implementations for the unsteady Taylor-Saffman instability is Tryggvason and Aref [31]. We studied the evolution of an interface perturbed by many waves for various viscosity ratios and value of the surface tension. Fig. 1 shows the evolution of an interface perturbed initially by random disturbances, as discussed in [31]. The top fluid is three times more viscous than the bottom one. The grid is 256^2 meshes, and resolution

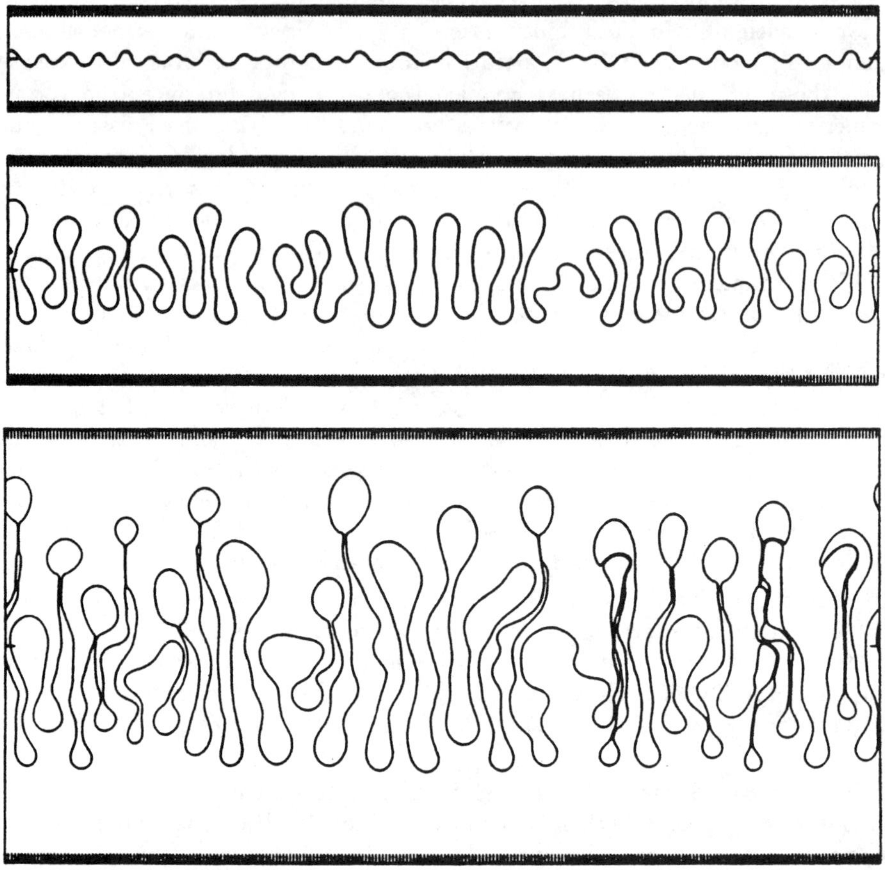

FIG. 1. The Taylor-Saffman instability. The top fluid is three times more viscous than the bottom one. The underlying grid is 256^2 meshes. Only the portion containing the interface is shown. Simulations of this type are discussed in [31].

test indicate that each of the initial waves is quite accurately represented. The simulations demonstrate clearly the importance of interactions mechanisms such as finger competition and merging. These large simulations showed that as the viscosity ratio increased, the competition between the fingers became stronger, and only for very high viscosity ratios do few very large fingers, as in the Saffman-Taylor experiment, emerge. This had been noted experimentally before, but the differences had sometimes been attributed to effects not explicitly accounted for in the Hele-Shaw equations [34]. Tryggvason and Aref [32] subsequently studied the selection mechanism in more detail. Fig. 2 shows the evolution of an interface perturbed by two waves. In addition to the most unstable wave there is a longer wave of much smaller amplitude present in the initial conditions. Here the long wave is three times longer than the most unstable wave and one tenth its amplitude. For simulations where the long wave is two times longer than the most unstable one see [32]. For the single fluid case the subharmonic instability sets in fairly quickly, and a single finger of the Taylor-Saffman type is generated. For smaller viscosity differences the growth of the subharmonic instability is much slower, and the fingers have shapes different from the Saffman-Taylor finger. Calculations of the evolution of only a single wave show considerable *insensitivity* to the viscosity ratio. Only by considering multimode initial conditions can the strong dependency observed in real systems be understood. Analytical calculations of this instability for the single fluid case have recently been done by Kessler and Levine [12].

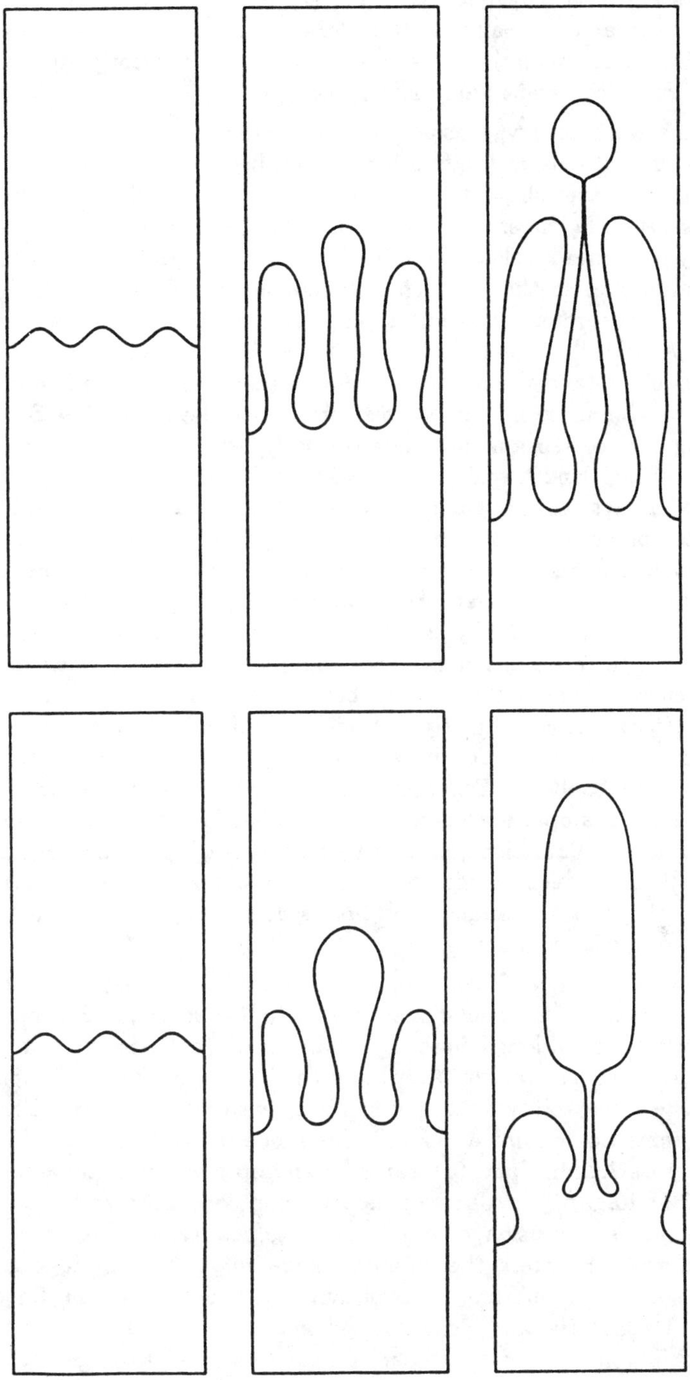

FIG. 2. The growth of a subharmonic instability. (a) $(\mu_1 - \mu_2)/(\mu_1 + \mu_2) = 0.7$. (b) $(\mu_1 - \mu_2)/(\mu_1 + \mu_2) = 1.0$. Here μ is the viscosity, and 1 refers to the top fluid and 2 to the bottom one. The amplitude of the long wave is ten percent of the short wave, which is the most unstable one. The underlying grid is 64 times 256 meshes.

When a single steady-state finger forms in the single fluid case, the numerically produced finger is the one calculated by Saffman and McLean at least for moderately large values of the surface tension. The simulations also demonstrated very clearly that it is quite stable, as has been confirmed by recent analytical investigation [4].

Following the work of Tryggvason and Aref several simulations have appeared that focus on the low surface tension limit. All these studies consider only the single fluid case, since Tryggvason and Aref showed that only in that case will the competition between interfacial waves generate fingers large compared to the surface tension length scale and also because only that case is related to solidification problems. These recent investigations all use boundary integral methods which although not as efficient as the VIC method are claimed to be more accurate. (It seems though difficult to do much better than fig. 4 in Tryggvason and Aref [32]). These studies are therefore confined to much smaller systems, and the goal is to understand the properties of a single finger, not the behavior of a complex large system. Several unanswered questions remain for the usual Hele-Shaw model in the limit of vanishing surface tension (or, more precisely, when the fingers are large compared with the surface tension length scale). It appears that the case of zero surface tension may be more tractable analytically, but being a singular perturbation to the full problem, it has properties that disappear with non-zero surface tension. For zero surface tension Meyer [22] found that the equations can generate a singularity in a finite time, in the form of a narrow spike with infinite curvature and velocity. Howison [11] has given a large class of solutions, which includes a solution found by Saffman [26] and several that generate a cusp in a finite time. How the surface tension modifies these solutions is not well understood at the moment, or in general, what is the relation between the full problem and the zero surface tension case. Surface tension definitely provides stabilizing effect and prevents "blow up" due to a singularity formation, but in view of [33] there is no reason to believe that it selects a single unique solution. Different initial conditions may well lead to considerable differences in the large amplitude stage, and the question is which one is most likely to be seen in initial value calculations, not so much how any particular one (say the Mclean Saffman solution) can be generated. For moderate values of the surface tension in the single fluid case, the Mclean-Saffman solution seems though to be an "attractor" for a fairly general set of initial conditions.

The stability of a single finger still is not completely clear. For sufficiently large surface tension the McLean-Saffman fingers are definitely stable, as has been demonstrated by several numerical simulations. Stability analyses of Taylor-Saffman fingers of half the channel's width, with surface tension (not quite the Mclean-Saffman solution, but close) [4], have found nonlinear instability with exponential sensitivity to noise. The most unstable modes are an asymmetric "hump" and a thinning mode (as found in [32]), but not a splitting mode as found by DeGregoria and Schwartz [9] and observed in many experiments. While the experimental finding may be due to pressure jump conditions more complex than those in the model equations (inclusion of more complete conditions make the fingers "fatter", and hence presumably counters the thinning instability), the findings of the numerical simulations are somewhat puzzling. To obtain the Mclean-Saffman finger for very low surface tension, DeGregoria and Schwartz [9] needed very careful redistribution of the computational points and found that coarser resolution lead to finger splitting. It is well known that lack of resolution generally acts as "artifical" surface tension in any boundary integral method, and it is possible that the fact that the fingers are "fat" is more important for the splitting mode than the stabilizing effect of inadequate resolution. Poorly resolved fingers were also "fat" in our VIC calculations but did not split. Almost all investigators of two-phase flow in Hele-Shaw cells have aired some concern over the appropriateness of usual boundary conditions at the interface. Homsy and collaborators [9] have studied the problem in some detail and stressed the significance of the terms commonly neglected. The required modification may be very important for delicate issues such as the stability of a

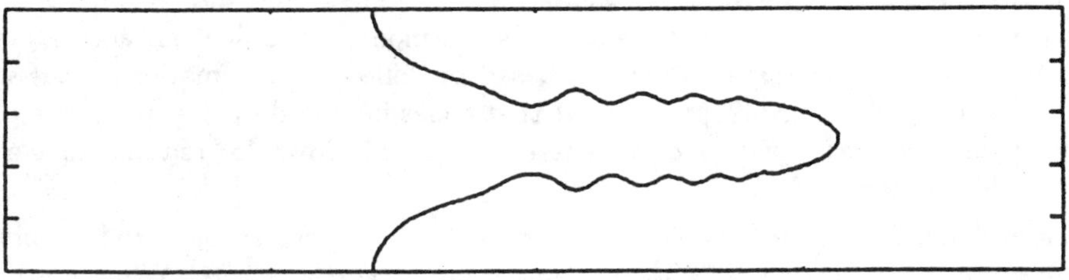

FIG. 3. The effect of slight small-scale anisotropy on a Taylor-Saffman finger of small surface tension. The grid is 64 times 256 meshes.

single finger but presumably less so for a rather "robust" phenomenon as the subharmonic instability of many fingers.

Several recent studies (e.g. [16]) have focused on the importance of anisotropy on the evolution of a single finger. A most interesting result of these studies have shown that a small amount of anisotropy and noise can affect the finger shape dramatically. Fig. 3 shows the large amplitude stage of a finger of small surface tension. The anisotropy in this case is triggered by slight small-scale anisotropy in the numerical method and disappears when this effect is corrected for. The wavelength along the finger is much larger than the scale of the numerical anisotropy and changing the resolution by a factor of two produces similar results. Although anisotropy is apparently necessary to trigger these waves, Howison [11] has found solutions with symmetric waves on the fingers' sides for no anisotropy and zero surface tension. It is likely that smaller anisotropy is required to generate these waves as the surface tension is reduced.

3. Inviscid Fluids, the Rayleigh-Taylor Instability. The model instability for the mixing of two fluids due to a density difference, where viscosity is thought to be unimportant, is the Rayleigh-Taylor instability. The original linear analysis goes back to Rayleigh, but in the late fifties G.I. Taylor renewed interest in this problem by pointing out the analogy with accelerated systems. Investigations of this problem have been well and alive ever since. The behavior of a single wavelength initial disturbance is relatively well established for moderately large amplitudes. For large density differences the heavy fluid falls into the lighter one in narrow spikes, while the light fluid rises in big round bubbles. For smaller density differences, vortices form on the sides of the spike, becoming larger and staying closer to the initial interface as the density difference becomes smaller. For the multimode Rayleigh-Taylor instability current knowledge is quite limited. The experiment is difficult to do; usually it consists of a tank accelerated downward for relatively a short distance. Detailed measurements are almost impossible, and photographic records are usually the only data obtained. Only quantities such as the overall width of the mixed region are easily obtainable. For strongly stratified flow, say air and water, subharmonic instability of the initial waves leads to strong bubble competition so only a few large bubbles grow to large amplitudes. (It is interesting, particularly when compared with the Taylor-Saffman instability, that no experimental evidence shows breakup of these large bubbles [24].) For smaller density differences an increase in mean length scale also takes place as the thickness of the mixed layer increases, but experiments give little or no indication of the mechanism for this.

Contrary to the Taylor-Saffman instability, which received relatively little attention by computational fluid dynamicsts, the Rayleigh-Taylor instability has continued to be a challenge for numerical studies. Since the first successful simulations of the Rayleigh-Taylor instability by Daly [8], several attempts have been made to achieve larger amplitudes and greater accuracy. Although some progress has been made, complete success has been rare.

Only for the special case of an interface between a dense fluid and a fluid of negligible density (the single fluid case) perturbed by one harmonic wave does an accurate solution exist [3,21]. For all other cases, methods based on following the motion of the interface have eventually failed. Sharp [27] points out that codes intended for this problems generally "..run up to aspects ratio of $\approx$ 1 or 2, where they break down for reasons that are often not thoroughly understood."

The problem for low density ratios is very similar to those encountered in simulations of roll-up of "ordinary" (that is, with no density jump) vortex sheets. When no stabilizing mechanism for high wavenumbers is included, the vortex sheet forms a curvature singularity in a finite time. Although this singularity has been the subject of both asymptotic and numerical studies recently (see e.g., [8,13]), the exact nature is still not completely understood. Real vortex sheets do not, of course, form a singularity. Viscosity will always smooth out the discontinuity in the velocity field and give rise to a smooth vorticity distribution. While viscosity cannot be included in boundary integral methods as ours, which depend on the fact that the interior flow remains irrotational, it is, in principle, possible to include surface tension to provide a small-scale stabilizing mechanism. However, it appears likely that surface tension is not sufficient to completely regularize the interface at large amplitudes.

Although it may be argued that the formation of a singularity in the vortex sheet model represents a break down of the model and that the correct solution is to go back to the full Navier-Stokes equation, the general belief is that there still exists a weak solution that may be obtained by the proper regularization (or desingularization) of the model. The hope is that the situation is somewhat analogous to shocks in inviscid incompressible gas dynamics where addition of "artifical viscosity" provides a physically relevant solution. Implied in this rather plausible conjecture (plausible, at least in two-dimension) therefore is that the smallest scales do not affect the large-scale motion strongly and that many reasonable regularizations would give a unique "weak solution" in the limit of small regularization. The failure to recognize explicitly these mathematical difficulties appears to be the main reason for the limited success in the past. For "ordinary" vortex sheets one of the more successful regularizations is the replacement of the point vortices used to model the sheet by vortex blobs [5]. Krasny [14] has recently used a vortex blob method to study in detail the large amplitude Kelvin-Helmholtz instability. We have repeated some of his calculations both by our own vortex blob method and also by the VIC method. Both methods give very similar results. Fig. 4 shows the large amplitude stage of a Kelvin-Helmholtz instability calculated by the vortex blob method for two blob sizes and the VIC method for two mesh sizes. The main effect of reducing the blob radius (or refining the grid) is to increase the roll up in the center of the vortex. The large scale structure is very much unaffected. We have tried to quantify this statement somewhat by comparing various moments of the vorticity distribution and indeed found that the low order moments are almost identical for the simulations in fig 4. The detailed comparison and a more thorough discussion will be published elsewhere. This behavior is also seen in the Rayleigh-Taylor instability. The extension of both methods to weakly stratified flow, the so called Boussinesq limit, is quite straight forward (see [1,28]) but for finite stratification the extension is more involved. Tryggvason [30] describes extensions of the VIC method, and recently we have successfully implemented a vortex blob method capable of regularizing the interface between fluids of any density ratio. Calculations with this new vortex blob method for three density differences are shown in fig. 5. Only the large amplitude stage is shown. Fig 5 (a) is the Boussinesq limit; (c) is the single fluid case where a singularity does not form, [3,21]; and (b) is for intermediate density ratio and is calculated by our new vortex blob method. Detailed description of both the methods and effect of the regularization is forthcoming. For similar calculations with the VIC method see [30]. The identification of

FIG. 4. Simulations of Kelvin-Helmholtz instability both by a vortex blob method, and by the Vortex In Cell method. (a) Vortex Blobs, ($\delta = 0.2$), 200 points. (b) Vortex In Cell, 32 grid spaces per wavelength. (c) Vortex Blobs, ($\delta = 0.1$), 200 points. For better resolution more points are needed in the center of the vortex. (b) Vortex In Cell, 64 grid spaces per wavelength. δ is a desingularization parameter. The VIC method used here has smoother core than the original method. (This modified method was also used for the simulations in fig. 2).

the difficulties encountered at large amplitudes and successful resolution of the problems have overcome many of the problems that have limited simulations in the past to somewhat small amplitudes. However, for long time calculations with small regularization, a large number of computational elements must be used. We have therefore used the VIC method for such studies. The VIC method has been extremely successful for relatively small density ratios and can repeat what other methods are able to do for any density ratios, at a fraction of the cost. Our recent experience with the vortex blob methods has suggested several improvements that are expected to make the VIC method equally successful for any density ratio.

Our simulations of multimode initial conditions have shown that, at least in two-dimension, the evolution for low density stratification is largely dominated by the inter-

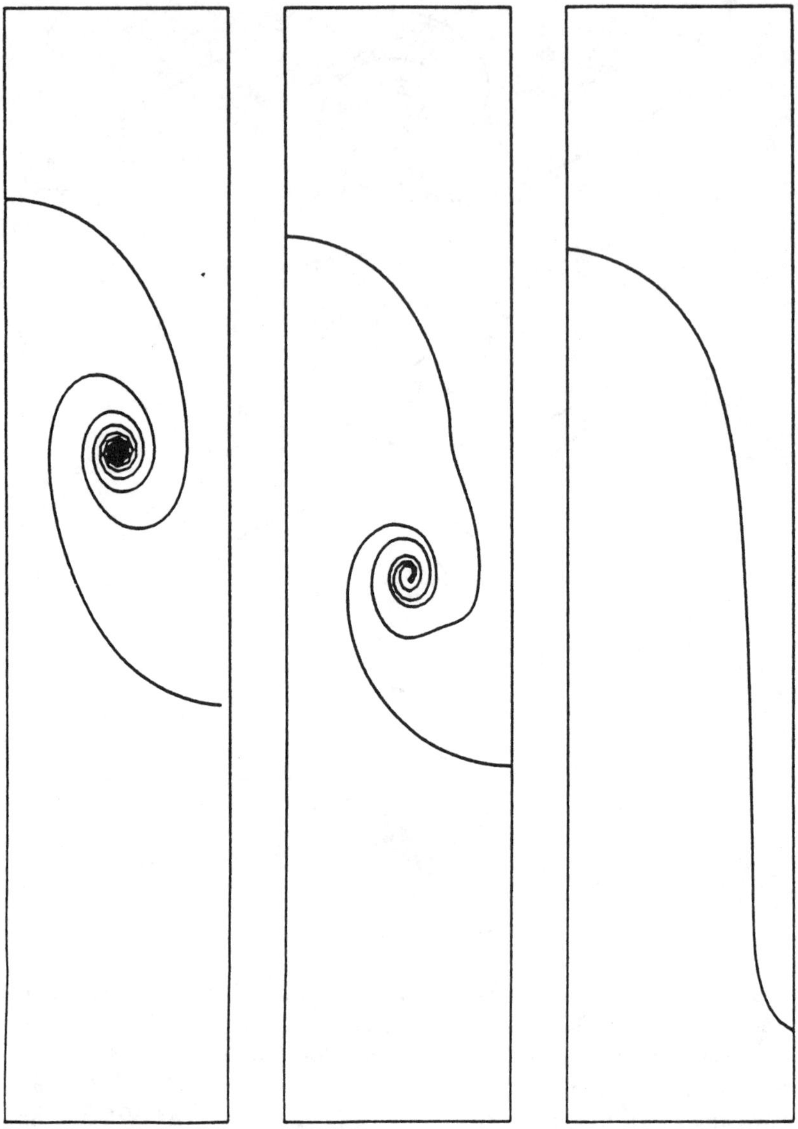

FIG. 5. Simulations of a single wave Rayleigh-Taylor instability by vortex blob method. (a) is the Boussinesq limit, (b) is for the top fluid three times denser than the bottom one, (c) is the single fluid case. For the calculations in (b) it is essential to us our new vortex blob method. Notice that even for 200 points in (a) and (b) the resolution in the center of the vortex is somewhat poor.

actions of the vortices that form in the early stage of the instability. There vortices are generally of alternating sign, and therefore interact somewhat differently than vortices of the same sign, often used to model interactions in shear layers. Fig. 6 shows the evolution of a few initial waves of random amplitude and phase for weakly stratified flow (the Boussinesq limit). Notice the formation of a dipole-like structure farthest to the right, rather early, and the strong interactions of the vortices closer to the center of the mixed region. It is quite obvious that the overall width of the mixed zone is more a result of the interactions of the vortices than the growth of each single wave. In the last frame the collective interactions have not only rearranged the vortices from their relatively regular early configuration, the vortices now interact sufficiently strongly so that some of them may actually pair. For stronger stratification the vortices are smaller and the dynamics is dominated by large bubbles.

The difficulties associated with full numerical simulations of the Rayleigh-Taylor instability have motivated the introduction of several simplified models aimed at capturing and explaining various aspects of the problem. Most such models have dealt with the single fluid case [15,27], but our simulations have led to one such model for weakly stratified flows. This model is based on the observation that the evolution is largely dominated by regions of relatively concentrated vorticity, and a governing equation for the net circulation of each region is easily determined in the weakly stratified limit. The model assumes that the evolution of the interface depends mostly on the first moment of the vorticity distribution, or equivalently, that there is a point vortex at the center of the roll up. At large amplitudes for a single periodic wave, higher moments do become important, but in multimode simulations the interactions between the vortices may be more important and the model valid for a longer time. Fig. 7 shows the evolution of a single periodic wave. To prevent infinite roll up around the point vortex a smooth core has been assumed.

In some cases the modelling can be even simpler. Consider for example an interface between fluids of very similar densities accelerated for a short period only such as by a passing shock. During the acceleration the interface will form concentrated vortices, so to model the subsequent evolution we can assume that the interface consists of few active vortices of constant, but uneven, strength (the shock has passed, so there is no acceleration) and connected by a passive interface. The motion of each point vortex depends now only on the other point vortices but is independent of the rest of the interface which is advected passively by the vortices. Since the vortex motion will in general be chaotic, quite complicated motion may be expected. Fig. 8 shows an example of such simulations. The points have not been redistributed during the simulation and are therefore quite far apart where the interface has undergone rapid stretching. A direct comparison with full simulation is quite possible but has not been done at the present time. Extensions of these models to include the effects of higher moments of the vorticity distribution appear to be possible. For such high order moment models for homogeneous flow see Melander, Zabusky, and Styczek [19]. The generation of simplified models that can be validated by direct comparison with full simulations may greatly enhance the understanding of the evolution of these large amplitude instabilities.

4. Conclusion. The ability to simulate complicated interface problems depends on the availability of a suitable numerical technique. For the calculations presented here it has been necessary to develop new methods more efficient than previously available ones. For the Rayleigh-Taylor instability a more traditional vortex blob method has also been implemented but with improvements that allow calculations where earlier methods have failed. The application of new methods to problems where previous results are few, or none, requires great care. We have therefore put considerable emphasis on testing and checking our methods. An example of such tests is figure 4 in [32] where the shape of a single finger in a Hele-Shaw cell is compared for resolutions differing by a factor of four, as well as with an analytic solution. Both numerical solutions are virtually identical to the analytic one. When it is necessary to use artificial regularization, as in the Rayleigh-Taylor problem, careful convergence studies such as those for the Kelvin-Helmholtz instability in fig. 5 are essential. Quite complete convergence tests have been done for the weakly stratified Rayleigh-Taylor instability, and comparison with the solution of Menikoff and Zemach [21] and Baker, Meiron, and Orszag [3] for the single fluid case may be found in [30]. Similar validation studies for the intermediate density ratios and comparison with our vortex blob method are underway. Detailed checks like these generally use rather

FIG. 6. Multimode Rayleigh-Taylor instability for weakly stratified fluids (the Boussinesq limit). Strong vortices form early in the evolution and interact strongly at larger amplitude. For stronger stratification the vortices are smaller, and the interactions are better described as bubble competition, [29].

FIG. 7. Model of the Rayleigh-Taylor instability for weakly stratified fluids (the Boussinesq limit). All the vorticity is assumed to be at the center of the roll-up. (Here a smooth core is used to prevent infinite roll-up.)

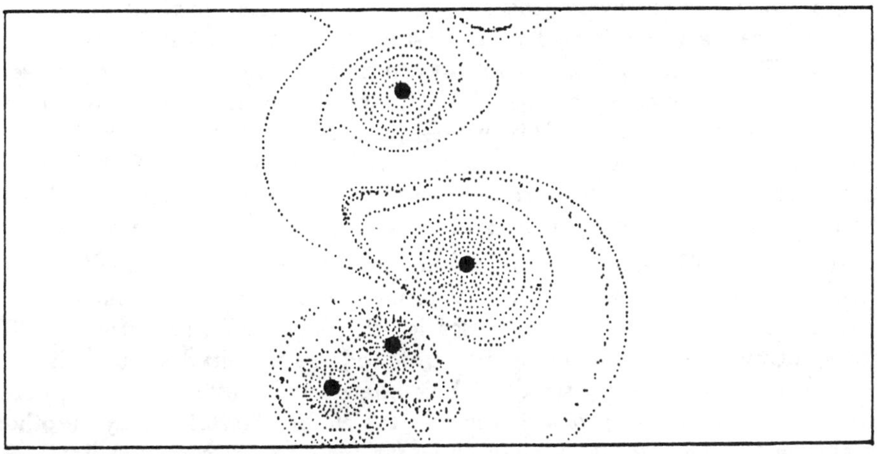

FIG. 8. Model of an instability of an impulsively accelerated interface. The "active" vortices are marked by dark dots. The initial condition was a vertical line across the box with active vortices of slightly different strength evenly spaced. The two vortices now at the top were initially in the middle. The box has periodic boundaries top and bottom. Due to rapid stretching of some parts of the interface the "passive" points have moved very far apart. In order to continue meaningful tracking of the passive interface they have to be redistributed or more points added.

simple initial conditions but establish the required resolution for simulations of interfaces perturbed by many waves of random amplitude and phase.

Currently the VIC method is the only way to simulate large systems of the type shown in figs. 1 and 6, but other vortex methods are rapidly being improved, both in speed and accuracy. An efficient vortex blob method has been introduced by Anderson [2] that is expected to be more accurate than the VIC method, and a clever implementation of "lumping" (treating many distant vortices as one) is said to reduce dramatically the cost of traditional vortex methods (Chorin, private communication). Some of these improvements may allow simulations of problems not easily done before; the lumping may, for example, be particularly suitable for flows where it is necessary to account for a large region of potential flows such as for many surface, or internal, waves where the VIC technique is not particularly suitable. Lumping can also be combined with our new vortex blob method. For simulations of more complex systems, where viscosity, diffusion, variable saturation, and so on, are important, vortex methods (or boundary integral methods in general) have only limited applicability. For such cases the finite difference methods rapidly being developed for compressible gas dynamics may provide a possible solution. These methods that may be referred to collectively as monotonic advection methods (several names, e.g., TVD and FCT, are in use for some subgroups of these methods) are particularly well suited for the advection of sharp interfaces. At the moment two-dimensionality is generally achieved by some splitting in the time stepping, and difficulties satisfying the incompressibility conditions are experienced. Working with the vorticity formulation may be a better alternative.

Although the systems discussed here differ considerably in the physics involved, they share some general morphological features. These generic properties are perhaps the most important findings of studies as ours and may provide guidelines as to what can be expected of systems that fall "between" those dealt with here. A understanding of the interactions of the "primary" structures at large amplitudes is essential to understanding bulk properties of the systems. Although in our simulations these interactions have generally been triggered by the presence of waves of wavelengths different from the most unstable one in the initial conditions, they can also be viewed as an instability of the initial primary structures' configuration. These instabilities, whereby the primary structures either merge or one structure grows at the expense of its neighbor, are the main mechanism that increase length scales in the mixed region. This increase in length scale is a prerequisite for other phenomena of much current interest such as the splitting of the fingers. Only in cases where the selection process leads to a big structure (or finger) relative to the initial length scale set by some stabilizing mechanism is the stability of the large structure a relevant question.

Detailed studies of large multimode systems is still very much in its infancy. It is not clear yet how they are best classified nor what properties are generic for a large class of such systems. There are several possible scenarios for the large amplitude growths. The simplest is for all or most of the initial waves to grow to large amplitudes with relatively mild interactions as fingers in a weakly stratified Hele-Shaw cell do. More complex subharmonic instabilities can lead to merging of structures or the suppression of one by an other and a rapid increase in the mean length scale such as for the strongly stratified Taylor-Saffman instability and the Rayleigh-Taylor instability. The large-scale structures may continue to grow until they themselves become unstable or they encounter boundaries. The presence or absence of boundaries is of considerable importance and can lead to different scaling laws as discussed in [31]. The interaction mechanisms may also be amplitude dependent such that competition at an early stage leads to few large structures far apart that all grow to large amplitude relatively independent of each other. In the case of the Rayleigh-Taylor instability, the large amplitude stage may be classified, for example into vortex dominated interactions when the stratification is weak and into bubble dominated interactions for strong stratification. How the complicated motion of a very contorted interface is best described in an statistical way is not yet well understood (except for rather low order statistics such as mean density profiles and so on). The motion is probably chaotic in many cases (well demonstrated by our model, Fig. 8) and the relation between the various length scales may sometime be best described by a fractal dimension. It is important to keep in mind that although the important interaction mechanisms can be studied by small simulations using only a few initial waves, it is not possible to access *a priori* the importance of each such possible interaction. Only by doing large simulations can the important interaction mechanisms be identified. Simulations of only one initial wave generally give only limited information about the behavior of the full system and may even miss completely the importance of some control parameters.

Acknowledgement: Much of the research described here has been done in collaboration with H. Aref. I also want to acknowledge constructive interactions with J. Glimm and N.J. Zabusky.

REFERENCES

[1] C.R. ANDERSON, A vortex method for flows with slight density variations. J. Comput. Phys. **61** (1985), 417-444.

[2] C.R. ANDERSON, A method of local corrections for computing the velocity field due to a distribution of vortex blobs. J. Comput. Phys. **62** (1986), 111-123.

[3] G.R. BAKER, D.I. MEIRON and S.A. ORSZAG, Vortex simulations of the Rayleigh-Taylor instability. Phys. Fluids **23** (1980), 1485-1490.

[4] D. BENSIMON, Stability of viscous fingering. Physical Review A, (1986), to appear.

[5] A.J. CHORIN and P.S. BERNARD, Discretization of a vortex sheet with an example of roll-up. J. Comput. Phys. **13** (1973), 423-429.

[6] J.P. CHRISTIANSEN, Numerical simulation of hydrodynamics by the method of point vortices. J. Comput. Phys. **13** (1973), 363-379.

[7] R.L. CHUOKE, P. VAN MEURS and C. VAN DER POEL, The instability of slow immiscible, viscous liquid-liquid displacements in permeable media. J. Petrol. Tech. **11** (1959), 64.

[8] B.J. DALY, Numerical study of two fluid Rayleigh-Taylor instability. Phys. Fluids **10** (1967), 297-307.

[9] A.J. DEGREGORIA and L.W. SCHWARTZ, A boundary integral method for two-phase displacement in Hele-Shaw cells. J. Fluid Mech. **164** (1986), 383-400.

[10] G. DE JOSSELIN DE JONG, Singularity distributions for the analysis of multiple-fluid flow through porous media. J. Geophys. Res. **65** (1960), 3739-3758.

[11] S.D. HOWISON, Fingering in Hele-Shaw cells. Preprint, (1986).

[12] D.A. KESSLER and H. LEVINE, Coalescence of Saffman-Taylor fingers: A new global instability. Physical Review A, **33** (1986), 3625-3627.

[13] R. KRASNY, A study of singularity formation in a vortex sheet by the point vortex approximation. J. Fluid Mech. (1986) (to appear).

[14] R. KRASNY, Desingularization of periodic vortex sheet roll-up. Preprint. (1986).

[15] D. LAYZER, On the instability of superposed fluids in a gravitational field. Astrophysical J. **122** (1955), 1-12.

[16] G. LI, D. KESSLER and L. SANDER, Sidebrancing of the Saffman-Taylor finger Preprint, (1986).

[17] J.W. MCLEAN and P.G. SAFFMAN, The effect of surface tension on the shape of fingers in a Hele-Shaw cell. J. Fluid Mech. **102** (1981), 455-469.

[18] D.I. MEIRON, G.R. BAKER and S.A. ORSZAG, Analytic structure of vortex sheet dynamics I. Kelvin-Helmholtz instability. J. Fluid Mech. **114** (1982), 283-298.

[19] M.V. MELANDER, N.J. ZABUSKY and A.S. STYCZEK, A moment model for vortex interactions of the two-dimensional Euler equations. I. Computational validation of an elliptical Hamiltonian representation. J. Fluid Mech. (1986) (to appear).

[20] J.C.S. MENG and J.A.L. THOMSON, Numerical studies of some nonlinear hydrodynamic problems by discrete vortex element method. J. Fluid Mech. **84** (1978), 433-453.

[21] R.C. MENIKOFF and C. ZEMACH, Rayleigh-Taylor instability and the use of conformal maps for ideal fluid flow. J. Comput. Phys. **51** (1983), 28-64.

[22] G.H. MEYER, Hele-Shaw flow with a cusping free boundary. J. Comput. Phys. **44** (1981), 262-276.

[23] C.-W. PARK and G.M. HOMSY, Two-phase displacement in Hele-Shaw cells: theory. J. Fluid Mech. **139** (1984), 291-308.

[24] K.I. READ, Experimental investigations of turbulent mixing by Rayleigh-Taylor instability. Physica D, **12** (1984), 45-58.

[25] P.G. SAFFMAN and G.I. TAYLOR, The penetration of a fluid into a porous medium or Hele-Shaw cell containing a more viscous liquid. Proc. R. Soc. Lond. A**245** (1958), 312-329.

[26] P.G. SAFFMAN, Exact solutions for the growth of fingers from a flat interface between two fluids in a porous medium or Hele-Shaw cell. Q. J. Mech. Appl. Maths. **12** (1959), 146-150.

[27] D.H. SHARP, An overview of Rayleigh-Taylor instability. Physica D, **12** (1984), 3-18.

[28] G. TRYGGVASON, Numerical experiments on slumping gravity currents. Woods Hole Oceanographic Institution, Tech. Rep. WHOI-83-41 (1983), 207-228.

[29] G. TRYGGVASON, A note on the simulation of vortex sheet roll-up by the vortex-in-cell method. Preprint, (1986).

[30] G. TRYGGVASON, Numerical simulation of the Rayleigh-Taylor instability. Submitted for publication, (1986).

[31] G. TRYGGVASON and H. AREF, Numerical experiments on Hele-Shaw flow with a sharp interface. J. Fluid Mech. **136** (1983), 1-30.

[32] G. TRYGGVASON and H. AREF, Finger interactions mechanisms in stratified Hele-Shaw flow. J. Fluid Mech. **154** (1985), 284-301.

[33] J.-M. VANDEN-BROECK, Fingers in a Hele-Shaw cell with surface tension. Phys. Fluids **26** (1983), 2033-2034.

[34] R.A. WOODING and H.J. MOREL-SEYTOUX, Multiphase fluid flow through porous media. Ann. Rev. Fluid Mech. **8** (1976), 233-274.

Multiphase Transport Phenomena: Matching Theory and Experiment

STEPHEN WHITAKER*

ABSTRACT

The matter of precise correspondence between theoretical dependent variables and the corresponding quantity measured in the laboratory is considered. It is shown that the details of a volume-averaged transport equation are altered by forcing an exact correspondence with experimental measurements, while the general form of the transport equation remains unchanged. Under certain circumstances, the closure problem is uneffected by the weighting functions used to describe the experimental measurements. For these conditions the closure scheme becomes a general tool that produces mapping functions that can be used both to compare theory and experiment and to predict system behavior in the volume-averaged sense.

1. INTRODUCTION

The multiphase transport equations for heat, mass, and momentum transfer are often generated in an intuitive manner (Hubbert, 1956) which leaves the dependent variables undefined. The problem is apparently resolved with the use of the method of volume averaging

---

* Department of Chemical Engineering
University of California
Davis, California 95616
This work was supported by NSF Grant CPE-8308461.

(Anderson and Jackson, 1967; Marle, 1967; Slattery, 1967; Whitaker, 1967) which provides defined dependent variables. For example, in the analysis of heat transfer in a packed bed catalytic reactor (Whitaker, 1986c), the approach used by Slattery and Whitaker would lead to a phase average temperature for the β-phase defined by

$$\langle T_\beta \rangle = \frac{1}{\mathcal{V}} \int_{V_\beta} T_\beta dV \tag{1-1}$$

Here T_β represents the "point" or continuum temperature of the β-phase, $\mathcal{V}$ represents the averaging volume shown in Fig. 1-1, and V_β is the volume of the β-phase contained within the averaging volume. To be more explicit we could write Eq. 1-1 as

$$\langle T_\beta \rangle \Big|_{\underset{\sim}{x}} = \frac{1}{\mathcal{V}} \int_{V_\beta} T_\beta(\underset{\sim}{x} + \underset{\sim}{y}) dV \quad , \quad dV = dy_1 dy_2 dy_3 \tag{1-2}$$

in order to clearly indicate that the average value is associated with the centroid located by $\underset{\sim}{x}$ and that the dummy variables of integration are the components of $\underset{\sim}{y}$, the position vector relative to the centroid of the averaging volume.

The approach favored by Anderson and Jackson, and used extensively by Marle (1967, 1982, 1984, 1985) would replace Eq. 1-1 with

$$\langle T_\beta \rangle = \int_{V_\beta^\infty} T_\beta m dV \tag{1-3}$$

where m is a weighting function normalized according to

$$\int_{\mathcal{V}_\beta^\infty} m \, dV = 1 \tag{1-4}$$

Here we have used V_β^∞ to indicate that integration is performed over an infinite region of the β-phase, and $\mathcal{V}^\infty$ is used to indicate that the integration in Eq. 1-4 is taken over an infinite domain containing both the β and σ phases. Marle (1967) pointed out that the weighting function can be chosen according to

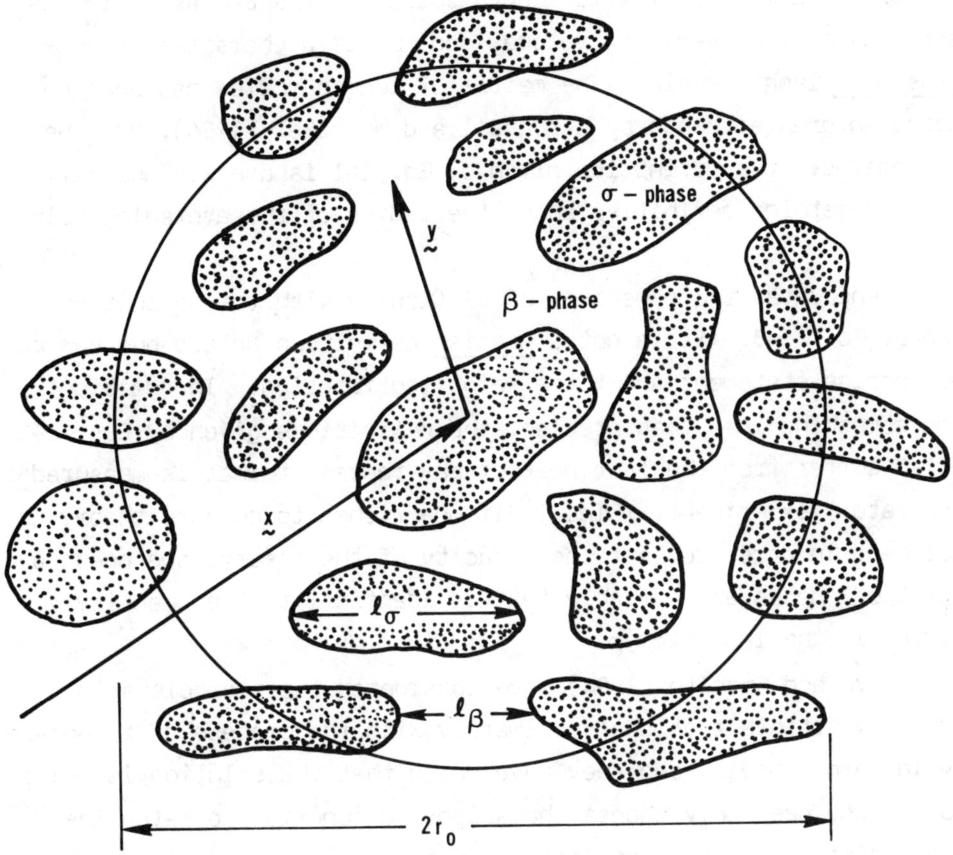

FIGURE 1-1

AVERAGING VOLUME FOR THE β-σ SYSTEM

(1-5) $$m(\underset{\sim}{y}) = \begin{cases} \frac{1}{V}, & 0 \leq |\underset{\sim}{y}| \leq r_o \\ 0, & r_o < |\underset{\sim}{y}| \end{cases}$$

so that the definition given by Eq. 1-3 reduces to that given by Eq. 1-1. In truth, Eq. 1-1 represents a precise definition only if one can say something definite about the averaging volume, V . What has been said about the averaging volume in the past is nothing more than a set of length scale constraints given by (Whitaker, 1969)

(1-6) $$\ell_\beta << r_o << L$$

Here ℓ_β is a characteristic length scale for the β-phase, r_o is the radius of the averaging volume, and L is a characteristic macroscopic length scale. The matter of length scales has been considered in greater depth by Carbonell and Whitaker (1984), but the fact remains: the definition given by Eq. 1-1 is precise only if we can say something definitive about the radius of the averaging volume, r_o.

We encounter the same type of difficulty with the definition given by Eq. 1-3. It is not a precise definition unless one can make some concise statement about the weighting function. In addition to being something less than precise, the definitions given by Eqs. 1-1 and 1-3 suffer from the fact neither may represent what is measured in a laboratory experiment. Clearly if one wishes to compare theory and experiment in order to test the veracity of the theory, one needs a theoretical dependent variable that is identical to the quantity measured in the laboratory.

Baveye and Sposito (1984) have considered these problems in a thoughtful article that deals primarily with the process of two-phase flow in porous media, and they have noted that the solution is quite simple. One need only choose the weighting function to match the characteristics of the experimental measurement in order to obtain: a) a precise correspondence between the theoretical dependent variable and the corresponding quantity measured in the laboratory, and b) a precise definition of a volume averaged quantity. The purpose of this article is to apply the approach of Baveye and Sposito to the problem of mass transport in the β-phase and to extend their development to include the influence of weighting functions on the closure process (Crapiste et al, 1985).

In order to simplify the nomenclature used with the weighting function approach to spatial smoothing, we note that the weighting functions will have non-zero values over a finite domain so that Eqs. 1-3 and 1-4 could be expressed as

$$\langle T_\beta \rangle \Big|_{\underset{\sim}{x}} = \int_{V_\beta} T_\beta(\underset{\sim}{x} + \underset{\sim}{y}) m(\underset{\sim}{y}) dV , \tag{1-7}$$

(1-8) $$\int_{\mathcal{V}} m(\underset{\sim}{y})dV = 1$$

(1-9) $$dV = dy_1 dy_2 dy_3$$

(1-10) $$m = 0 \quad , \quad |\underset{\sim}{y}| > r_o$$

Here it must be understood that the size of the averaging volume depends on the weighting function, and we have not explicitly indicated this in order to keep the nomenclature as simple as possible. Throughout the remainder of this paper will express Eqs. 1-7 through 1-10 as

(1-11) $$\langle T_\beta \rangle = \int_{V_\beta} T_\beta \; m \; dV$$

(1-12) $$\int_{\mathcal{V}} m \; dV = 1$$

where it is understood that $\langle T_\beta \rangle$ is associated with the point $\underset{\sim}{x}$ and that V_β and $\mathcal{V}$ are functions of m .

In certain cases the use of a weighting function is not required, and one can revert to the traditional approach of Slattery and Whitaker by the use of Eq. 1-5 and the constraint that $\mathcal{V}$ is independent of the particular weighting function. For example, in the case of heat conduction in porous media under the constraint of local thermal equilibrium (Whitaker, Sec. 3.4, 1980), Nozad et al (1985) have shown that the point temperature is a suitable representation of the volume averaged temperature. This results from the fact that changes in the local spatial deviation of the temperature are small compared to the measured changes in the temperature for typical experimental studies of heat conduction in porous media. A similar situation occurs for diffusion and reaction in porous catalysts (Ryan et al, 1979 and 1981) and for the pressure field during single phase flow in porous media (Whitaker, 1986a). On the other hand, the measurement of the concentration on a distillation tray is likely to be highly dependent on the measuring device, and in that case the weighting function approach is essential if one seeks a precise comparison between theory and experiment. By "precise comparison" I have in mind that the coefficients in the volume-averaged transport equation are determined by theoretical means and the comparison between theory and experiment will be carried out to test the closure scheme. This is quite different from using model equations to extract experimental values of effective transport coefficients which may be site-specific and therefore of little use in analysis and scale-up. While no general closure scheme currently exists for the flow and mass transport that occurs in complex systems such as distillation trays or trickle bed reactors (Saez and Carbonell, 1986), there is considerable interest in, and effort directed toward, the development of methods of closure for multiphase flows. For two-phase Stokes flow in porous media a method of closure is available (Whitaker, 1986b).

2. THEORY

To illustrate the role that the weighting function plays in the process of spatial smoothing and the subsequent closure, we consider a two-phase system in which diffusion, convection and reaction occur, and we describe the process in terms of the following governing equations and interfacial boundary conditions.

$$\frac{\partial c_\beta}{\partial t} + \nabla\cdot(\mathbf{v}_\beta c_\beta) = \nabla\cdot(\mathcal{D}_\beta \nabla c_\beta) \quad , \quad \text{in the } \beta\text{-phase} \tag{2-1}$$

$$- \mathbf{n}_{\beta\sigma}\cdot\mathcal{D}_\beta \nabla c_\beta = - \mathbf{n}_{\beta\sigma}\cdot\mathcal{D}_\sigma \nabla c_\sigma \quad , \quad \text{at the } \beta\text{-}\sigma \text{ interface} \tag{2-2}$$

$$c_\beta = K_{eq} c_\sigma \quad , \quad \text{at the } \beta\text{-}\sigma \text{ interface} \tag{2-2}$$

$$\frac{\partial c_\sigma}{\partial t} = \nabla\cdot(\mathcal{D}_\sigma \nabla c_\sigma) - k c_\sigma \quad , \quad \text{in the } \sigma\text{-phase} \tag{2-2}$$

We consider the σ-phase to be rigid and assume that the no-slip condition applies at the β-σ interface. Our analysis will be restricted entirely to the β-phase since our objective is only to illustrate a set of ideas that are important in the analysis of multiphase transport phenomena. The system under consideration is illustrated in Fig. 2-1, and in order to compare theory and experiment we require that the concentration, velocity, and volume fraction of the β-phase

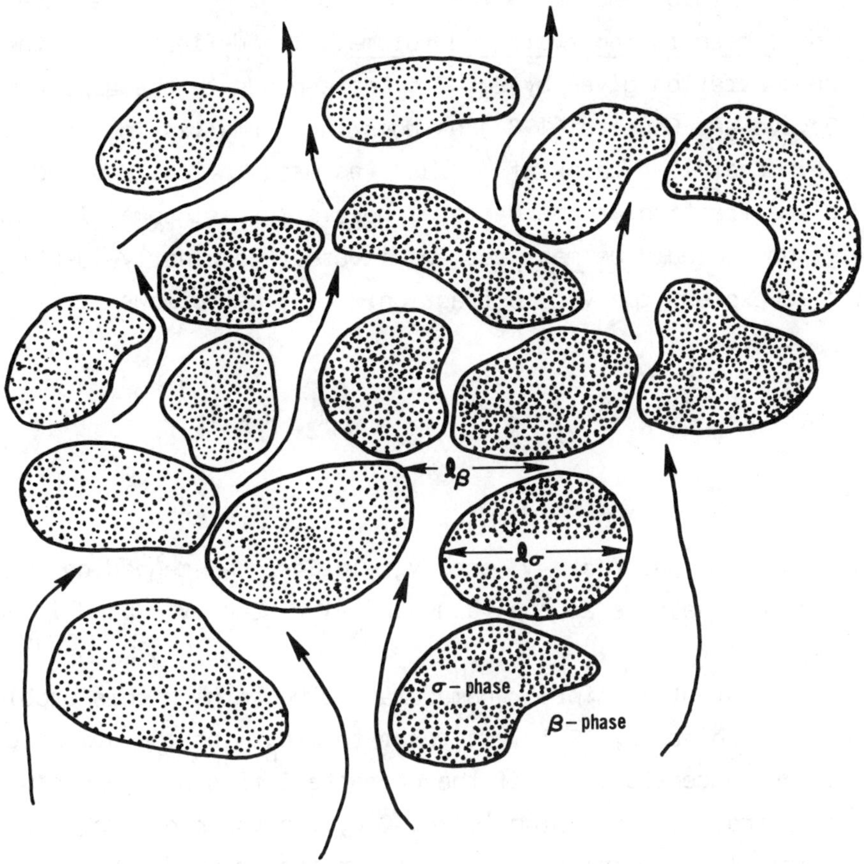

FIGURE 2-1

MASS TRANSPORT IN A TWO-PHASE SYSTEM

will be measured by devices which provide some average value. This measured value can be expressed (for the concentration) as

$$\langle c_\beta \rangle = \int_{V_\beta} c_\beta m_1 \, dV \tag{2-5}$$

where m_1 is the <u>concentration</u> <u>weighting</u> <u>function</u>. The weighting function is normalized according to

$$\int_V m_1 \, dV = 1 \tag{2-6}$$

in which $\mathcal{V}$ is large enough so that $m_1 = 0$ for points that lie outside of the <u>concentration averaging volume</u>. The definition of the average concentration given by Eq. 2-5 corresponds to the approach used by Baveye and Sposito (1984) as does the normalization indicated by Eq. 2-6. One might think of $V_\beta/\mathcal{V}$ (as used in Eqs. 2-5 and 2-6) as the volume fraction of the β-phase; however, this geometrical parameter will undoubtedly be <u>measured</u> and in order to clearly identify this fact we express the volume fraction, ε_β , as follows

$$(2\text{-}7) \qquad \varepsilon_\beta = \int_{V_\beta} m_2 \, dV \quad , \quad \int_{\mathcal{V}} m_2 \, dV = 1$$

Here one must remember that $\mathcal{V}$ and V_β in Eqs. 2-7 depend on m_2 and need not be the same as the values of $\mathcal{V}$ and V_β used in Eqs. 2-5 and 2-6.

To be consistent with prior nomenclature concerning this problem (Carbonell and Whitaker, 1983), we refer to $\langle c_\beta \rangle$ as the weighted phase average concentration. If the characteristics of the measuring device are normalized according to Eq. 2-6, the value of $\langle c_\beta \rangle$ can be computed directly from the knowledge of the measuring device. However, $\langle c_\beta \rangle$ is generally not the dependent variable of choice since $\langle c_\beta \rangle$ is not equal to c_β when the latter is a constant. The preferred dependent variable, referred to as the weighted intrinsic phase average concentration, has this characteristic and is defined by

$$(2\text{-}8) \qquad \langle c_\beta \rangle^\beta = \int_{V_\beta} c_\beta m_1 \, dV \Big/ \int_{V_\beta} m_1 \, dV$$

One can use Eqs. 2-5 through 2-8 to obtain

$$(2\text{-}9) \qquad \langle c_\beta \rangle = \varepsilon_\beta \, \langle c_\beta \rangle^\beta \zeta$$

where the coefficient ζ is given by

$$(2\text{-}10) \qquad \zeta = \int_{V_\beta} m_1 \, dV \Big/ \int_{V_\beta} m_2 \, dV$$

In the traditional method of volume averaging (Gray, 1975) one finds ζ to be unity and Eq. 2-9 reduces to the usual relation between phase average and intrinsic phase averages quantities.

In order to obtain a transport equation for $\langle c_\beta \rangle$, and subsequently $\langle c_\beta \rangle^\beta$, one multiplies Eq. 2-1 by the weighting function m_1 and integrates over the β-phase to obtain

$$\int_{V_\beta} \left(\frac{\partial c_\beta}{\partial t}\right) m_1 \, dV + \int_{V_\beta} \underset{\sim}{\nabla} \cdot (\underset{\sim}{v}_\beta c_\beta) m_1 \, dV = \int_{V_\beta} \underset{\sim}{\nabla} \cdot (\mathcal{D}_\beta \underset{\sim}{\nabla} c_\beta) m_1 \, dV \tag{2-11}$$

Since the σ-phase is rigid we have

$$\int_{V_\beta} \frac{\partial c_\beta}{\partial t} m_1 dV = \frac{\partial}{\partial t} \int_{V_\beta} c_\beta m_1 dV = \frac{\partial \langle c_\beta \rangle}{\partial t} \tag{2-12}$$

provided m_1 is not a function of time. The weighting function form of the averaging theorem (Anderson and Jackson, 1967; Marle, 1967) provides

$$\int_{V_\beta} \underset{\sim}{\nabla} \cdot (\underset{\sim}{v}_\beta c_\beta) m_1 dV = \underset{\sim}{\nabla} \cdot \int_{V_\beta} \underset{\sim}{v}_\beta c_\beta m_1 dV + \int_{A_{\beta\sigma}} \underset{\sim}{n}_{\beta\sigma} \cdot \underset{\sim}{v}_\beta c_\beta m_1 dA \tag{2-13}$$

along with a similar result for the right hand side of Eq. 2-11. Use of these relations in Eq. 2-11 leads to

$$\frac{\partial \langle c_\beta \rangle}{\partial t} + \underset{\sim}{\nabla} \cdot \langle \underset{\sim}{v}_\beta c_\beta \rangle = \underset{\sim}{\nabla} \cdot \langle \mathcal{D}_\beta \underset{\sim}{\nabla} c_\beta \rangle + \int_{A_{\beta\sigma}} \underset{\sim}{n}_{\beta\sigma} \cdot \mathcal{D}_\beta \underset{\sim}{\nabla} c_\beta m_1 dA \tag{2-14}$$

in which we have made use of the no-slip condition. It is permissible to neglect the variations of $\mathcal{D}_\beta$ within the averaging volume and write

$$\langle \mathcal{D}_\beta \underset{\sim}{\nabla} c_\beta \rangle = \mathcal{D}_\beta \langle \underset{\sim}{\nabla} c_\beta \rangle = \mathcal{D}_\beta \left[\underset{\sim}{\nabla} \langle c_\beta \rangle + \int_{A_{\beta\sigma}} \underset{\sim}{n}_{\beta\sigma} c_\beta m_1 dA \right] \tag{2-15}$$

Use of this relation in Eq. 2-14 leads to

$$\frac{\partial \langle c_\beta \rangle}{\partial t} + \nabla \cdot \langle \mathbf{v}_\beta c_\beta \rangle = \nabla \cdot \left\{ \mathcal{D}_\beta \left[\nabla \langle c_\beta \rangle + \int_{A_{\beta\sigma}} \mathbf{n}_{\beta\sigma} c_\beta m_1 dA \right] \right\} + \int_{A_{\beta\sigma}} \mathbf{n}_{\beta\sigma} \cdot \mathcal{D}_\beta \nabla c_\beta m_1 dA \tag{2-16}$$

At this point we recall that we are interested in deriving an equation for $\langle c_\beta \rangle^\beta$, and we can represent Eq. 2-16 in terms of the weighted intrinsic phase average concentration by the use of Eq. 2-9. Provided ε_β and ζ are independent of time, this leads to

$$\varepsilon_\beta \zeta \frac{\partial \langle c_\beta \rangle^\beta}{\partial t} + \nabla \cdot \langle \mathbf{v}_\beta c_\beta \rangle = \nabla \cdot \left\{ \mathcal{D}_\beta \left[\varepsilon_\beta \zeta \nabla \langle c_\beta \rangle^\beta + \langle c_\beta \rangle^\beta \nabla (\varepsilon_\beta \zeta) + \int_{A_{\beta\sigma}} \mathbf{n}_{\beta\sigma} c_\beta m_1 dA \right] \right\} + \int_{A_{\beta\sigma}} \mathbf{n}_{\beta\sigma} \cdot \mathcal{D}_\beta \nabla c_\beta m_1 dA \tag{2-17}$$

In order to deal with the convective transport term, we define a weighted intrinsic average velocity by

$$\langle \mathbf{v}_\beta \rangle^\beta = \int_{V_\beta} \mathbf{v}_\beta m_3 dV \Big/ \int_{V_\beta} m_3 dV \tag{2-18}$$

and then propose the decompositions defined by

$$c_\beta = \langle c_\beta \rangle^\beta + \tilde{c}_\beta \quad , \quad \mathbf{v}_\beta = \langle \mathbf{v}_\beta \rangle^\beta + \tilde{\mathbf{v}}_\beta \tag{2-19}$$

It is reasonable to treat averaged quantities as constants with respect to local volume averaging so that

$$\int_{V_\beta} c_\beta m_1 dV = \langle c_\beta \rangle^\beta \int_{V_\beta} m_1 dV + \int_{V_\beta} \tilde{c}_\beta m_1 dV \tag{2-20}$$

From Eq. 9 we see that this leads to

$$\int_{V_\beta} \tilde{c}_\beta m_1 dV = 0 \tag{2-21}$$

and we can use this result to simplify the convective transport according to

$$\begin{aligned} \langle \mathbf{v}_\beta c_\beta \rangle &= \int_{V_\beta} \left(\langle \mathbf{v}_\beta \rangle^\beta \langle c_\beta \rangle^\beta + \langle \mathbf{v}_\beta \rangle^\beta \tilde{c}_\beta + \tilde{\mathbf{v}}_\beta \langle c_\beta \rangle^\beta + \tilde{\mathbf{v}}_\beta \tilde{c}_\beta \right) m_1 dV \\ &= \langle \mathbf{v}_\beta \rangle^\beta \langle c_\beta \rangle^\beta \int_{V_\beta} m_1 dV + \int_{V_\beta} \tilde{\mathbf{v}}_\beta m_1 dV \, \langle c_\beta \rangle^\beta + \langle \tilde{\mathbf{v}}_\beta \tilde{c}_\beta \rangle \end{aligned} \tag{2-22}$$

We can use Eqs. 2-7 and 2-10 to express this result as

$$\langle \mathbf{v}_\beta c_\beta \rangle = \varepsilon_\beta \zeta \langle \mathbf{v}_\beta \rangle^\beta \langle c_\beta \rangle^\beta + \underbrace{\langle \tilde{\mathbf{v}}_\beta \rangle \langle c_\beta \rangle^\beta}_{\text{added convective transport}} + \underbrace{\langle \tilde{\mathbf{v}}_\beta \tilde{c}_\beta \rangle}_{\text{dispersive transport}} \tag{2-23}$$

Note that the analogous form of Eq. 2-21 for the velocity is given by

$$\int_{V_\beta} \tilde{\mathbf{v}}_\beta m_3 dV = 0 \tag{2-24}$$

and if $m_3 = m_1$ the "added" convective transport in Eq. 2-23 is zero because

$$\langle \tilde{\mathbf{v}}_\beta \rangle = 0 \,, \text{ if } m_3 = m_1 \tag{2-25}$$

Use of Eq. 2-23 in Eq. 2-17 brings us closer to the final form of an equation for $\langle c_\beta \rangle^\beta$ which is given by

$$\varepsilon_\beta \zeta \frac{\partial \langle c_\beta \rangle^\beta}{\partial t} + \nabla \cdot \left(\varepsilon_\beta \zeta \langle \mathbf{v}_\beta \rangle^\beta \langle c_\beta \rangle^\beta \right) + \nabla \cdot \left(\langle \tilde{\mathbf{v}}_\beta \rangle \langle c_\beta \rangle^\beta \right)$$

$$+ \nabla \cdot \langle \tilde{\mathbf{v}}_\beta \tilde{c}_\beta \rangle = \nabla \cdot \left\{ \mathcal{D}_\beta \left[\varepsilon_\beta \zeta \nabla \langle c_\beta \rangle^\beta + \langle c_\beta \rangle^\beta \nabla(\varepsilon_\beta \zeta) \right. \right. \tag{2-26}$$

$$\left. \left. + \int_{A_{\beta\sigma}} \mathbf{n}_{\beta\sigma} c_\beta m_1 \, dA \right] \right\} + \int_{A_{\beta\sigma}} \mathbf{n}_{\beta\sigma} \cdot \mathcal{D}_\beta \nabla c_\beta m_1 \, dA$$

At this point we use the first of Eqs. 2-19 to obtain

$$\int_{A_{\beta\sigma}} \mathbf{n}_{\beta\sigma} c_\beta m_1 \, dA = \int_{A_{\beta\sigma}} \mathbf{n}_{\beta\sigma} m_1 \, dA \ \langle c_\beta \rangle^\beta + \int_{A_{\beta\sigma}} \mathbf{n}_{\beta\sigma} \tilde{c}_\beta m_1 \, dA \tag{2-27}$$

and note that the weighting function from of the averaging theorem can be used to obtain

$$\int_{A_{\beta\sigma}} \mathbf{n}_{\beta\sigma} m_1 \, dA = - \nabla(\varepsilon_\beta \zeta) \tag{2-28}$$

Substitution of Eqs. 2-27 and 2-28 into Eq. 2-26 yields

$$\varepsilon_\beta \zeta \frac{\partial \langle c_\beta \rangle^\beta}{\partial t} + \nabla \cdot (\varepsilon_\beta \zeta \langle \mathbf{v}_\beta \rangle^\beta \langle c_\beta \rangle^\beta) + \nabla \cdot (\langle \tilde{\mathbf{v}}_\beta \rangle \langle c_\beta \rangle^\beta)$$

$$+ \nabla \cdot \langle \tilde{\mathbf{v}}_\beta \tilde{c}_\beta \rangle = \nabla \cdot \left\{ \mathcal{D}_\beta \left[\varepsilon_\beta \zeta \nabla \langle c_\beta \rangle^\beta + \int_{A_{\beta\sigma}} \mathbf{n}_{\beta\sigma} \tilde{c}_\beta m_1 \, dA \right] \right\} \tag{2-29}$$

$$+ \int_{A_{\beta\sigma}} \mathbf{n}_{\beta\sigma} \cdot \mathcal{D}_\beta \nabla c_\beta m_1 \, dA$$

At this point we have derived a governing partial differential equation for a concentration $\langle c_\beta \rangle^\beta$ that is directly related to the experimental measurements of $\langle c_\beta \rangle^\beta$, $\langle \mathbf{v}_\beta \rangle^\beta$ and ε_β . Before we can use Eq. 2-29 in a comparison of theory and experiment, we need to consider the spatial deviations $\tilde{\mathbf{v}}_\beta$ and $\tilde{c}_\beta$ and to see how the

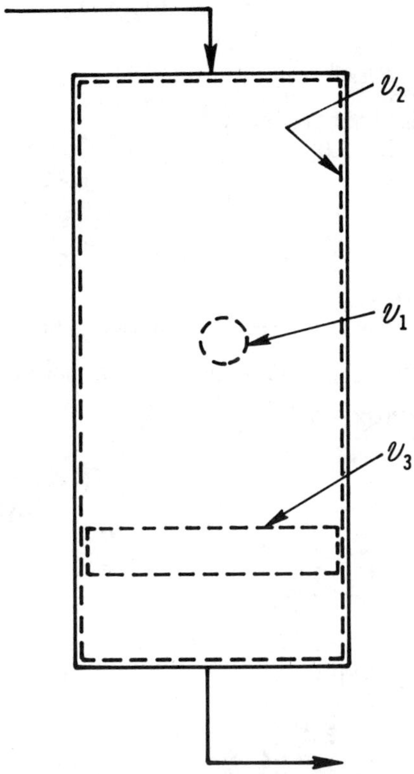

FIGURE 2-2
AVERAGING VOLUMES FOR AN EXPERIMENTAL STUDY OF MASS TRANSPORT AND REACTION IN A PACKED BED

method of closure is influenced by the use of the weighting functions, m_1 , m_2 and m_3 . Before moving on to the closure problem, it will be worthwhile to spend some time thinking about the three weighting functions that have been introduced in this analysis.

In Fig. 2-2 we have illustrated the volumes $(\mathcal{V}_1, \mathcal{V}_2, \mathcal{V}_3)$ associated with typical experimental measurements of the concentration, void fraction, and velocity for the process of mass transport and reaction in a packed bed. The thought here is that some type of a probe would be used to measure the concentration in the β-phase, the volume fraction would most likely be determined gravimetrically for the entire bed, and the velocity would be measured in terms of the volumetric flow rate and the cross-sectional area of the bed.

For design purposes, the interfacial flux per unit volume is often the key quantity of interest. This quantity is usually thought of in terms of V_1 as illustrated in Fig. 2-2, and we can express this quantity as

$$\text{(2-30)} \quad \left\{\begin{matrix}\text{traditional definition of} \\ \text{the interfacial flux} \\ \text{per unit volume}\end{matrix}\right\} = \frac{1}{V_1} \int_{A_{\beta\sigma}} \underset{\sim}{n}_{\beta\sigma} \cdot \mathcal{D}_\beta \underset{\sim}{\nabla} c_\beta \, dA$$

Here $A_{\beta\sigma}$ is the interfacial area contained within V_1 . If $A_{\beta\sigma}$ represents the total interfacial area in the packed bed, we express the total interfacial flux as

$$\text{(2-31)} \quad \left\{\begin{matrix}\text{total} \\ \text{interfacial} \\ \text{flux}\end{matrix}\right\} = \int_{A_{\beta\sigma}} \underset{\sim}{n}_{\beta\sigma} \cdot \mathcal{D}_\beta \underset{\sim}{\nabla} c_\beta \, dA$$

It is generally <u>assumed</u> that;

$$\text{(2-32)} \quad \int_{A_{\beta\sigma}} \underset{\sim}{n}_{\beta\sigma} \cdot \mathcal{D}_\beta \underset{\sim}{\nabla} c_\beta \, dA \overset{?}{=} \int_{V_2} \left\{ \frac{1}{V_1} \int_{A_\beta} \underset{\sim}{n}_{\beta\sigma} \cdot \mathcal{D}_\beta \underset{\sim}{\nabla} c_\beta \, dA \right\} dV$$

however, the error associated with this assumption does not appear to have been explored. The nature of Eq. 2-29 suggests that we should consider an alternate representation for the interfacial flux given by

$$\text{(2-33)} \quad \left\{\begin{matrix}\text{operational definition} \\ \text{of the interfacial flux} \\ \text{per unit volume}\end{matrix}\right\} = \int_{A_{\beta\sigma}} \underset{\sim}{n}_{\beta\sigma} \cdot \mathcal{D}_\beta \underset{\sim}{\nabla} c_\beta \, m_1 \, dA$$

The analog of Eq. 2-32 is expressed as

$$\text{(2-34)} \quad \int_{A_{\beta\sigma}} \underset{\sim}{n}_{\beta\sigma} \cdot \mathcal{D}_\beta \underset{\sim}{\nabla} c_\beta \, dA \overset{?}{=} \int_{V_2} \left\{ \int_{A_{\beta\sigma}} \underset{\sim}{n}_{\beta\sigma} \cdot \mathcal{D}_\beta \nabla \underset{\sim}{c}_\beta m_1 \, dA \right\} dV$$

and one is forced to consider the possibility that this result is either more or less accurate than Eq. 2-32. Two special cases are of interest. If m_1 is a Dirac delta function, Eq. 2-34 is <u>exact</u>, while if m_1 is given by Eq. 1-5 with r_0 the radius of V_1 one can see that Eq. 2-34 is identical to Eq. 2-32. A little thought will lead to the conclusion that there is no real preference for the traditional definition of the interfacial flux per unit volume, and the form of Eq. 2-29 is quite satisfactory in its present state. In order to obtain an equation that can be compared with experiment or used for predictive purposes, we must represent the spatial deviations in terms of volume averaged quantities and this matter is taken up in the next section.

3. CLOSURE - THE PASSIVE CASE

The determination of $\tilde{\mathbf{v}}_\beta$ is relatively straightforward for flow in porous media (Eidsath et al, 1983), and in this section we will be concerned only with the determination of $\tilde{c}_\beta$. In order to simplify the discussion, we consider the case of passive dispersion so that the boundary value problem given by Eqs. 2-1 through 2-4 is simplified to

$$(3\text{-}1) \qquad \frac{\partial c_\beta}{\partial t} + \nabla\cdot(\mathbf{v}_\beta c_\beta) = \nabla\cdot(\mathcal{D}_\beta \nabla c_\beta), \text{ in the } \beta\text{-phase}$$

$$(3\text{-}2) \qquad \text{B.C.1} \qquad -\mathbf{n}_{\beta\sigma}\cdot\mathcal{D}_\beta\nabla c_\beta = 0, \text{ at the } \beta\text{-}\sigma \text{ interface}$$

$$(3\text{-}3) \qquad \text{B.C.2} \qquad c_\beta = F(\mathbf{r},t), \text{ at } A_{\beta e}$$

Here we have used $A_{\beta e}$ to represent the macroscopic area of entrances and exits of the β-phase. In order to obtain the boundary value problem for $\tilde{c}_\beta$ we make use of the decompositions given earlier by Eqs. 2-19

$$(3\text{-}4) \qquad c_\beta = \langle c_\beta\rangle^\beta + \tilde{c}_\beta \quad , \quad \mathbf{v}_\beta = \langle \mathbf{v}_\beta\rangle^\beta + \tilde{\mathbf{v}}_\beta$$

in Eq. 3-1 in order to obtain

$$(3\text{-}5) \qquad \frac{\partial \tilde{c}_\beta}{\partial t} + \nabla\cdot(\mathbf{v}_\beta\tilde{c}_\beta + \tilde{\mathbf{v}}_\beta\langle c_\beta\rangle^\beta) - \nabla\cdot({}_\beta\nabla\tilde{c}_\beta) = - \left[\frac{\partial\langle c_\beta\rangle^\beta}{\partial t} + \nabla\cdot(\langle c_\beta\rangle^\beta\langle \mathbf{v}_\beta\rangle^\beta) - \nabla\cdot({}_\beta\nabla\langle c_\beta\rangle^\beta)\right]$$

Since the right hand side of this result can be treated as a constant when $(r_0/L)^2 \ll 1$ (Carbonell and Whitaker, 1984), we can form the integral over V_β and follow the approach of Crapiste et al (1985) to arrive at

$$\frac{\partial \tilde{c}_\beta}{\partial t} + \nabla\cdot(\mathbf{v}_\beta \tilde{c}_\beta + \tilde{\mathbf{v}}_\beta \langle c_\beta\rangle^\beta) - \nabla\cdot(\mathcal{D}_\beta \nabla \tilde{c}_\beta) = \tag{3-6}$$

$$\frac{1}{V_\beta}\int_{V_\beta} \left[\frac{\partial \tilde{c}_\beta}{\partial t} + \nabla\cdot(\mathbf{v}_\beta \tilde{c}_\beta + \tilde{\mathbf{v}}_\beta \langle c_\beta\rangle^\beta) - \nabla\cdot(\mathcal{D}_\beta \nabla \tilde{c}_\beta)\right] dV$$

Use of the first of Eqs. 3-4 in Eq. 3-2 yields

$$\text{B.C.1} \quad -\mathbf{n}_{\beta\sigma}\cdot \mathcal{D}_\beta \nabla \tilde{c}_\beta = \mathbf{n}_{\beta\sigma}\cdot \mathcal{D}_\beta \nabla \langle c_\beta\rangle^\beta, \text{ at the } \beta\text{-}\sigma \text{ interface} \tag{3-7}$$

and at the macroscopic boundaries we specify $\tilde{c}_\beta$ as

$$\text{B.C.2} \quad \tilde{c}_\beta = (\mathbf{r},t), \text{ at } A_{\beta e} \tag{3-8}$$

While the inclusion of this boundary condition suggests that the closure problem is to be solved for the macroscopic region, this is not the case since the $\tilde{c}_\beta$- field will be dominated by Eq. 3-6 and the boundary condition given by Eq. 3-7. Because of this, the closure problem is a local problem and Eq. 3-8 can be replaced with a periodic boundary condition (Brenner, 1980).

In general the $\tilde{c}_\beta$-field can be treated as quasi-steady since the constraint

$$\mathcal{D}_\beta t \,/\, \ell_\beta^2 \gg 1 \tag{3-9}$$

is usually satisfied. In addition, we can impose the constraint

$$\nabla \mathcal{D}_\beta \cdot \nabla \tilde{c}_\beta \ll \mathcal{D}_\beta \nabla^2 \tilde{c}_\beta \tag{3-10}$$

in order to express the local problem as

$$\nabla\cdot(\mathbf{v}_\beta \tilde{c}_\beta + \tilde{\mathbf{v}}_\beta \langle c_\beta\rangle^\beta) - \mathcal{D}_\beta \nabla^2 \tilde{c}_\beta$$

(3-11)

$$= \frac{1}{V_\beta} \int_{V_\beta} \left[\nabla \cdot (\mathbf{v}_\beta \tilde{c}_\beta + \tilde{\mathbf{v}}_\beta \langle c_\beta \rangle^\beta) - \mathcal{D}_\beta \nabla^2 \tilde{c}_\beta \right] dV$$

(3-12) B.C.1 $$-\mathbf{n}_{\beta\sigma} \cdot \nabla \tilde{c}_\beta = \mathbf{n}_{\beta\sigma} \cdot \nabla \langle c_\beta \rangle^\beta , \quad \text{at } A_{\beta\sigma}$$

(3-13) B.C.2 $$\tilde{c}_\beta(\mathbf{r} + \boldsymbol{\ell}_i) = \tilde{c}_\beta(\mathbf{r}) , \quad i = 1, 2, 3$$

Here $\boldsymbol{\ell}_i$ represents the three non-unique lattice vectors that are used to describe a spatially periodic porous medium. In addition to the governing differential equation and boundary conditions for $\tilde{c}_\beta$, we also require that the average of $\tilde{c}_\beta$ be zero, i.e.

(3-14) $$\langle \tilde{c}_\beta \rangle^\beta = 0$$

Some simplification of Eq. 3-11 can be achieved by use of the continuity equation

(3-15) $$\nabla \cdot \mathbf{v}_\beta = 0$$

and the volume averaged form of this result which is given by

(3-16) $$\nabla \cdot \langle \mathbf{v}_\beta \rangle = 0$$

Here $\langle \mathbf{v}_\beta \rangle$ is defined by

(3-17) $$\langle \mathbf{v}_\beta \rangle = \int_{V_\beta} \mathbf{v}_\beta m_3 dV$$

and we can use Eqs. 2-7 and 2-18 in order to express Eq. 3-16 as

(3-18) $$\nabla \cdot (\varepsilon_\beta \xi \langle \mathbf{v}_\beta \rangle^\beta) = 0$$

The parameter ξ is analogous to ζ and is given by

$$\xi = \int_{V_\beta} m_3 dV \bigg/ \int_{V_\beta} m_2 dV \tag{3-19}$$

For the particular case illustrated in Fig. 2-2, the porosity is constant and Eq. 3-18 reduces to

$$\nabla \cdot (\xi \langle \mathbf{v}_\beta \rangle^\beta) = 0 \tag{3-20}$$

When this result is used in conjunction with Eqs. 2-19 and 3-15, we find

$$\nabla \cdot \tilde{\mathbf{v}}_\beta = \xi^{-1} (\nabla \xi) \cdot \langle \mathbf{v}_\beta \rangle^\beta \tag{3-21}$$

We can now use this result and Eq. 3-15 to express the governing differential equation for $\tilde{c}_\beta$ as

$$\begin{aligned} &\mathbf{v}_\beta \cdot \nabla \tilde{c}_\beta + \tilde{\mathbf{v}}_\beta \cdot \nabla \langle c_\beta \rangle^\beta + \xi^{-1} (\nabla \xi) \cdot (\langle \mathbf{v}_\beta \rangle^\beta \langle c_\beta \rangle^\beta) \\ &- \mathcal{D}_\beta \nabla^2 \tilde{c}_\beta = \frac{1}{V_\beta} \int_{V_\beta} \Big[\mathbf{v}_\beta \cdot \nabla \tilde{c}_\beta + \tilde{\mathbf{v}}_\beta \cdot \nabla \langle c_\beta \rangle^\beta + \\ &\xi^{-1} (\nabla \xi) \cdot (\langle \mathbf{v}_\beta \rangle^\beta \langle c_\beta \rangle^\beta) - \mathcal{D}_\beta \nabla^2 \tilde{c}_\beta \Big] dV \end{aligned} \tag{3-22}$$

Arguments have been put forth elsewhere (Eidsath et al, 1983; Carbonell and Whitaker, 1984) indicating that the right hand side of Eq. 3-22 can be set equal to zero. Under these circumstances the boundary value problem for $\tilde{c}_\beta$ takes the form

$$\tilde{\mathbf{v}}_\beta \cdot \nabla \langle c_\beta \rangle^\beta + \xi^{-1} (\nabla \xi) \cdot (\langle \mathbf{v}_\beta \rangle^\beta \langle c_\beta \rangle^\beta) + \mathbf{v}_\beta \cdot \nabla \tilde{c}_\beta = \mathcal{D}_\beta \nabla^2 \tilde{c}_\beta \tag{3-23}$$

$$\text{B.C.1} \quad -\mathbf{n}_{\beta\sigma} \cdot \nabla \tilde{c}_\beta = \mathbf{n}_{\beta\sigma} \cdot \nabla \langle c_\beta \rangle^\beta , \quad \text{at } A_{\beta\sigma} \tag{3-24}$$

$$\text{B.C.2} \quad \tilde{c}_\beta (\mathbf{r} + \boldsymbol{\ell}_i) = \tilde{c}_\beta (\mathbf{r}) , \quad i = 1,2,3 \tag{3-25}$$

Here we can see three "sources" for the $\tilde{c}_\beta$ -field: a surface source given by $\mathbf{n}_{\beta\sigma}\cdot\nabla\langle c_\beta\rangle^\beta$, a volumetric source given by $\tilde{\mathbf{v}}_\beta\cdot\nabla\langle c_\beta\rangle^\beta$, and a third source term which is proportional to $\nabla\xi$. When these source terms are zero one can prove that the solution of Eqs. 3-23 through 3-25 leads to $\tilde{c}_\beta = 0$.

For the particular case under investigation illustrated in Fig. 2-2, it is easy to show that $\langle\mathbf{v}_\beta\rangle^\beta$ and $\nabla\xi$ can have only axial components and if the bed is homogeneous in the axial direction we have $\nabla\xi = 0$. Under these circum-stances, the boundary value problem for $\tilde{c}_\beta$ simplifies to

$$\tilde{\mathbf{v}}_\beta\cdot\nabla\langle c_\beta\rangle^\beta + \mathbf{v}_\beta\cdot\nabla\tilde{c}_\beta = \mathcal{D}_\beta\nabla^2\tilde{c}_\beta \tag{3-26}$$

$$\text{B.C.1}\qquad -\mathbf{n}_{\beta\sigma}\cdot\nabla\tilde{c}_\beta = \mathbf{n}_{\beta\sigma}\cdot\nabla\langle c_\beta\rangle^\beta \ , \quad \text{at } A_{\beta\sigma} \tag{3-27}$$

$$\text{B.C.2}\qquad \tilde{c}_\beta(\mathbf{r} + \boldsymbol{\ell}_i) = \tilde{c}_\beta(\mathbf{r}) \ , \quad i = 1,2,3 \tag{3-28}$$

This particular problem has been solved by Eidsath et al (1983), and they have shown that $\tilde{c}_\beta$ can be expressed as

$$\tilde{c}_\beta = \mathbf{f}\cdot\nabla\langle c_\beta\rangle^\beta \tag{3-29}$$

where the vector $\mathbf{f}$ is given by

$$\tilde{\mathbf{v}}_\beta + \mathbf{v}_\beta\cdot\nabla\mathbf{f} = \mathcal{D}_\beta\nabla^2\mathbf{f} \tag{3-30}$$

$$\text{B.C.1}\qquad -\mathbf{n}_{\beta\sigma}\cdot\nabla\mathbf{f} = \mathbf{n}_{\beta\sigma} \ , \quad \text{at } A_{\beta\sigma} \tag{3-31}$$

$$\text{B.C.2}\qquad \mathbf{f}(\mathbf{r} + \boldsymbol{\ell}_i) = \mathbf{f}(\mathbf{r}) \ , \quad i = 1,2,3 \tag{3-32}$$

It should be clear that $\mathbf{v}_\beta$ will depend on the geometry of the unit cell as will $\mathbf{f}$ through the solution of eqs. 3-27 through 3-29. In addition, $\mathbf{v}_\beta$ will depend upon the weighting function m_3 since $\mathbf{v}_\beta$ is defined by

$$\tilde{\mathbf{v}}_\beta = \mathbf{v}_\beta - \int_{V_\beta} \mathbf{v}_\beta m_3 \, dV \Big/ \int_{V_\beta} m_3 \, dV \tag{3-33}$$

It is important to note that $\mathbf{f}$ is independent of the weighting functions m_1 and m_2, and if the functional dependence of $\tilde{\mathbf{v}}_\beta$ on m_3 is weak we can think of the closure problem as being independent of the weighting functions used in the averaging process described in Sec. 2.

Use of Eq. 3-29 in Eq. 2-29 leads to the closed form of the volume-averaged transport equation, and this is given by

$$\begin{aligned}\varepsilon_\beta \zeta \frac{\partial \langle c_\beta \rangle^\beta}{\partial t} &+ \nabla \cdot (\varepsilon_\beta \zeta \langle \mathbf{v}_\beta \rangle^\beta \langle c_\beta \rangle^\beta) + \nabla \cdot (\langle \tilde{\mathbf{v}}_\beta \rangle \langle c_\beta \rangle^\beta) \\ &= \nabla \cdot \left\{ \varepsilon_\beta \zeta \mathcal{D}_\beta \left[\mathbf{I} + (\varepsilon_\beta \zeta)^{-1} \int_{A_{\beta\sigma}} \mathbf{n}_{\beta\sigma} \mathbf{f} \, m_1 \, dA \right] \cdot \nabla \langle c_\beta \rangle^\beta \right\} \\ &+ \nabla \cdot \left\{ \left[- \int_{V_\beta} \tilde{\mathbf{v}}_\beta \mathbf{f} \, m_1 \, dV \right] \cdot \nabla \langle c_\beta \rangle^\beta \right\}\end{aligned} \tag{3-34}$$

We can define an effective diffusivity tensor, $\mathcal{D}_{\text{eff}}$, and a dispersion tensor, $\mathbf{D}$, in the obvious manner in order to write Eq. 3-31 as

$$\begin{aligned}\varepsilon_\beta \zeta \frac{\partial \langle c_\beta \rangle^\beta}{\partial t} &+ \nabla \cdot (\varepsilon_\beta \zeta \langle \mathbf{v}_\beta \rangle^\beta \langle c_\beta \rangle^\beta) + \nabla \cdot (\langle \mathbf{v}_\beta \rangle \langle c_\beta \rangle^\beta) \\ &= \nabla \cdot \left[(\mathcal{D}_{\text{eff}} + \mathbf{D}) \cdot \nabla \langle c_\beta \rangle^\beta \right]\end{aligned} \tag{3-35}$$

Here it is important to remember that $\mathcal{D}_{\text{eff}}$ depends on m_1 through the appearance of m_1 in Eq. 3-34 and the fact that

$$\varepsilon_\beta \zeta = \int_{V_\beta} m_1 \, dV \tag{3-36}$$

and that $\mathbf{D}$ depends on m_1 and may depend upon m_3 via Eq. 3-33.

The magnitude of this influence of the weighting functions on effective transport coefficients needs to be explored in more detail; however, the traditional scatter encountered in the measurement of effective diffusivities and dispersion coefficients suggests that the influence of the weighting functions should not be ignored.

CONCLUSIONS

Weighting functions have been used to derive a volume-averaged transport equation for mass transport in a packed bed. The void fraction, velocity, and concentration that appear in the volume-averaged equation are defined in exactly the manner in which they are measured experimentally, and this gives rise to effective transport coefficients that depend upon the weighting functions.

ACKNOWLEDGMENT

This work was supported by NSF Grant CPE-8308461.

REFERENCES

[1] Anderson, T.B. and Jackson, R., A fluid mechanical description of fluidized beds, Ind. Eng. Chem. Fundam. 6, 527-538, 1967.

[2] Baveye, P. and Sposito, G., The operational significance of the continum hypothesis in the theory of water movement through soils and aquifers. Water Resour. Res. 20, 521-530, 1984.

[3] Brenner, H., Dispersion resulting from flow through spatially periodic porous media. Trans. Roy. Soc. (London) 297, 81-133, 1980.

[4] Carbonell, R.G. and Whitaker, S., Heat and mass transport in porous media, in J. Bear and M. Y. Corapcioglu (eds.), Fundamentals of Transport Phenomena in Porous Media, Martinus Nijhoff, Dordrecht, pp. 121-198, 1984.

[5] Crapiste, G.H., Rotstein, E., and Whitaker, S., A general closure scheme for the method of volume averaging. Chem. Engng. Sci. 41, 227-235, 1985.

[6] Eidsath, A., Carbonell, R.G. and Whitaker, S., Dispersion in pulsed systems--III. Comparison between theory and experiments for packed beds. Chem. Engng. Sci. 38, 1803-1816, 1983.

[7] Gray, W.G., A derivation of the equations for multiphase transport. Chem. Engng. Sci. 30, 229-233, 1975.

[8] Hubbert, K.M., Darcys law and the field equations of the flow of underground fluids, A.I.M.E. Petrol. Trans. 207, 222-257, 1956.

[9] Marle, C.-M., Ecoulements monophasiques en milieu poreux. Rev. Inst. Francais du Petrole 13, 1471-1509, 1967.

[10] Marle, C.-M., On macroscopic equations governing multiphase flow with diffusion and chemical reactions in porous media. Int. J. Engng. Sci. 20, 643-662, 1982.

[11] Marle, C.-M., Les ecoulements polyphasiques en milieu poreux: de l'echelle des pores a l'echelle macroscopique. Annales des Mines, No. 5-6, 51-56, 1984.

[12] Marle, C.-M., Equations macroscopiques pour l'equilibre et l'ecoulement de deux fluides non miscibles en milieu poreux, a temperature non uniforme. Euromech 194, Simultaneous Heat and Mass Transfer in Porous Media, Nancy, France, 173-179, 1985.

[13] Nozad, I., Carbonell, R.G., and Whitaker, S., Heat conduction in multiphase systems: I Theory and experiment for two-phase systems, Chem. Engng. Sci. 40, 843-855, 1985.

[14] Ryan, D., Carbonell, R.G., and Whitaker, S., A theory of diffusion and reaction in porous media, AIChE Symposium Series, edited by P. Stroeve and W.J. Ward, #202, No. 77, 1981.

[15] Saez, E., and Carbonell, R.G., Hydrodynamics of trickle bed reactors, submitted to A.I.Ch.E. Journal, 1986.

[16] Whitaker, S., Diffusion and dispersion in porous media, AIChE J., 13, 420-427, 1967.

[17] Whitaker, S., Advances in the theory of fluid motion in porous media, Ind. Eng. Chem. 12, 14-28, 1969.

[18] Whitaker, S., Advances in Drying (edited by Mujumdar A.S.) Vol. 1, Hemisphere, New York, 1980.

[19] Whitaker, S., Flow in porous media I: A theoretical derivation of Darcy's law. Transport in Porous Media 1, 3-25, 1986a.

[20] Whitaker, S., Flow in porous media II: The governing equations for immiscible two-phase flow. To be published in Transport in Porous Media, 1986b.

[21] Whitaker, S., Local thermal equilibrium: An application to packed bed catalytic reactor design. To be published in Chem. Engng. Sci, 1986c.